THE KRAKEN AND THE
COLOSSAL
OCTOPUS

From mermaids to living islands, Dr Bernard Heuvelmans, the founder of cryptozoology and author of other works such as *On the Track of Unknown Animals* and *The Great Sea Serpents*, here dives into the blue to dispel myths and to explore the vast ocean and its mysterious and astounding inhabitants. Often the facts rival fiction as he discusses tales of battle between sperm whales and giant squid, between sharks and octopus, octopus and man. He takes on Melville and Verne's seemingly farfetched stories and hypothesises on what could have really plausibly happened. Distributed within the text are over 140 illustrations: line drawings, artist renditions and photographs. The reader can see photographed scars from a sperm whale-squid battle, the comparison of a manatee to a mermaid; compelling evidence as to what sailors of centuries ago really saw, and various comparison figures of oceanic giants to other species and to humans so that the reader can truly comprehend their Goliath proportions. Entertaining and informative, Kegan Paul is pleased to present another volume of Dr Heuvelmans's groundbreaking discoveries.

www.keganpaul.com

Also by the author

On the Track of Unknown Animals

THE KRAKEN AND THE COLOSSAL OCTOPUS

In the Wake of Sea-Monsters

BERNARD HEUVELMANS, D.S.C.

Routledge
Taylor & Francis Group

LONDON AND NEW YORK

First published 2006 by
Kegan Paul Limited

2 Park Square, Milton Park, Abingdon, Oxon OX14 4RN
711 Third Avenue, New York, NY 10017, USA

*Routledge is an imprint of the Taylor & Francis Group,
an informa business*

First issued in paperback 2016

British Library Cataloguing in Publication Data
Heuvelmans, Bernard
The kraken and the colossal octopus : in the wake of sea-monsters
1. Sea monsters 2. Cryptozoology
I. Title
591.7'7

ISBN 978-1-138-99295-5 (pbk)
ISBN 978-0-7103-0870-2 (hbk)

TO THE MEMORY OF

PIERRE DENYS de MONTFORT,
outlawed naturalist,
and to all those who, in good faith,
have had the courage to bear witness to barely credible facts.

B.H.

CONTENTS

———

ILLUSTRATIONS IN THE TEXT

ILLUSTRATIONS IN THE TEXT

xi

ILLUSTRATIONS IN THE TEXT

ILLUSTRATIONS IN THE TEXT

Part One

HERE, EVERYTHING
IS POSSIBLE

————

"As the earth produces among many other admirable things monsters of various sorts, one should also not doubt that in the sea – which is the greater in area and full of an infinite number of fish and other beasts - would arise things monstrous and of a strange kind..."

PIERRE BELON,
La Nature et Diversité des Poissons avec leurs pourtraicts représentés au plus près du naturel
1555

Chapter One

BELOW THE WAVY CURTAIN

———

It is a rather facile truism to repeat, with the Greek poet Oppian, that "the sea hides many things". However, I shall not fear to repeat what may be seen as a platitude, especially since most people nowadays no longer believe in the truth of this statement.

The impression that our planet is thoroughly known down to its most intimate recesses steadily grew through the XIXth century until it reached a state of heady confidence at the end of that period. By that time, carried away by the wave of fascinating inventions which were to revolutionize daily life - electric light, automobiles, wireless communication, aviation, etc, - people imagined that nothing was beyond the power of Science, which had by then been raised to the level of a divinity. Scientists, high priests of this new religion, were deemed capable of knowing everything and indeed were already thought to know most everything, but for a few trivial details. A few scientists were even lulled into actually believing this illusion. In such an atmosphere of euphoria and naive optimism, imbued with a quasi-religious fanaticism, there was no longer any room for ignorance, doubt, uncertainty, the unexplainable, the irrational. Some of the scientists and philosophers of that time behaved like those young students or those curious laymen who, through their readings, discover the vast sum of knowledge accumulated since the beginning of civilization, as well as the apparently inexhaustible potential of technology, and imagine that they know everything and can do anything. Theirs was the arrogance of the self-taught who think they know all the answers and rule on any question with a tone which brooks no objection, while the specialist must often, in similar conversations, tread timidly, watch his words, express doubts and even withdraw into a prudent silence, often ending up looking like a poor imbecile.

The XIXth century was truly a period of simple-minded scientific thought and deserves in that respect the label of "stupid" bestowed upon it by Léon Daudet.

We have since then ventured into the vastness of our ignorance, in a perspective already perceived by the pre-socratic philosophers. We have recognized the *a priori* fragile and preliminary character of our knowledge, capable of attaining, in the words of Henri Poincaré, at best convenience, but hardly the truth. However, we are still paying for the consequences of the past century. The aura of excessive positivism advocated by the high priests of triumphant science still hovers over us. This situation is particularly acute among the general public, which always responds with some lag to the progress of thought, and often indulges in an irksome tendency to over-generalize.

The public harbours the most naive illusions about marine science. For example, since it was already said eighty years ago that the ocean had been criss-crossed in every which direction and that we had discovered even the smallest islets, doesn't it follow that we must know the ocean even better today? Motorized navigation has freed us from the limitations imposed by winds and currents. Since then, divers and submarines have penetrated beneath the undulating curtain of the waves to explore the depths of the sea. Modern fishing fleets exploit the rich resources of the world ocean and draw from great depths using sophisticated gear. Oceanographic expeditions have reached beyond, all the way to the bottom; using samplers and trawls, they have touched the bottom of supposedly unfathomable abysses. The shape of submerged continents has been traced using acoustic sounding. People, in bathyscaphes and deep submersibles, have dived deep into the eternal night of the abyss, which they have pierced with the light of their cold lanterns. The seas have been sieved, and if there remains any undiscovered creature, it cannot be very large. Undiscovered sardines perhaps, but certainly not mysterious man-sized animals and even less likely real giant creatures. That is what many people think. In practice, the situation is quite different.

In another book (*On the Track of Unknown Animals*, 1958), I have pointed out the imperfect and fragmentary nature of our knowledge of the continents, on which people can nevertheless move about with ease. Our ignorance of the marine world is much greater still. It is such that I do not hesitate to claim that, in the ocean, *everything* is still possible! Faced with the immensity of Neptune's realm, a certain degree of gullibility is preferable to blind incredulity. If it was claimed tomorrow that a real mermaid had been captured - not just an ugly manatee, but a creature boasting Marilyn Monroe's bust and the tail of a coelacanth - the attitude of the zoologist who wished to see it would be much more scientifically justifiable that that of his colleague who would merely shrug the news away.

Just think! The sea covers more than three fifths of the surface of the globe, but our ships cross it only along rather narrow and fixed paths. Already at the end of the last century, Frank Bullen wrote:

"It would at first sight appear strange that, in view of the enormous traffic of steamships through Malacca Straits, so easily "gallied" a creature as the cachalot should care to frequent its waters; indeed, I should certainly think that a great reduction in the numbers of whales found there must have taken place. But it must also be remembered, that in modern steam navigation certain well-defined courses are laid down, which vessels follow from point to point with hardly any deviation there from, and that consequently little disturbance of the sea by their panting propellers takes place, except upon these marine pathways; as, for instance in the Red Sea, where the examination of thousands of log-books proved conclusively that, except for straight lines drawn point to point between Suez and Perim, the sea is practically unused to-day."

Far from having broadened the scope of our marine wanderings, the advent of motorized navigation has constrained it. A propeller driven vessel needs no longer be concerned with favorable but fickle winds, and can steam against currents: its route is thus much more regular. At sea, travellers have no time to waste. Even when they travel for their pleasure, they make no detours to admire the scenery or exceptional beauties: marine horizons are despairingly monotonous. Even if the shortest way between two ports is not always a great circle, it is always the same path. Much of the art of navigation consists in not deviating from the assigned route.

Our knowledge of the marine life at the *surface of the ocean* is thus comparable to that of the naturalist who would know of the fauna of a country only that which he had been able to observe from its main highways.

Furthermore, while the fauna of dry lands occupies an essentially two-dimensional world, of rather negligible thickness[1], that of the sea roams through a truly three-dimensional world. In some areas, its depth exceeds 10,000 metres; on the average, the ocean is over 3,000 metres deep.

The exploration of the third dimension of the marine world has just begun. SCUBA and even hard-hat divers hardly ever reach below 100 metres, and almost always over the continental shelf, usually quite near the coast. They can never know more than a small fraction of the marine fauna, since marine habitats are as sharply differentiated as terrestrial ones, and each one possesses its characteristic inhabitants.

A few areas of the deep ocean have been tentatively explored with tethered bathyscaphes and free diving submersibles. The focus of exploration has been on mid-ocean ridges, where unexpected new life forms have been found near hydrothermal vents. We readily forget, in the euphoria of these discoveries, that only a minute fraction of the deep ocean has been examined. Results obtained so far are so promising that many countries are investing vast sums of money in constructing submersibles capable of reaching greater depths and surveying larger areas of the abyss. In the conquest of this New World, which lies below the wavy curtain of the sea surface, the pioneers Barton and Beebe have played a role similar to that of the Viking adventurers, who long before Columbus landed in Vinland; Auguste Piccard has been the Amerigo Vespucci, the guiding light of this epic venture; commander Houot, the engineer Willm, and oceanographer Ballard have been its first conquistadors. However only the first steps have been taken.

Even oceanographic expeditions, devoted though they are to systematic investigation, have but scratched the surface of the deep. There have been quite a few of them since that of the Challenger, which showed the way in 1872, but even the most refined methods at their disposal for surveying the contents of the ocean remain quite rudimentary. With net and trawls, one cannot hope to capture large animals, or creatures rapid or clever enough to avoid such blind devices. Unfortunately, it is the most interesting "monsters" which are also best protected from our scrutiny.

---

[1] *We need really not take into account the aerial domain, whose inhabitants are always visible from the ground and are tied to the latter, where they come to rest: the air is for them only a medium of passage. In practical terms, the thickness of the biosphere, at least of its macroscopic inhabitants, is limited below by the depth reached by burrowing animals and above by the height of the tallest trees. On dry land, the thickness of the biosphere is thus at most a few tens of metres, and on the average much less.*

THE KRAKEN AND THE COLOSSAL OCTOPUS

One should not also imagine that the oceans are constantly criss-crossed by fishing fleets and thus systematically surveyed, at least near the surface. Fishermen rapidly learn where the fishing areas are and waste no time on "surveying" unprofitable grounds. In 1949, at a scientific conference on Conservation and Utilization of Resources, sponsored by the United Nations, international experts established that 98% of the fish are caught in the northern hemisphere, mostly north of the Tropic of Cancer. In that hemisphere, only 60% of the surface area is covered by water, while this proportion reaches 80% in the southern hemisphere. South hemisphere fisheries have since expanded considerably, and have reached into many unexploited areas; nevertheless, fishing activities remain focused in a few, mainly coastal areas, and the vastness of the Southern Ocean remains only lightly touched.

Because water is not our natural element, we cannot compare the manner of collecting zoological specimens on land and at sea. Methods differ fundamentally: fishing is basically a blind chase, a manner of trapping. In this respect, underwater spear hunting, although severely limited in range, is quite a revolutionary advance.

Overall, our methods of investigating the oceanic fauna have so far consisted in nearly random, blind sampling. In the words of the great naturalist writer A. Hyatt Verrill, in his book *Strange Prehistoric Animals*:

"If inhabitants of another planet were to fly far above ours and slowly drag their nets on its surface, what could they learn about our fauna? They would catch insects, perhaps a few birds, frogs, lizards and turtles; occasionally a cat, a dog, a marmot or some other small mammal, even a sheep, a pig, or chickens. But they would not stand a chance in a million of picking up larger species, or human beings. Scientists from Mars or Mercury could never imagine in examining their catch that men and women, elephants and giraffes, rhinoceros and moose, and a hundred other beasts live on the Earth. However, it is through the use of such simple-minded and inefficient devices that we have acquired what little knowledge we have about the depths of the ocean."

In surveying the marine fauna, we have most often been served by random chance. Nothing has contributed as much to our learning about large aquatic animals as accidental strandings. Carcasses thrown onto the beach or onto rocky shores have been for us like meteorites fallen from the sky, messengers from another world which we can hardly grasp. It is surprising to realize how many marine species have first been identified in this fashion, and how many are known in no other way.

It is almost always a stroke of luck that sets us on the right path, directing the course of future research.

If, on 22 Dec 1938, Captain Goosen, returning from his usual fishing grounds in South-African waters, had not impulsively cast his nets at a depth of 65 metres off the mouth of the Chalumna River, we might to this day remain ignorant of the survival of a group of fish reputedly extinct for seventy million years. He also had to notice the unusual appearance of that large blue armoured fish which the net had brought up, together with a ton and a half of edible fish and two tons of sharks. What had undoubtedly attracted the attention of the crew was that the blue fish, with its small stump-like fins, had not been knocked out by the weight of the catch, as normally happens, but had savagely tried to bite the hand of the captain. Finally, the strange fish had to arrive in some reasonable state - in the middle of the southern summer - in the hands of a competent person, Miss Courtenay-Latimer, of the East London Museum, a young woman sufficiently curious of natural phenomena to alert a specialist, Prof. J.L.B.Smith, who could recognize it as a coelacanth. After a nightmarish *tête-à-tête* with the escapee from prehistory, the eminent ichtyologist had to accept the reality of the evidence. To make it more concrete and serious, he even gave it a Latin name: *Latimeria chalumnae*.

Thus a series of exceptional circumstances were necessary to trigger this event, the most stupendous of the century in the field of zoology. It was soon learned that this was not the first time that such a bizarre fish had been caught. In the Comoro Archipelago, the natives occasionally caught it and ate it salted; they called it *Kombessa*. A trivial detail: the spiny scales of the "living fossil" were commonly used by those people in scratching the surface of punctured bicycle inner tubes before applying patches.

Indeed, as soon as the range of distribution of the coelacanth had been deduced by Prof. Smith, and that the capture of a second specimen near the island of Anjouan had confirmed his conclusions, the Institute of Scientific Research of Madagascar undertook systematic investigations, with the help of local fishermen, and seven specimens were captured in the space of two years.

There is one important point here that one should grasp fully: if the first coelacanth had been immediately eaten by gourmets eager for new taste sensations; if it had just been thrown back to sea by casual fishermen; or if it had completely decomposed before reaching a competent observer, so that only a good, detailed description had reached zoologists, hardly anyone would have believed its existence.

History has shown that the incredulity of scientists can go much further. Even while the precious *Latimeria* was conserved in a laboratory, where a professional ichtyologist had identified it, there were still professionals who denied the whole thing as much too improbable.

In his fascinating history of the discovery of the coelacanth (*Old Fourlegs*), my friend Prof. J.L.B. Smith relates how a government scientist, who had known him for years, came to see him in his office at the university soon after the announcement of the extraordinary catch. He put both hands on his shoulders and said, in a tone of deep gravity:

"Doc, what has made you do this thing? It is terrible to see you ruin all your scientific reputation in this way."

"I asked him what thing?" said Smith. *He replied:*
"Calling this fish a coelacanth."
"I said it was *a coelacanth". He shook his head in sorrow.*

"No, man, he said, I have just been talking to X [a scientist} and he says you are crazy, that it is only a Rock Cod with a mutilated and regenerated tail."

THE KRAKEN AND THE COLOSSAL OCTOPUS

As if to find an excuse for their doubting-Thomas attitude, a number of zoologists spread the rumour that *Latimeria* was an inhabitant of the "inaccessible depths of the ocean". Only that could explain, in their eyes, how it could have remained unknown for so long. This attitude was particularly pronounced among paleontologists, for whom the survival of a coelacanth was a terrible blow, a stark contradiction of laborious reconstructions of the past, based on the most fragile conjectures. They had repeatedly stated that the coelacanth group had been completely extinct since the end of the Cretaceous, since not a trace of a fossil had been found in younger sediments. This accursed fish had showed up to taunt them with its frightening smile.

Nevertheless, the skipper of the trawler had been clear: the fish had been caught in water of 40 fathoms depth, about 65 m. Well, one could easily imagine that it was a lost wanderer....

Even the least astute of zoologists that explanation did not hold water: everything in the appearance of the coelacanth spoke against an abyssal origin. Its thick armour of large spiny scales, the bony plates that covered its head, and its robust fins indicated an animal built to live among rocks and coral reefs, able to hide in a flash in a crevice without risk of injury. Abyssal fish are as a rule small or elongated, almost thread-like, usually delicate and fragile. *Latimeria*, weighing 40 kilograms and 1.25 metres long, had nothing in common with them; its fearsome jaws, powerfully muscled, did not look as if they were built for flimsy prey. Finally, its metallic grey colour, like that of the spring of a watch, distinguished it further from abyssal fish, usually black, grey or dark brown. A bluish being would not fit in the austere abyssal world.

Further, while it was already extraordinary, even for a surface fish, to have survived under a load of several tons when Capt. Goosen net had suddenly been emptied its contents, such sturdy vitality was unthinkable on the part of an abyssal dweller, usually destroyed by decompression before arriving at the surface.

Prof Smith's repeated categorical statements did not suffice to counter general opinion. A photograph of a live coelacanth taken at a depth of 15 metres by a diver of an Italian expedition in the summer of 1963 was claimed to be a fake. A report by a spear fishing diver, G.F. Cartwright, of Salisbury, was brushed aside. While enjoying his favorite sport in October and November 1952 in Malindi, Kenya, he saw, and even tried to harpoon on the coral reef, a fish which looked, in all respects, like *Latimeria*.

In vain. When it was announced that according to local fishermen the second specimen had been caught at a depth of 32 metres, the press, especially in France, exaggerated this depth without hesitation. A Danish expedition even went off to fish for a coelacanth in the abyss. Coelacanth number 8 was captured alive by Comoro Islanders, this time from a depth of 252 metres; it could be kept alive only for a few hours, upon which Prof. J.Millot expressed this rather imprudent diagnostic:

"There can be no doubt that death was brought about by decompression, combined with a rise in temperature."

BELOW THE WAVY CURTAIN

To this, Prof. Smith, an experienced line fisherman, had this to say:

"Professor Millot and his collaborators are possibly not aware of the experience that large fishes taken alive after a struggle on a line, even with no visible laceration, rarely live long after, certainly not in aquaria, and even when liberated may die very shortly."

That the coelacanth is truly a "living fossil" *par excellence*, a creature almost unchanged after three hundred million years, should have been enough to convince naturalists of the unlikelihood of a presumed abyssal habitat. To invade the abyss, an animal must be the very opposite of an archaic form, or "living fossil". It must start from a young type, undifferentiated, still malleable, capable of adapting to the terrible conditions prevailing at depth: the crushing pressure, the intense cold, a complete absence of light and hence of plant life, and thus the necessity of an exclusively carnivorous existence. As Rachel Carson describes it in her masterpiece *The Sea Around Us*:

"The terms of existence in these deep waters are far too uncompromising to support life unless that life is plastic, molding itself constantly to the harsh conditions, seizing every advantage that makes possible the survival of living protoplasm in a world only a little less hostile than the black reaches of interplanetary space."

Because of its hellishly inhospitable character, it is clear that of all habitats, the abyss is the last to have been conquered by the barbaric hordes of life. Although the discovery of chemotrophic life forms near hydrothermal vents has re-opened the question of where life originated on Earth, no one is suggesting that the pelagic abyssal fauna might be related to these newly discovered organisms. The abyss is also that habitat in which survivors from the past are least likely to be found, in spite of what the oceanographers of the last century were still imagining. I have shown, in *On the Track of Unknown Animals*, that there exist on our planet surviving representatives of all geological ages, and that it is thus always possible to discover new ones, unknown to this day. Fossils live among us. Nowhere is this more true than in the vast ocean, where conditions have changed much less over the ages than on dry land. Species and groups have a much longer life in the sea than on land. Water preserves, one might say...

Even though it is quite plausible for the curtain of waves to hide creatures from some other age, it is not from the depth of the abyss that they will rise, like the Beast of the Apocalypse.

But, one might object, doesn't this mean that the portion of the marine world most likely to hide remarkable unknown animals, either through their size or their archaic form, is now simply reduced to two dimensions?

Certainly not! Even if that were the case, one would have to recognize that even such a restricted field of investigation is much less well known than the surface of the earth, which is much smaller and still fertile in unresolved zoological mysteries. Actually, the pelagic zone which overlies the abyss is a few hundreds of meters thick. Beyond the coastal fringe where frogmen dive, this zone, the most populous of all, remains out of sight. Its contents are revealed only by the random sampling mentioned above.

On the high seas, one may expect to glimpse only those animals, which for some reason or other occasionally pierce the wavy curtain of the surface. Among those, we must count those fish that hunt near the surface, such as the sharks, but also mainly lunged vertebrates, reptiles and mammals, which *must* return at regular intervals to renew their air supply. (No need to mention at this point birds, which are never exclusively aquatic, or amphibians, whose habitat is now restricted to freshwater.)

However, it is not because an animal has to pierce the surface of the ocean that it becomes readily visible. While large baleen and toothed whales are noticed by their powerful blow when they come up to breathe - reaching up to 15 metres in height and lasting up to 5 seconds in some species - this is not so for all cetaceans. Many of them, even some with body lengths of about ten metres, blow only briefly and quietly: this is the case for orcas, pilot whales, and many other small whales. Smaller cetaceans do not blow at all. Other marine mammals - pinnipeds and sirenians - also come up for air very discretely; often their nostrils barely graze the surface of the water. That is also the case for reptiles adapted to aquatic life: crocodiles, turtles and snakes. Even the least discrete of cetaceans reveal only a small portion of their back during their most ostentatious blows.

Of course, many cetaceans - even among the largest - are in the habit of making prodigious leaps above the surface for reasons which are as varied as they are poorly understood: play, mating display, or frenetic spasms provoked by the bites of external parasites. Few people have had the opportunity to witness such displays, except of course in the case of smaller species, such as porpoises and dolphins, which naturally travel by leaps and bounds and often follow ships.

Most oceanic creatures have not the slightest reason to make spectacular appearances at the surface. *A Happy Life is A Hidden Life* is a universal law of the animal world.

Thus, without exception, the inhabitants of the pelagic zone hardly ever show themselves; when they do it is only for a fleeting glimpse. Animals careless, or rather unlucky enough, to exhibit themselves to man are hunted without quarters: some because we eat them; other because they eat us; others because their fur, ivory, or shell flatter our elegance; others still because they provide us with precious resources, oil in particular; others, alas! because we find it amusing to kill them... All those species - tuna, swordfish, sharks, whales, sperm whales, seals, sea-lions, walruses, turtles, etc... are generally well known[2]. But nothing proves that we know them all, or that there may not exist other species, entirely different in shape and habits. Undoubtedly, it is *precisely* because they may be different that we cannot know them: we are caught in a vicious circle.

Does this mean that there does exist very rare animals? The idea of

---

[2] *Everything is relative, of course. Ivan T. Sanderson, who has carefully examined our knowledge of Cetaceans, did not hesitate to write in his book* Follow the Whale *(1956): "After at least ten thousand years in pursuit of the whale, we still know very little about him..."*

rarity, from a zoological point view, is rather subjective. No animal is truly rare within its biotope, unless it be on the eve of extinction. We call an animal rare because it is unknown to us, or difficult of access. There are undoubtedly as many, if not more, aardvarks than zebras in Africa, but the former are nocturnal diggers, while the latter run freely through the open savanna. There is not a visitor to black Africa who has not seen the frolicsome horses in black and white pajamas; however how many can boast of having seen the "earth pig", or aardvark, of the Boers?

The animals that we call rare are those which we observe purely by chance. This is not a condition which is likely to make them better known, or even accepted. "Seeing is believing" is not an acceptable rule for the majority of zoologists, for whom merely seeing is an insufficient criterion for the existence of an animal, especially when dealing with creatures apparently difficult to fit among known genera or even recent groups. Scientists want to touch: they insist on concrete evidence. A tooth, a vertebra, a scale, has more value in their eye than a careful description of the whole animal. One can of course understand that zoologists would normally demand some material evidence before proceeding with the scientific naming of an animal: this practice is to be respected. However, when they deny without any further trial the existence of animal under the pretext that no anatomical fragment is available, their professional bent becomes alarmingly distorted.

To prove that there may, and indeed that there *must* still exist in the sea a number of unknown animals, I need only limit my demonstration to the largest creatures. For smaller animals, there is in any case no argument: everyone accepts without argument that there may be found each year many new species of fish, crustaceans or molluscs small enough to hold in the palm of one's hand.

It is not that large animals have any more difficulty hiding: a rhinoceros may be more easily missed than a macaw. But nothing impresses people so much as size.

"Size is important, very important," noted Prof. J.L.B. Smith, "An elephant is far more exciting than even a rare *Peripatus* ... If the original coelacanth had been only 5 inches [15 centimetres] long, it would not have stirred public imagination a fraction as much."

There is little risk of being proven wrong in predicting that there are still many unknown man-sized fish.

Perhaps other species of coelacanths will some day be caught in seas remote from the Indian Ocean. We already know for sure that there exists in the Gulf of Mexico a large fish of a primitive type, still unknown to science. One of its scales, nearly 4 centimetres in diameter, has been preserved since 1949 at the National Museum in Washington.

This priceless piece was sent for identification by a woman from Tampa, Florida, who creates ornamental souvenirs from sea shells, fish scales and other marine objects. She had bought a full barrel of these scales from fishermen and their strange appearance had puzzled her so much that she had sent a sample to the Museum. The scale was examined by an eminent ichtyologist, Dr. Isaac Ginsburg, who declared that:

THE KRAKEN AND THE COLOSSAL OCTOPUS

"This scale is like no other fish scale I have ever seen!" It appeared to be of a primitive structure. It was not impossible, according to Dr. Ginsburg, that it might come from a species of coelacanth. Unfortunately, it has not been possible to obtain from the Tampa crafts-woman any further information on the origin of the scales and on the man from whom she had purchased them.

In any case, we will hardly pause in this work for creatures merely of the size of *Latimeria*. We are after much larger game. We discuss here only those marine creatures described, because of their enormous size, as "monsters".

Of course, we recognize that this term is ambiguous and imprecise; Voltaire was right when he saw in this word one of the most difficult to define. In this case however no other word seems more appropriate.

"*Monstrum*, according to Armand Landrin, describes all that is strange, incredible, extraordinary, bizarre, hideous, stunning, excessive in its kind, of unheard wildness, fabulous."

From this point of view, there are few animal species that do not deserve in some way the label "monstrous", since they are all distinguished by something strange or extraordinary - at least from our point of view. In the ocean, the shark is a monster because of its gluttony, the whale because of its size, the sea-horse because of its shape and posture, the starfish because of its bizarre radial structure, the angler-fish because of its ugliness, the coelacanth because of its anachronistic nature.

To confuse things further, we also call "monstrous" any individual that is anomalous with respect to the other members of its species: a five-legged sheep; a bearded woman; or a white sperm-whale like Herman Melville's Moby Dick.

This is however not a treatise on Teratology, the science of abnormally shaped individuals, but on Zoology, with a focus on marine animals of monstrous size. In this respect, I would have preferred to use in this book the old French term "bellue" (from the Latin *bellua*, wild beast), but it has long gone out of usage and would have been incomprehensible to most. The recently introduced term "cryptid", a hidden animal, would apply only to those "bellues" still unrecognized by Science. In any case, the only oceanic creatures of striking size invoked here are those whose mysterious character has made into legends. Thus, the only creatures discussed here are indeed those said to be strange, incredible, extraordinary, bizarre, hideous, stunning, excessive in its kind, of unheard wildness (in some cases at least), and fabulous: monsters *par excellence*. The choice of the term "monster " is thus fully justified.

Nothing would be more convincing to those who will deny our ignorance of the marine fauna that the discovery of an unknown animal as large as a whale. This is of course not to say that it is the largest animals that are the most interesting zoologically. On land, the capture of a *Peripatus*, a kind of false caterpillar intermediate between worms and arthropods, is more significant than that of an elephant; at sea, fishing a coelacanth is more sensational than finding a new kind of whale. One of these days, we may discover in the ocean animals far more precious than the famous sea-serpent itself. Catching a living trilobite, no larger than a penny, would be more interesting for zoologists than finding a giant new species of crocodiles, cetaceans or pinnipeds. It would then be possible to study the internal anatomy, the physiology and the behaviour of an animal which was part of a formerly cosmopolitan group extinct for hundred of millions of years. That is of course the principal interest in the discovery of a live coelacanth. *Latimeria* fulfills, in a small way, the dream of Wells' Time Machine. It is doubtful however whether the capture of living trilobites would be the cause for sensational newspaper headlines.

This point was nicely demonstrated in the 1930's by the discovery of a marine animal which caused a complete rearrangement of the classification of Molluscs. This creature, a bridge between chitons (segmented molluscs) and limpets (primitive molluscs with a conical shell), was caught in 1912 in the Pacific, off the coast of Costa Rica, by the Danish oceanographic vessel *Galathea*. Upon examination, astounded scientists discovered it to be a late representative of a group extinct since the Primary Era, the Monoplacophora. According to paleontologists, *Neopilina galathea*, as Dr. Henning Lemche named it in 1957, was thought to have disappeared 280 million years ago! The discovery of a live specimen was thus much more extraordinary than that of the coelacanth, thought to have been extinct for a mere 70 million years. However, this discovery was unnoticed by the general public because this strange mollusc is about the size of a one-franc coin...

Equally astounding was the 1957 discovery by Dr. Howard L.Sanders of Yale University of what is undoubtedly the most ancient known crustacean. This tiny animal, named *Hutchinsoniella*, is so small that it could fit through the eye of a needle; however, it is a Cephalocarid, namely a kind of marine arthropod having retained some of the traits of trilobites and thus a missing link between the five principal groups of contemporary crustaceans. This inestimable find did not come from the abyss, but from the mud at the bottom of Long Island Sound, near New York City, in a depth of only 8 to 22 metres.

What would truly raise unanimous enthusiasm, in the public as well as among specialists, would be the appearance, in our era, of a giant of days gone by. That's a lot to ask, one might think. Nevertheless, it is within the realm of possibilities. The story of the coelacanth, of the *Neopilinia* and of *Hutchinsoniella* has shown - or rather confirmed, since these are but the latest in long series of "living fossils" - that animals from the most remote eras have survived in the shadows to this day. It remains to be shown that not only small, but gigantic animals may also have remained incognito for such long periods.

As islands, because of their small area, favour the development of dwarf species, so is the vast, unbounded ocean, the Kingdom of Giants *par excellence*.

Marine animals are generally larger than their terrestrial relatives. That is true from one end to the other of the ladder of living beings. The noctiluca which cause marine phosphorescence are among the rare protozoans visible to the naked eye. River sponges are ridiculously small compared to those of the sea, which sometimes reach a metre in diameter. Among coelenterates, fresh water hydra are nearly microscopic, while some medusae in the sea grow to incredible sizes: the gelatinous dome of a *Cyanea capillata* found off Massachusetts and described by Alexander Agassiz, measured 2.40 metres in diameter and its fishing tentacles hung down 40 metres. The few land molluscs - slugs and snails- are puny besides their marine equivalents, especially the giant cephalopods. The longest of all worms, Sowerby's *Lineus longissimus* is a marine nemertean (or ribbon worm); some specimens measure up to 15 metres, even double that length[3]. Terrestrial arthropods (Insects, Myriapods and Arachnids) cannot rival the size of marine arthropods (Crustaceans): a 35 centimetre long Atlas beetle (*Coscinocera hercules*) looks puny compared to a giant spider crab (*Macrocheirus kämpferi*) which, legs outstreched, may span 4 metres.

We find the same pattern among vertebrates. Starting with fish, what fresh water champions can rival with the largest sharks, the whale shark (*Rhineodon typus*), for example? This fish exceeds 15 metres in length, and might reach up to 20 metres according to some estimates; it weighs tens of tons. One specimen 12 m long was weighed at 12 tonnes.

Among reptilians, the largest individual ever taken is said to have been of the species *Crocodylus porosus*, the only crocodile which ventures to sea. Shot a hundred years ago by two hunters from Luzon, Philippines, it was according to them 8.85 m long. However, on the basis of a measurement of its skull, preserved in the U.S.A., Karl Schmidt of the Chicago Museum of Natural History, estimates that the monster did not exceed 6.85 m in length. Alligators (*Alligator mississipiensis*) and the Orinoco crocodile (*Crocodylus intermedius*) can both exceptionally reach a length of 7 metres. Marine turtles on the other hand are certainly the largest: the leatherback turtle (*Dermochelys coriacea*) sometimes reaches 2.75 m and weighs up to 700 kilos. The giant tortoises of the Galapagos Islands are much less impressive, hardly ever tipping the scales at more than 300 kilos.

Finally, among mammals, elephants, the heavy-weight champions on land, look puny compared to the largest cetaceans. The blue whale (*Balaenoptera musculus*) can weigh as much as a herd of thirty adult elephants. Even *Baluchiterium*, the most gigantic land mammal of all times - a kind of rhinoceros taller at the shoulder than the horns of a giraffe - must have weighed ten times less than a large whale, tipping the scales at about twice as much as a new-born baby whale. Just as an example, an 8 metre long, 7 ton foetus was once taken from the body of a 24 metre long female blue whale. A baby baleen whale is already heavier before its birth than the largest adult elephant!

One should not imagine that the largest inhabitants of the sea were necessarily the first to be found. It is chance above all, as we have said, which usually guides discovery. The most persistent will has little say in the matter. The ocean is reminiscent of those small remote-controlled cranes found in amusement arcades, with which one tries to catch objects of various degrees of attraction; it seems quite easy to catch a large, prominently placed and desirable prize such as a camera, but in spite of the most persistent efforts, one never catches anything but a few mint candies or some useless trinkets.

A few details regarding the discovery of some marine giants will prove

---

[3] *It's somewhat difficult to be very precise about the size of worms, which are so elastic. A 7 metre individual can in the interval of only a few seconds stretch up to a length of 30 metres!*

instructive. This book relates the progressive, and extremely slow growth of our knowledge of the largest known invertebrate, a creature with a mass exceeding that of the largest land animals. It is only at the end of the last century that we began to believe in its existence...

The largest of all fish, the whale shark *(Rhineodon typus)* was only discovered in 1828. A specimen had been harpooned in Table Bay, South Africa: chance had it that it could be examined by Dr. Andrew Smith, a military surgeon as well as an expert in South-African fishes. It took another twenty years before a description of the giant reached western science. It has thus been only a century that we have known of a fish larger than any boastful fisherman ever dreamed of.

Our acquaintance with the largest ray, the Manta Ray, or Great Sea Devil *(Manta birostris)*, dates from the same period. It was first described scientifically by Dr. Edward Bancroft, after a specimen was captured the previous year in the port of Kingston, Jamaica, by Major General Sir John Keane. Although that terrifying monster, with a wingspan of up to 8 metres, was already known, its nature had up to then remained mysterious. In his *Relacion Historica del viaje a la America meridional* (1758), don Antonio de Ulloa, lieutenant general of the Spanich navy, had already alluded to a monstrous fish, named "manta" which attacked divers between Panama and Guayaquil: " as long and wide as a bed-spread, it wraps in its fins any creature within its range and chokes it immediately". Later a colonel Montagu had already tentatively suggested that the manta might be a ray. We shall see later that in spite of Dr. Bancroft's meticulous description, a well-informed traveller could still, forty years later, mistake a manta for some kind of white-armed siren.

The largest specimen of *Manta* whose measurements could be verified barely exceeded 5 metres in wingspan. Harpooned in 1919 at Bimini, in the Bahamas, it weighed 1,500 kilos. These monstrous rays are totally harmless. The only danger comes from the habit which they have of leaping above the surface. The loud noise which they produce when falling flat on the sea surface is quite terrifying. It has also happened that wounded mantas have fallen on the boats of fishermen careless enough to attack them. A ton of angry muscle can be quite damaging!

Most significant is our lack of knowledge of cetaceans, among which one finds the most enormous animals ever seen on our planet. Contrary to popular opinion, the most gigantic dinosaur of the Jurassic age never got close to the weight of today's largest whales: the largest baleen whales and sperm whales. Estimates suggest that the largest reptiles of prehistory did not exceed 50 tonnes in weight, while today's whales can reach 150 tonnes.

Among whales, the giants, exploited by man on an industrial scale, especially for their oil, have by now become familiar. However, might there not exist others? It would be rather bold to deny it, since many of the larger cetaceans have been discovered only in the last century and that many others are still very poorly known.

No other family of ceataceans remains as mysterious in our eyes as that of the Ziphiids, or Beaked Whales[4]. Of its six genera, only one, the genus *Hyperoodon* is rather common: it was named by Lacépède in 1804 and was the only one known at the beginning of the last century. There seems to exist several forms of *Hyperoodon,* with a length approaching 10 metres, but the most distinctive one is known practically only from skulls found on the shores of the southern seas. No *Hyperoodon* has ever been caught outside the North Atlantic.

Cuvier's whale, *Ziphius*, which gave its name to the whole family, was first thought to be an extinct species. All that was known about it was a skull in the process of lithification discovered in 1804 on a Mediterranean shore; Cuvier described it in 1823 as a fossil. Nearly half a century after the harvest of this first piece of anatomical evidence, another, just like it, was found stranded at the same place. It was only later that there was found in New Zealand a live cetacean with an identical skull. Curiously enough, the skull nearly lithifies in older individuals, literally beginning to fossilize. The animal also appeared in other respects as a joke played on zoologists. White above and dark below, contrary to all rules of marine animal coloring, it was difficult to accept as a real animal.

The genus *Mesoplodon* is perhaps the most enigmatic of all. It was first discovered through a carcass cast onto the coast of Elgin, Scotland, which the naturalist Sowerby had the good fortune to examine. It was a brownish animal, with a strangely curved lower jaw carrying only two teeth. In 1825, a similar animal, but without any teeth at all, was stranded alive on the beach at Le Havre. It remained alive for two days and attracted quite a crowd. The onlookers, drawing on a rather personal conception of the diet of a whale, offered it bread soaked in water and other similarly unorthodox food. To express its anguish, the poor beast emitted a kind of cavernous cry resembling the mooing of a cow. Because of its lack of teeth, it received the name *Aodon dalei*. Some specialists claimed that it might have been an aged specimen of Sowerby's whale, toothless as is appropriate to the very old. It could also have been a female, as these were later often found to be toothless.

In 1850, the same whale received from Paul Gervais the name *Mesoplodon bidens*, which it has since officially kept. Since then, a number of *Mesoplodon* species have been noted, but the descriptions are based on such poor material that it is difficult to tell whether there are 10 or 15 distinct species. One of the rarest is Gervais' Whale (*Mesoplodon europaeus*), known through only six specimens. The first of these was found floating in the English Channel in 1840; however, the next three stranded on the coast of New Jersey, in the USA (in 1889, 1933 and 1935) and the last two, a female and its young, were cast ashore in Jamaica, in 1953... a rather embarrassing event for an animal baptized "European". Blainville's Whale *(M. densirostris)* has an even more fantastic geographic distribution. Only seven specimen have been found, but they were caught in the most disparate parts of the globe: in the Seychelles; at Lord Howe Island, south of Africa; near Massachusetts; in Madeira; and off New Jersey. *M.stejnegeri* is known only from two specimens from the Pacific coast of North America; *M. hectori* from two specimens from New Zealand. The description of

---

[4] *These whales are really dolphins, since the name whale ought strictly be reserved for baleen whales.*

M.bowdoini is also based on a pair of New Zealand specimens, but these were only skeleta, so that we know nothing about the external appearance of the animal.

Needless to say, we know next to nothing about the behaviour of the various *Mesoplodon* species, some of which sometimes exceed 5 metres in length.

Little more is known about the two species of the genus *Berardius*: Arnoux's and Baird's whales, described respectively by Duvernoy in 1851 and Stejneger in 1883; these can reach almost 13 metres. The dentition of Baird's whale is such that no zoologist would have believed it, had it been described by a non-specialist: its teeth are encased in a cartilaginous sleeve from which it seems that they may be extruded at will to act as hooks.

It is expected that new species of Ziphiids continue to be discovered from time to time. In 1937, Oliver had to create a clearly distinct genus, *Tasmacetus*, following the stranding of three beaked whales of a completely new type on the coasts of New Zealand. These were not small fry, measuring from 7 to 9 metres in length! All this happened, let us remind ourselves, not in the Middle Ages or in the Renaissance, but just before the Second World War. More recently, in 1958, *Mesoplodon ginkgodens* was described, from Japanese waters; another beaked whale, was discovered in California in 1966: *Mesoplodon carlhubbsi*; and yet another, *Mesoplodon peruvianus*, off the coast of Peru, in 1991. Nevertheless, on still hears in cocktail conversations cocky assertions that the sea cannot possibly hide any more large unknown animals.

Given the rather confused situation which still prevails among the beaked whales, it is not surprising that among the dolphins, cetaceans of a lower calibre, many species should remain known only through a few rare specimens, even sometimes a single one.

There are, for example, many species of the genus *Lagenorhynchus*, also known as whiteside dolphins, whose external appearance remains quite poorly known. Some seem uncatchable, for example, Wilson's hourglass dolphin (*Lagenorhynchus wilsoni*):

"This kind, wrote Francis C.Fraser in his classical work on the Giants of the Sea, written in collaboration with the ichtyologist John R.Norman, was frequently seen just north of the ice pack in Antarctica, during the Discovery expedition as well as during that of the Terra Nova. None has ever been captured."

About the spotted Dolphin of the genus *Prodelphinus*, named as early as 1889 by Paul Gervais, Fraser adds that:

"A host of species has been described, about none of which is there much information concerning habits. Many of the forms are known only from skulls..."

The sperm whale family (*Physeteridae*) includes an extremely rare and peculiar genus, the dwarf sperm whale, (*Kogia*), described as late as 1846. This 3 metre long animal has never been seen alive. Only a few random strandings have revealed its existence. We don't know to this day the number of species included within the genus *Kogia*; half a dozen have been described, but its not clear, given the paucity of the evidence, whether these might not be individual variations within the same species.

The dwarf of the baleen whales has been known only since 1864. It is a strange animal whose ribs are so spread out that its chest looks like a barrel. Even today, it is extremely rare to glimpse a dwarf whale (*Neobalaena*), although this "midget" can be as long as 18 metres!

There is thus an impressive number of rather large cetaceans about which we know very little or which we are just beginning to discover. Some are known only by sight. How many more might there exist whose very existence is still unknown? How many more marine animals are there, from various groups, which because of their anatomy, physiology or life style are unlikely to end up stranded on a beach? Most animals do not float, once dead; unless they live near shore, they are not very likely to be cast onto the beach or onto the rocks.

In spite of the widely recognized incompleteness of our census of the oceanic fauna, the bitterest skepticism surfaces as soon as the subject of sea "monsters" is brought up. This is because these are always described in the most fanciful terms. Rather than adopting a purely negative attitude, one should perhaps ask whether there aren't *a priori* reasons for such descriptions.

First of all, marine creatures always seem more or less fantastic because they are adapted to a medium which is different from ours: they appear odd from *our* point of view. To quote Montaigne: "What we call monstrous is not so to God, Who sees in the vastness of His work the infinity of forms which he has created."

We have also emphasized fanciful aspect of marine denizens by giving them inappropriate names. Lacking reference points in our own world, we have seen in the creatures hiding behind the wavy curtain the counterparts of every terrestrial and celestial object imaginable. We have peopled the ocean with spiders and scorpions, mice and hare, calf, cows and pigs, cats and dogs, lions and tigers, wolves and bears, elephants and horses, men and women; we have planted in the sea anemones, lilies and nettles; we have grown in it grapes and cucumbers; we have scattered in its midst stars, moons and suns. Then, trapped by own words, we have been surprised to see the stars and anemones eating each other, the cucumbers crawling on the bottom, the grapes giving birth to tentacled devils, the hare moving more slowly than the turtle, women and cows fitted with a fish tail.

It is to the poverty of our vocabulary, to the very mechanisms of our language - which introduces, through words, artificial discontinuities in a continuous universe - that marine monsters owe much of their incredible appearance. We are incapable of describing an unknown animal except by taking it apart piece-by-piece to compare them to those of known creatures: this method is bound to yield heterogeneous monsters. Such a description allows us to recognize the animal when we see it, but it provides only a vague idea of its appearance before we are familiar with it. Language is sometimes a remote abstraction of reality and cannot replace direct sensory observation. Nothing is more satisfying than a concrete representation, a sketch. Unfortunately, there is rarely an

Figure 1. The fabulous mermaid and its prototype: a Sirenien (manatee)

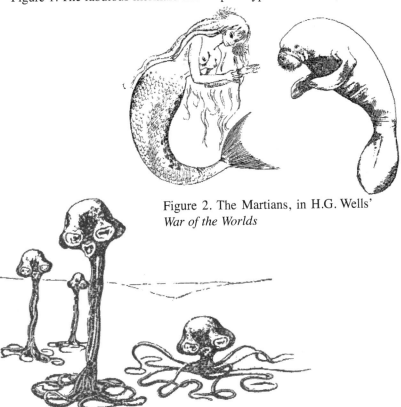

Figure 2. The Martians, in H.G. Wells'
War of the Worlds

Figure 3. The octopus, depicted as a fish with legs by Sebastian Munster, 1556

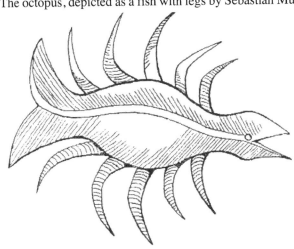

Figure 4. The coelacanth, a refugee from the Devonian (Photo Le Cuziat-Atlas)

Figure 5. Prof. J.L.P. Smith (1897-1968) with the author in 1967 (Photo B. Heuvelmans)

Figure 6. Whiteside dolphin (*Lagenorhynchus*)

Figure7. Whale-shark

Figure 8. The blue whale, compared to a Baluchiterium and to a man

Figure 9. A 13 m long whale shark, weighing eight tonnes (Photo *New York Zoological Society)*

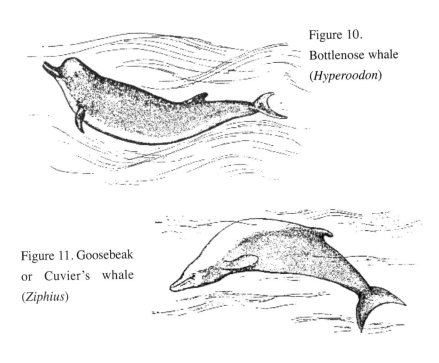

Figure 10.
Bottlenose whale
(*Hyperoodon*)

Figure 11. Goosebeak
or Cuvier's whale
(*Ziphius*)

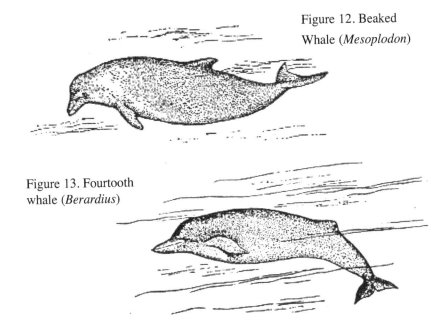

Figure 12. Beaked
Whale (*Mesoplodon*)

Figure 13. Fourtooth
whale (*Berardius*)

Figure 14. Hercules, in battle with the hydra of Lerna, on a Greek amphora of the Louvre Museum Paris
(Photo Chuzeville)

Figure 15. *Neopilina* (Photo Henning Lemche)

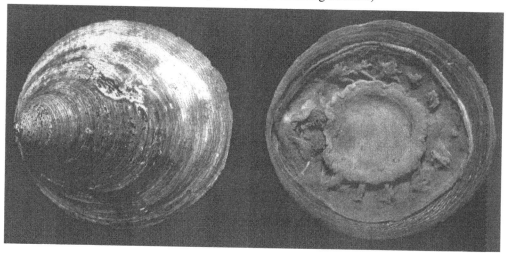

Figure 16. Internal anatomy of the octopus

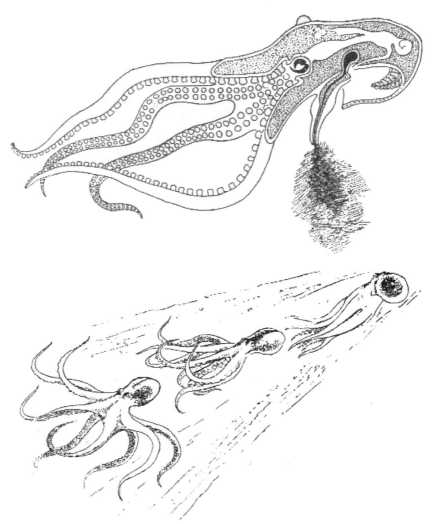

Figure 17. Swimming motions of the octopus

Above: Figure 18. The author, at the Ile du Levant, struggling with an octopus with a 1.20 m span (Photo L. Demol)

Below: Figure 19. A duel between an octopus and a shark (Film R.K.O.)

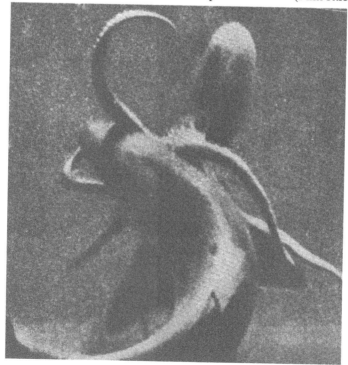

illustrator or a talented painter on hand at the appearance of each new animal. The description of the unknown is usually through a verbal portrait. That is where the worm of confusion crawls into the zoological apple, for this process is deeply flawed; the more vague and coarse the original description, the least faithful the results which it yields.

Describe a walrus as a large seal with elephant tusks and a good artist will be able to create a reasonably satisfying, and certainly recognizable, likeness on the basis of this simple portrait. If however, in the manner of some ancient naturalist, you speak of the octopus as an eight-legged fish, you are likely not to recognize the animal when you see the drawing. It is a fact that very few people are reminded of an octopus when they perceive, on a map of Sebastian Munster's *Cosmographia Universalis* (1552), where are assembled all the monsters of the northern seas, a kind of large scaly fish with ten spider-like legs. However, this sketch of a rather surprising creature was taken by the cosmographer from Ryff's (1545) German translation of the work of Albert the Great, which undoubtedly relates to the octopus.

Can we blame the artist who illustrated this work in the vernacular, meant for a broad public, of having misrepresented the mollusc? Certainly not, because the author himself referred to it as a marine fish with eight legs. And can we blame Albert of Bollstadt, the most pretigious scientist of the Middle Ages, for having defined the octopus in this fashion, when in his time every kind of marine animal, be it jellyfish, starfish or whale, was called a fish?

Distortions arising from unfortunate mis-interpretation by the illustrator can have far-reaching effects; they are often caused by some excessive care on his part. It is not surprising to see marine monsters dressed up in a coat of scaly mail. Fish are the most characteristic, certainly the most striking of marine denizens and they are almost always covered with scales. So, as soon as a description mentioned the marine nature of an animal, the keen illustrator never failed to cover its skin with scales. Even in those cases where the artist might be doubtful about the presence of scales, it was a symbolic fashion of explaining the marine nature of the animal. One might think that all that needed to be done was to draw the animal bathing in its natural environment. Yes, of course, but water is not easy to draw. That is the crux of the problem. A similar convention affected the illustration of the first water closet, from the sixteenth century, where the water tank is drawn full of fish[5]. Obviously, no one ever thought of using as an aquarium a container made to be flushed regularly down the sewer drain. But how else to show clearly the original aspects of the invention?

Everything is conventional in pictorial representation. Even in a colour photograph, which might seem the epitome of fidelity, the actual size of the objects is left entirely to the judgment of the viewer, as are their relief and distance. At each period, in each civilization, people have known how to apply to illustrations the appropriate mental filters, in conformity with the prevailing

---

[5] *This drawing is to be found in a book called* A New Discourse on a Stale Subject, Called the Metamorphosis of Ajax *(1556) due to the pen of the inventor of the device, Sir John Harrington, nephew of Queen Elizabeth.*

conventions of the time. All that is required is to know them!

Thus, an experienced zoologist of an earlier time was not fooled by the distortions and even the most outrageous embellishments brought in by observers and illustrators. Take for example the rather subtle commentaries expressed by Guillaume Rondelet in his *Histoire entière des Poissons* (1558) on the representation of a "leonine sea-monster":

"The monster here portrayed is a perfect animal without any parts appropriate for swimming. That is why I have often doubted that it was a marine monster, but I was assured in Rome that such a beast was caught in the sea shortly before the death of Pope Paul III, and as they vouched for its veracity, I had it portrayed as shown[6]. It was of the shape and size of a lion, with four well formed feet, without webbing between its toes as one finds in beavers or river ducks, but divided in fingers with claws, its long tail hairy to the very end, large ears and scales all over its body. It did not survive long outside its natural element. Although this description was made to me by well informed people, and in good faith, it seems to me, that the artist may have added some improvements of his own which have taken away from nature: the feet are much longer than they normally are in marine beasts; he may also have forgotten about the webbing between the toes. The large ears are not in keeping with aquatic beings, and scales are shown instead of rough and horny skin such as covers the feet of sea turtles; for all beasts which breathe through lungs and are held together by bones do not have scales. With many other marine beasts and monsters artists add and remove much, as may be noticed with the whales painted on the northern charts of the Cosmographia of Munster, as may also be noted with the sea calf, the orca and others."

What this "leonine sea monster" was, it is difficult to tell. But there is a good chance that it might have been some maned sea-lion, extraordinarily lost in the Mediterranean. At first sight described as a sea lion, by which name we still know it, the animal was probably soon described, second hand, as an authentic lion. But, after depriving it of the webbing between its toes, and of its imperfect feet, what was left of its marine origin? To restore its marine nature and its prestige as a monster, it had to be decorated with scales. That's all there was to it!

If marine creatures are often described in some fanciful manner, which immediately provokes skepticism, it is because they are almost always observed fleetingly and imperfectly. They are either glimpsed in a flash, as they pierce the water surface, or discovered stranded on a beach, mutilated or decomposed.

---

[6] *In the Appendix to* The Book of Monsters *by Ambroise Paré, one finds more detailed information on this animal:*

"This monster in the shape of a scale-covered lion was caught in the Thyrrenian Sea, near the town of Castres, and was presented to Marcel, then bishop, who later succeeded Pope Paul III after the latter's death. This lion had a voice similar to that of a man and was much admired in the town, where he died, having been taken from its natural element as is related by Philippe Forestus, in Book III of his chronicles."

BELOW THE WAVY CURTAIN

It is easy to chuckle at the naivety and extravagance of ancient animal illustrations and even of the zoological descriptions of yesteryears. One should not conclude that artists of the time were less skillful than those of today - who have never surpassed Dürer and Pisanello, or some ancient oriental artists- or that naturalists then were not as good observers as they are today. In fact it is simply that the information then available was much more meager so that imagination had to fill in the gaps in missing knowledge.

These days, when an intelligent and well-informed person apprehends at sea some unknown animal, he is still likely to depict it in a most misguided fashion. In 1866, at a time when the *Manta* had already been known to zoologists for nearly forty years, don Enrique Onffroy de Thoron, an eminent South American philologist, met up with one of these gigantic rays in the Gulf of Ancon de Sardinas, between Esmeraldas and the Rio Mira, off the coast of Ecuador. In order to understand the dramatic description which he left of the incident, it is important to recall that, in the manta, two appendices of the pectoral fins appear on either side of the mouth as long flexible, slightly bent horns. This prominent feature explains various names that have been given to this ray: Cephaloptere, giant horned ray, Sea Devil and finally its official name, *Manta birostris*. Among the largest specimens, these horns are nearly a metre long. Viscount Onffroy de Thoron, completely unaware of the existence of such strange appendices in a marine animal, saw in them much more familiar organs:

"Suddenly there rose from the bottom of the ocean a marine monster, remarkable by its strangeness; it came to lie so close to our whaler than we could have hit it with an oar. To escape it, I would have needed two more rowers; the two at my disposal were already exhausted by their work. Under the circumstances, I decided to act with prudence, not knowing whether this visitor had peaceful or hostile intentions, and I had the oars pulled in so as not to excite it. We stayed on the spot. The pilot, upon seeing the beast, said to me: "Sir, it is a Manta; take your machete and if it tries to grip the boat, cut its hand." I saw that we were dealing with an amphibian. With raised arm, and machete in hand, I was ready to strike the monster parked alongside. However, not willing to initiate the attack, I undertook a detailed examination of the animal, in as much as this was possible while I watched its movements to make sure we were not caught by surprise and made to capsize."

"The manta had real human arms; they were white and about a metre and a half long, but very thin compared to the length and width of the body; in addition, they articulated as do ours, at the wrist, in the middle - but in a more rounded manner than our elbow - and at the shoulder. Its hands were small and slightly curved and the colour of old parchment rather than white, which gave them the appearance of being dirty. Its fingers, sharp and poorly separated, were probably webbed or cartilaginous; I couldn't tell for sure because at that time they seemed glued to each other. The head of the manta was extremely flattened in the horizontal; it was triangular in shape and broadened towards the shoulders; at its base, it was more than two feet (60 cm) wide and its mouth, which it kept closed, was fully as wide as the head. Its body was rather thin (only a few centimetres), but no less than four feet (120 cm) wide in the horizontal; its back was flat and

uniformly wide and that part of the animal visible above the surface was about three metres long and without fins. The size of the rest of the body, which continued under the water at an angle of 20°, could not be estimated. Its flesh or its skin was white, and in the middle of its back there was a line of spots as seen on the sea-calf or on leopards. Was its skin of the nature of a frog's or of a sea-calf's? That, I don't know. I saw no appearance of hair or scales on its arms, its head or its body. The spots seen along its vertebral line would be no proof of the contrary, since amphibians and fish almost always have markings on their back. In any case, my pilot assured me that he had seen mantas which were completely white; however, in order to stick to what I actually saw and not to be led into error, I did not wish to interrogate him. This is why I cannot tell whether the manta is simply a quadruped, like the frog, or whether its body ends with fins."

After having interpreted in this most romantic manner that small part of the animal, which he had seen at the surface, don Enrique launched into a rather bold hypothesis on the nature of the monster:

"Had I perhaps found in the manta a living chirotherium, a frog reaching 12 to 15 feet (3 to 5 metres) in length, which geologists have found as a fossil? For me, there is no doubt that both are varieties of the same family of batrachians."

Actually, for paleontologists, it was rather doubtful whether *Chirotherium*, the beast with hands, was any kind of amphibian. At that time, all that was known about it was a set of fossilized tracks found in 1833 near Hildburghausen, in Germany. It has been determined more recently, following paleontological discoveries in North America, that it was more likely to be a reptile, a kind of dinosaurian.

To confirm that today's illustrators are no less naive nor more faithful than in the past, one should see the engraving which illustrates don Enrique's story, quoted in the 1870 French edition of B.H. Révoil's *Shooting and Fishing in the Rivers, Prairies and Backwoods of North America*. The *Manta* is shown as an exquisite maiden with outstretched and inviting arms, truly a rival of the most charming sirens of legend.

This modern story is at least the equal of the eight-legged fish, the armoured whales and the scaly lion of the Renaissance.

There are objective, as well as subjective, reasons behind the fanciful descriptions of the unknown animals of the sea. It is important never to be entirely set back by what seems most incredible in this matter. The essential is to hold back judgment, ready to adjust and rectify later as required. With time, everything falls into place; the dimensions of poorly seen animals shrink down; the deformations introduced by illustrators are corrected, the embellishments fall away.

It matters little to the objective existence of whales that they were formerly described as fully armored as jousting horses, or embellished with the horns of triceratops and armed with teeth such as to frighten the bravest, and that their size was exaggerated to unbelievable proportions. Their industrial exploitation has lowered them to the trivial level of meat cattle. When we now speak of them as monsters, we only refer to their size. The halo of terror and strangeness that surrounded them has forever vanished.

This book and its companion volume *In the Wake of the Sea-Serpents*, are about the history of those sea monsters which have not quite lost all of their mystery. They are divided in two parts: the first deals with the tentacular monsters known to legend as the Kraken and the Colossal Octopus; the second collects all the fables and reports pertaining to the Great Sea-Serpent. Although these two stories are in practice independent, they are curiously parallel in their development.

More than two thousand years ago, Aristotle had this to say about animals which did not fit into the corpus of knowledge of his day:

"Beyond the animals studied to this point, there are also some in the sea that one cannot place in categories because they are too rare. Among experienced fishermen, there are some who pretend to have seen in the sea animals similar to beams, black, round and everywhere of the same thickness. Some other animals resemble a shield; they are reddish in colour and have numerous fins."

These are, in my view, clear allusions to the Great Sea Serpent on the first hand and to the Giant Squid on the other. Briefly juxtaposed in this terse statement, they later received a fuller treatment by the Scandinavian bishop Pontopiddan, who was the first to research them thoroughly. Nourished as twins at the breast of the erudite prelate, the Kraken and the Great Sea-Serpent might well be thought of as milk brothers.

Both of these monsters have followed a similar evolution since the remotest antiquity. Both were first known as fabulous creatures through rumours, heroic accounts and superstitions. Described later with somewhat more sobriety and attention to detail in a few isolated reports, which gradually multiplied into a body of convincing evidence, they have nevertheless both continued to be denied and ridiculed, called improbable, and judged to be impossible. In the case of the Kraken, even the undeniable evidence of anatomical fragments of gigantic cephalopods did not suffice for a long time to sway general public incredulity. Let's not be surprised then if the Great Sea- Serpent, about which the collected evidence still remains rather meager, continues to be an object of ridicule. The imaginary character of the animal has become the butt of classical jokes, its very name synonymous with hoax.

It is precisely because these two monsters, the most famous of the oceanic fauna, have until recently had parallel careers that I will endeavour to relate in some detail in this book the fascinating and very poorly known history of our knowledge of giant cephalopods. This first story has a happy ending: the mystery was finally solved, at least in its broad lines. Why shouldn't then the other story, that of the Great Sea Serpent, similar in so many respects, which I have treated in another book, not meet with a similar ending?

Chance alone has dictated that the mystery of the giant cephalopods be the first elucidated, at least partly. The giant squid today finds it place in zoology textbooks; it has a Latin name which one may express without fear of ridicule in academic surroundings. Had the dice of nature landed in a different manner, perhaps the Great Sea-Serpent would appear today as a legitimate entry in learned tomes under a variety of scientific names while the Kraken would still remain an item of folkloric legend.

Part Two

ANIMALS WITH FEET
ON THEIR HEAD

"It is quite necessary that someone should restate, with emphasis, following in the steps of Tolstoi, that besides their famous masterpieces, geniuses have also written inferior works, insipid, nonsensical and barely readable; it should also be proclaimed that the majority of the masterpieces of the human spirit are much over-rated. It is such an important truth that a writer shouldn't die without having expressed it.

"Whatever the consequences, which may be quite serious. Imagine for example, expressing the truth about Moliere in France. Or, in their own countries, about Shakespeare, Cervantes, Goethe! One's safety, even the safety of one's private life, will suffer..."

" The absence of critical sense, the absence of any sense at all, the lack of appreciation for the truth, the cowardice (to state it somewhat differently), snobism, professional bias: all these contribute to the uncompromising adoration of the "masters".

"Forced admiration is counterbalanced by forced injustice. Only one thing matters, which is nothing short of heroic: to see things as they really are."

HENRY DE MONTHERLANT
Carnets

CLASSIFICATION OF THE CEPHALOPODS
(This list of families comprises only those which include species mentioned in the text)

Sub-class PROTEROCEPHALOPODA

       Order Nautiloidea   (nautilus)

       *Order Ammonoidea   (ammonites; all fossil)

Sub-class METACEPHALOPODA

       Order Decapoda   (with ten arms)

         *Sub-order Belemnoidea   (belemnites, all fossil)

         Sub-order Sepioidea   (cuttle-fish)
             Family Spirulidae   (with coiled shell)
             Family Sepiidae   (cuttle-fish proper)
             Family Sepiolidae   (with lateral fins)

         Sub-order Teuthoidea   (calamaries or squids)
             Family Loliginidae   (coastal squids)
             Family Ommastrephidae (giant squids)
             Family Architeuthidae   (super-giant squids)
             Family Onychoteuthidae (clawed squids)
             Family Enoploteuthidae   (armed squids)
             Family Lepidoteuthidae   (scaly squids)
             etc, etc

       Order Vampyromorpha   (ten-armed ciliated octopus)

          Family Vampyrotheuthidae

       Order Octopoda   (with eight arms)

         Sub-order Cirroteuthoidea   (ciliated octopus)
             Family Cirroteuthidae
             Family Opisthoteuthidae
             etc..

         *Sub-order Paleoctopoidea   (fossil octopus)

         Sub-order Octopoidae   (non-ciliated octopus)
             Family Octopodidae   (ordinary octopus)
             Family Argonautidae   (argonauts)
             etc, etc...   *   Extinct   order   or   sub-order

Chapter 2

ON THE AGRESSIVITY OF THE OCTOPUS, IN LITERATURE AND IN THE OCEAN.

There exist fabulous monsters whose existence is well established in fact, but from which the most striking proof will never remove the aura of legend. The horror which they inspire is so intense, their structure so weird from a human perspective, and their size so large that, in spite of scientific acceptance, they will forever remain stuck in the mire of the fantastic and the surreal.

The *kraken* of Scandinavian fishermen, the heir of the awful Skylla of Homeric legends, is one such creature. Even after removing the exaggerations, reducing it to its real proportions, and putting aside naive embellishments, it will continue to be a creature of nightmare. The terror which it inspires has communicated itself to the more modest representatives of its group: the octopus, the smaller squids, and other many tentacled molluscs which science has grouped under the collective name of cephalopods: those who have feet on their head. Hieronymus Bosch could not have dreamed up a stranger anatomy for the creatures inhabiting his scenes of Hell. Wells clearly used them as a prototype when he wished to make his Martians as frightful as possible.

Not so long ago, navigators still feared that these giants of the infernal hordes might threaten the safety of their ships. The days of superstitious terror, born out of the first contacts with an unknown world, have now passed. The skeptics have gained the upper hand. After all, they reasoned, are these bogeymen of the sea as dangerous as all that? Exaggerating in the opposite direction, they have come to deny that they would even have the audacity of attacking a lone person. Some modern scuba divers go as far as thinking of them as playmates. What prosaic decadence, what lackluster fall for these oceanic demons!

Nevertheless, some general level of terror does remain. There are good reasons to be alarmed, and those who would deny with such confidence the agressivity of the octopus might well be behaving like scared little children, who reassure themselves by whistling in the dark...

Are cephalopods voracious ogres or timid phantoms? On this point, the scientific literature is split. It will thus not be without some difficulty that we shall reach the truth, through thickets of contested facts and contradictory opinions. A truth "à la Pirandello", with as many faces as the octopus has arms.

Today, nobody can imagine that a giant cephalopod could drag a large ship into the abyss. However, if one is to believe the films or contemporary novels where the action takes place on the sea bottom, it would appear to be impossible for a diver to withdraw sunken treasure from a wreck, or to return with a prize pearl, without having first given battle to some giant octopus. Even in those stories with some pretension of authenticity, such a struggle seems necessary: without it a documentary would be too boring for the usual thrill-seeking audience. What do people want? Today, just as yesterday, they like to see their hero - Hercules, Gilliatt, Captain Nemo or Superman - triumph over the seven-headed or eight-legged Hydra, the secret personification of Evil.

There are many instances of such duels. To plunge the reader in the climate of horror traditional to these events, I have selected a single example, from one of the most popular accounts. My choice of a chapter from *Treasure Hunter* (1945) by Lieutenant Harry E.Rieseberg is motivated by his claim of relating actual facts, the personal reminiscences of a famous American hard-hat diver, specialized in the salvage of sunken treasure.

A few years before the event, a Spanish schooner, loaded with silver bars, had hit a reef and sunk off Malpelo Island, Columbia. Seven times already divers had tried to recover the precious cargo, but none had returned to the surface. A curse seemed to hang on the wreck, partly buried in the sand at a depth of 65 metres. Completely undaunted, and obviously lured by the prospect of a handsome profit, fearless Rieseberg did not hesitate to dive and try his luck.

After quickly locating the wreck, our diver first stumbles upon the skeleton of one his predecessors, still wearing his hard copper hat and a torn diving suit (`which looked like a pair of pants that a puppy had chewed'). However, he has to rush back to the surface, having discovered that his air hose has been mysteriously cut.

In spite of this warning, Rieseberg dives again, two days later. His courage is rewarded: he discovers the treasure, as well as a bronze idol at the feet of which are piled up human remains. This gives him something to mull over!

"Caught in my thoughts, I had the strange and sinister feeling of being watched. So strong was the impression of some presence near me that I turned around and pointed the beam of my light around the room. And then... Great God! From behind the rather vague contours of the great bronze statue, a gigantic form rose before my eyes. As I sought to pierce the shadows, a shiver ran through my spine, for there, shifting its bulk across the door... and cutting any hope of retreat, was a creature which could have featured in the worst nightmares of a demented mind or of a hashish smoker. A disgusting body, covered with warts, was oscillating slowly, twisting from one side to the other, continuously shivering and writhing. The enormous monster had a span of at least 4.5 m, with a rounded barrel-like body over a meter across. Its long viscous arms, or tentacles, were covered with suckers as large as a saucer, of which there appeared to be hundreds.

"It looked like a creature out of hell. Its enormous swollen body was mottled with patches, which slowly shifted from brown to dirty yellow, to clear marroon and then to grey and white. Its eyes were those of a demon, which seemed to watch my slightest movements, like a vampire.

"It was horrible!" concluded our treasure hunter, and one could hardly disagree with him.

There then begins a terrible struggle, during which Rieseberg manages to sever with his knife three of the monster's tentacles. What happens is that the latter is stupid enough - in spite of our hero's reference to its "diabolic cunning" -

ON THE AGRESSIVITY OF THE OCTOPUS,
IN LITERATURE AND IN THE OCEANS

to attack his opponent with only one arm at the time, in the fashion of a fencer: it's hardly worth having eight of them! When the beast finally decides to behave like a normal octopus and to finish up with our submarine Tarzan, he plunges his steel blade in the "only vulnerable part of the octopus' body: the jugular vein [*sic*]."

Before expiring, the beast still has the strength to shake up his ennemy like a child's rattle and to tear up his diving suit as well as some of his skin. Bleeding and suffocating, Rieseberg looses consciousness while thinking about the horde of sharks which his blood is bound to attract.

Will he be spared no horror?   Well, yes. He wakes up in the ship's decompression chamber. It turns out that his associate on board had sent down two native divers who untangled him from the grasp of the dead monster, turned off the air valves of his ripped diving suit, and cut his safety line, entangled in the wreck.

Rieseberg doesn't wax lyrical about the rescue operation, carried out by un-assisted divers at a depth of *64 metres*. *That* is undoubtedly the most extraordinary part of the whole adventure, so extraordinary in fact that one is justified in wondering if the whole story, sprinkled as it is with a number of suspicious details, might not just be one very elaborate and boastful fabrication.

Never mind! I am not interested here in assessing the authenticity of a report on the private life of the octopus and on its relationships with people. What I am trying to depict is how people perceive these relationships. If few people even think of casting doubt on the authenticity of the above story, it is because of its classical character: after all, who would doubt what is normal and usual?   The incident is characteristic of the literature on submarine adventures and faithfully reflects public opinion on a creature which is often referred to as *devil-fish*.

In the mythology of our time, the giant octopus, guardian of sunken treasures, has replaced the dragon of medieval times.

Two books are particularly responsible for the popularity of the octopus as an underwater scarecrow: *The Toilers of the Sea* by Victor Hugo, and *Twenty Thousand Leagues Under the Sea*, by Jules Verne. Moreover, it is these purely fictional stories which, either directly or through their derivatives, provide the basis of the public's knowledge about the anatomy and behaviour of cephalopods.

Strange reference works these are! The first, from an undoubtedly prestigious poet whose verbal excesses and immoderate taste for pathos often led to the worst fabrications; the other from the adventure novelist who charmed the youth of many generations, but whose scientific culture was entirely superficial flim-flam. The broad public, quite naturally, would rather read novels than scientific magazines, so often presented in unintelligible jargon. So, in the first of these books, the man in the street heard about the fundamentally evil nature of the octopus, while he discovered in the second just how enormous these animals could be. It would be difficult to be more badly misinformed!

Let's first have a look at these immortal masterpieces, starting with the first.

The famous battle between Gilliatt, the courageous Guernsey diver, and the octopus of the Douvres rocks, an episode to which Hugo devotes no less than three chapters of *The Toilers of the Sea*, is perhaps one of the most perfect examples of what we may call `literature', in the most pejorative sense of the word. Here, words are no longer at the service of ideas (we too often forget that this is indeed their natural role): verbal associations - the falsest, the most extravagant and absurd - are the seed of cascades of hyperboles and of comparisons as treacherous as they are admirable; a whole philosophy is built up on premises of pure fantasy. Readily fascinated by the kaleidoscope of rich imagery, the reader is gradually lulled by the sound of assonances; his critical sense is put to sleep by the magic of verbal mirages, flattened by the barbaric gallop of sonorous harmonies; the word is king, even tyrant! All is melodious garglings and verbal legerdemains. One finishes out of breadth, stunned, amazed. Just as after a laser-light show!

Let's wake up, shake off our sleepiness and return to Earth! Let's try to analyse cold-bloodedly what the Poet-Illusionist has told us.

"To believe in the octopus, one must have seen it", writes Hugo. To believe in what he describes to us, one would have never to have seen one. The father of the *Miserables* nevertheless invites us to a regular lecture on natural history and has no hesitation to quote along the way, to emphasize the seriousness of his words, those of Bory Saint-Vincent and Denys-Montfort, to praise Bonnet, to criticize Buffon and refute Lamarck. His preamble is certainly gripping:

"Compared with it, the hydras of old were insignificant"...
 "At certain times one is tempted to imagine that the mysterious forms which float in dreams meet, in the realms of the possible, with magnetic points to which their lineaments adhere, and from these mysterious centres of a dream beings are created."...
 "The Unknown has the marvelous at His disposal, and out of it He makes the monster."...
 "Orpheus, Homer and Hesiod were only able to imagine the Chimera' God made the devil-fish."...
 "When God wills it, he excels in the creation of the execrable."...
 "The reason for this is to strike the religious thinker with awe."...
 "All ideals being admitted, if terror be the object, the devil-fish is the masterpiece."

If, by that point, one is not already hypnotized, one notices that there follows a long series of antitheses where the author tries to display his deep zoological erudition, but merely succeeds in revealing his complete ignorance of the anatomy of the octopus; here are a few illustrative samples:

"...the jaracara [a variety of fer-de-lance pit viper] makes a hissing sound, the devil-fish is silent [...]; the buthus (a kind of scorpion) has claws, the devil-fish has no claws; the howler monkey has a prehensile tail, the devil-fish has no tail [...]; the vampire-bat has wings with claws, the devil-fish has no wings [...]; the torpedo [cramp-fish] has an electric shock, the devil-fish has no electric shock; the toad has a virus, the devil-fish has no virus; the viper has venom, the devil-fish

ON THE AGRESSIVITY OF THE OCTOPUS,
IN LITERATURE AND IN THE OCEANS

has no venom; the bearded vulture has a beak, the devil-fish has no beak, etc..."

One might readily forgive Mr. Hugo for being unaware that the octopus is venomous; although Oppian had already said so eighteen centuries earlier, this fact was experimentally verified only a few years ago[1]. The actual manner in which the animal uses this venom also remains rather mysterious.

For quite a few years already, people had wondered how the octopus managed to defeat enemies much larger than itself, more powerful as well as better armed. One day, at the Naples Aquarium, Lo Bianco noticed that crabs and lobsters were paralyzed at a distance as soon as they were immersed in the same vat as the octopus[2]. Did the frightening mollusc, whose eyes have horizontal, rectangular pupils, perhaps hypnotize its preys?

Such an explanation might have satisfied a romantic poet; it could certainly not satisfy scientists. Still in Naples, Krauss and Baglioni, after minutious observations, found the key to the mystery. They noticed that when an octopus caught a prey, it always brought it first to its mouth without touching it, just as a human gourmet will delight in first smelling a delicate dish before eating it. It was then observed that if, at that moment, the prey was taken away, it would soon die without showing the slightest wound. It seemed to have been poisoned!

Krauss then collected the salivary secretion found on the octopus's thick tongue and readily demonstrated its venomous nature. When poured into an aquarium, the saliva paralyzed all the animals living in it; injected to a rabbit, it killed it instantly.

Today, we know that the venom of a few species of octopus is dangerous, even for people. In 1947, Don Simpson was collecting live specimens for the Steinhart Aquarium in San Francisco; he had the unfortunate idea to pose for a photograph holding on his hand a baby octopus of the species *Paroctopus apollyon,* native to the Pacific Ocean. The little devil took advantage of the fact that its captor was showing his best side to bite the back of his hand, which began to bleed profusely. After a few minutes, Simpson felt a strong burning sensation, and during the night, his hand swelled up like a balloon. It took four weeks for the swelling to subside.

In another well-authenticated case, a person who was fishing for octopus among Hawaii's coral reefs was bitten in the palm of his hand by its prey. Seized with vertigo, he had to take to bed for two days. His flesh became enflamed and swollen around the two tiny holes created by the octopus's beak.

Recently, in the USA, a physician and a squid expert, Dr. Bruce W.Halstead and S.Stillman have made a careful study of six cases of this kind of bite. Here are their conclusions:

"Symptoms consisted of a sharp pain upon contact (described as similar to a bee

---

[1] *"There are some whose bites are not benign; such as those of the crawling octopus and of the sepia, which hold within themselves in small quantity a liquor, or ink, which is dangerous and noxious." (Halieutica, Chap II)*
[2] *Pliny was thus quoting careful observations when he related this apparently fantastic comment: "The lobster is so afraid of the octopus that it dies as soon as it comes near it."*

sting), tingling, throbbing, redness, swelling, and in one case, abnormally profuse bleeding. Symptoms seem to vary considerably, depending upon the size and possibly the species of the octopus, the size of the wound and doubtless the amount of venom injected. Octopus bites are of the puncture-wound variety and with the smaller animals commonly handled re relatively minor in nature. The venom is secreted by the anterior and posterior salivary glands."

That is however not always the case. On September 18, 1954, a twenty-one year old scuba hunter, Kirke Dyson-Holland, was practicing his favorite sport off East Point, near Darwin, Australia. He was accompanied by his friend John Baylis, like him a member of the *Arafura Skindivers' Club*. While swimming back to the beach, the latter saw a small (15 centimetres across) blue octopus swimming near him. With long-practiced skill, he grabbed it and allowed it to run around his arms and neck. He then jokingly threw it onto his friend's shoulders. The small beast crawled on his back, stopped for a moment near his neck and slipped off into the sea. As soon as they arrived on the beach, Dyson-Holland complained of a dry throat and of some difficulty in swallowing. He said nothing about being bitten, but Baylis noticed a droplet of blood at the spot where the small octopus had stopped on his upper back. Soon the unfortunate young man was subject to violent nausea and vertigo and collapsed. Baylis rushed him to the hospital in Darwin, three kilometres away. The victim moaned: "It's the small octopus...the small octopus." Half-way to the hospital he lost consciousness. When he reached it, he was blue and had stopped breathing. Adrenalin injections and the use of an iron lung did not revive him. Fifteen minutes later, only two hours after having been bitten, Dyson-Holland was dead.

At the autopsy, nothing was found, not even a trace of the bite. It was, however, discovered that the young man had been slightly asthmatic, a common allergy. The two physicians who studied his case, Dr. H. Flecker and Dr. Bernard C.Cotton suggested the following explanation in the *Medical Journal of Australia*:

"Although his symptoms are not typical of allergic response, the possibility of some hypersensitivity reaction exists."

A lesser or greater sensitivity to venoms is always a personal matter: some people have died after touching a jellyfish or being stung by a wasp; others are not seriously affected even by the bite of a cobra or a viper.

I must add that octopus living in our region (the eastern North Atlantic and the Mediterranean) do not tend to bite when they are manipulated; I have also been able to verify that when they do, their bite is not poisonous[3].

---

[3] *The American specialist MacGinitie, who has handled thousands of octopus from the Gulf of California, says that he has never been bitten, in spite of doing everything possible to incite some of them to do so. MacGinitie recognizes that he has seen some octopus use their beak in fighting. One may well ask, naively, how would octopus ever eat if they never bit anything?*

ON THE AGRESSIVITY OF THE OCTOPUS,
IN LITERATURE AND IN THE OCEANS

In June 1968, at the Ile du Levant, I caught with my naked hands an octopus spanning 1.20 m, so as to examine it and to study its crawling motion on dry land. In order to prevent it from escaping my grasp as I was swimming ashore, I held it firmly against my left shoulder. He quickly undertook to carve me up with its beak. As I tried, in reaction to the pain, to push it off me, I moved it a number of times, which allowed it each time to get to work on a different part of my skin. I carried sucker marks on my skin for 7 to 8 weeks: the scars from its bites showed for many years. I shall be forgiven for disagreeing with the specialists who claim that Mediterranean octopus never bite!

When Victor Hugo claims that the octopus has no beak, he is guilty of a much graver fault and has in this case no excuse at all. All he had to do was to handle one at a fish market to discover his mistake.

All cephalopods have, in the middle of the ring of their arms, a horned beak very similar to that of the parrot, but upside down. It is a powerful tool, with sharp edges, with which they can shear the flesh of their preys or crush the shell of crustaceans, or even the hard shell of bivalves on which many of them feed. Drawn in when it is at rest, this beak may not be very prominent, but it is nevertheless there, believe me!

Our impetuous romantic poet has indeed such a peculiar idea of how octopus feed themselves that, according to him, they don't need a beak at all, sucking the blood of their victims directly through their suckers! [4]

"No grasp equals the embrace of the cephalopod."...

"It is the sucking apparatus attacking one. One has to combat with a vacuum furnished with claws, which make an indescribable scar, although they neither scratch nor bite. A bite is formidable, but no less than suction. A claw is nothing in comparison with the cupping-glass. The claw means the beast entering into your flesh, the cupping-glass means yourself entering into the beast.

"The muscles swell, the fibres stretch, the skin cracks under the loathsome weight, blood spurts forth and mixes frightfully with the lymph of the mollusc. The creature fastens itself upon one by a thousand dreadful mouths; the hydra incorporates itself in the man. The man is amalgamated with the hydra. They form but one. This idea haunts one. The tiger can but devour one; the devilfish, oh, horror! Sucks one in. It draws one into it, and, bound, glued, and powerless, one feels one's self slowly absorbed into that frightful sack, which is the monster itself.

---

[4] *Others before Hugo had made the same mistake, following an erroneous interpretation of a text of Pliny:* 'Luctatur enim complexu, et sorbet acetabulis, ac numeroso suctu, dum trahit, cum in naufragos urinantesve impetum cepit.` *Guéroult (1802) among others, translated this as: 'When it graps a swimmer or a diver, he binds him and squeezes him in his arms, whereupon, dragging him towards itself he sucks all his blood through that multitude of suckers along its tentacles.` In 1558, Guillaume Rondelet already knew that "octopus don't suck you dry any more than cupping glasses do". The application of cupping glasses has never been thought to leave anyone sucked dry of blood!*

THE KRAKEN AND THE COLOSSAL OCTOPUS

"Beyond the terror of being eaten alive is the inexpressible terror of being imbibed alive."

Besides this most advanced sucking apparatus, Hugo endows the octopus with one more, a thousand and first mouth, even `fouler' than all the others:

"It has a single orifice in its centre. Is this one opening, the anus or the mouth? It is both."

"The same aperture fulfils both functions. The place of entrance is also the exit."

The poor monster must have terribly bad breath. In fact, in all molluscs, particularly among non-literary cephalopods, the anal orifice is always quite distinct from the mouth. This has been known since the days of Aristoteles's careful observations. In the octopus, for example, the mouth is located at the junction of the eight arms, while the anus opens under the "mantle". The latter is a kind of a fold of the body which extends at the back and tucks underneath it to create a kind of a bag. The lower part of the mantle floats, as it were, against the body so that the space between them (the mantle cavity) is in contact with the exterior through a transversal slit. In addition, this cavity communicates with the outside through a siphon, called the locomotory funnel, for reasons which will become clear below.

Ambient water penetrates freely into the mantle cavity through the transversal slit to bathe and oxygenate the gills which project within it. This water circulation is not only for the purpose of respiration. When the octopus has had enough of crawling on the bottom in the manner of a crab, he presses the mantle tightly against its body, closing the transversal slit; then through a sudden contraction of his whole body, he expels through his siphon the water held in the cavity. Since the siphon points in the same direction as its arms, this action pushes it backwards. At the end of its motion, it relaxes and opens the transversal slit, refills the mantle cavity, closes the slit anew and repeats the cycle. A series of rhythmic contractions push it backwards by jerks in a true jet propulsion system. Should the animal wish to rush, tentacles first, towards some prey, he can curve his locomotory funnel backwards; in practice it is quite rare to see an octopus perform this `forward' motion[5].

The products of excretion are also ejected within the mantle cavity, to be expelled into the surrounding waters by the siphon. Victor Hugo's description is thus entirely incorrect.

The body of the octopus has the shape of a bag. This bag is actually made of two layers of very different thicknesses; one of them is the actual body wall, while the other is only a membrane, the mantle. In the fleshy part of the inner bag are located most of the animal's vital organs: its genitals; the digestive tract, including the liver; the kidney; a complex circulatory system, with heart,

---

[5] *There exists however a very rare species of octopus,* Opisthoteuthis, *in which the locomotory funnel normally points backwards. It is the only cephalopod for which forward jetting motion is the usual mode of locomotion.*

veins, arteries and gills. The latter project into the mantle cavity and they may be exposed by lifting up the edge of the mantle, or "hood" as fishermen call it. That is actually what they do when they wish to overwhelm the octopus: exposing the gills to the air hastens asphyxia and leads to fairly rapid death[6].

Mr Hugo's octopus has a much simpler anatomy than the real animal; so simple in fact that it is possible to turn the animal completely inside out, just like an old jacket:

"It has no bones, no blood, no flesh. It is flabby. There is nothing within it. It is a skin. One can turn its tentacles inside out, like the fingers of a glove."

Sober and attentive readers will wonder what exquisite taste this emptiness may well have that Mediterranean gourmets are so insistent in including octopus in their bouillabaisse. Ah! Ha! The octopus which Gilliatt encountered in the mysterious cave must have been only a rubber model, an empty toy of the kind sold in bazaars to amuse children in their bathtub. It would be easy to imagine that the hardy Normand fisherman will burst it with a single pin-prick....

Well, not exactly! In fact, the struggle is epic!

The unfortunate hero, in water up to his waist, is suddenly whipped, bound, garotted and choked by five of the tentacles of the pneumatic monster, each armed - note the high degree of precision - with fifty suckers. Actually, in the common octopus, there are usually 240 suckers per tentacle; 1920 in total. The octopus of The Toilers of the Sea must have been a very special kind of animal.

"Gilliatt had two hundred and fifty suckers upon him. A combination of agony and disgust results from being thus pressed by an enormous fist, whose elastic fingers are nearly a metre in length, and the interior surface of which is full of living pustules which penetrate the flesh."

What to do? Gilliatt has his knife in hand, blade drawn, but can this weapon help him?

"The tentacles of the devil-fish cannot be cut; they are leathery and impossible to sever, and slip from under the blade; moreover they are so placed that a cut into these tentacles would make an incision into one's own flesh."

As one can notice, the octopus of Mr Hugo's day were not as soft as today's. Fortunately, one reads with relief, they nevertheless did have a weak point. "There is way to use the knife" of which Gilliatt was not ignorant. We learn that the trick is to wait for the right moment, "when the devil-fish thrusts forth its head - a fleeting moment; he who fails to strike at that juncture is lost".

---

[6] *Pliny the Elder was already aware of this procedure. He wrote:* "si invertatur, elanguescit vis", *i.e.* "when it is turned inside out, its strength fades". *The translators of the erudite Roman were often not as well informed as he was. Thus Gueroult (1802) renders the above as:* "When the octopus is put on its back, it looses its strength." *No wonder that Pliny acquired a reputation for telling ridiculous fables!*

THE KRAKEN AND THE COLOSSAL OCTOPUS

The man watches the beast, and...
"It was like the conflict between two bolts of lighting.
"Gilliatt plunged the point of his knife in the flat, viscous mass, and with a gyratory movement, like the twist of a whip-lash, describing a circle around the two eyes, he tore out the head as one wrenches out a tooth."

I don't know where Victor Hugo learned the secret of this astounding thrust, but I would not advise anyone finding themselves in the poor Gilliatt's situation to waste their time trying to pull the same trick. Better use a simpler method. According to E.G. Boulenger, a well-known specialist of marine fauna, it is enough to squeeze the animal strongly between the head and the body to force it to let go; this is supposed to work even for large octopus. However I doubt whether this would always work. The surest method is to stick a knife between the animal's eyes, piercing the brain. Death is instantaneous and the animal relaxes immediately. For lack of a pointed object, one may imitate Polynesian divers and take a full bite of the animal between its eyes. This method may seem rather unappetizing, but there are circumstances when one shouldn't be too fussy! I must say that this is easier said than done. Once, when there was no weapon at hand, I tried to finish off in this manner a 1.50 m octopus which a hunter had wounded, torn apart and abandoned. Even though I could use both hands to hold the poor animal, it took me more than 10 minutes to crush his cartilagineous skull, which kept avoiding my bite within its viscous bag.

Where the devil did Victor Hugo pick up his information on the anatomy and biology of the octopus?

At the time when he was writing *The Toilers*, published in 1866, there were a number of excellent works on cephalopods available to the poet: his only problem was an overabundance of choice.

It is of course possible that he might not have known about the *Manuel de l'histoire naturelle des mollusques et de leur coquilles* (1829) by Karl Sander Rang, or the handsome monograph by Jean-Baptiste Vérany, of Nice, entitled *Mollusques Méditerranéens* (1851), or also the volume on cephalopods (1862) in Gwen Jeffrey's *British Conchology*. It is however unthinkable that he would not have known about the remarkable *Mémoires pour servir à l'histoire des Mollusques* (1817), by Georges Cuvier, the *Manuel de malacologie et de conchyliologie* (1825-27) by Ducrotay de Blainville, and especially about the treatise to which modern cephalopod science owes so much: *l'Histoire naturelle, générale et particulière des céphalopodes acétabulifères* (1835-1848), by Alcide d'Orbigny and the baron de Ferussac. Without having to consult the work of specialists, old man Hugo could have obtained a good background from the *Encyclopédie d'Histoire Naturelle* of Dr.Chenu, whose volume on molluscs had appeared in 1858. He could even have spared himself all intellectual effort by consulting a popular work, *Le Monde de la Mer*, published in 1865, just at the time when he was writing his marine novel.

The author of this book was an unknown by the name of A.Fredol; soon after its publication, it became generally known that this was a pseudonym for Alfred Moquin-Tandon, an eminent zoologist, author among other works, of a voluminous *Histoire Naturelle des Mollusques de France* (1855).

Instead of drawing upon that selection of solid references, Hugo seemed

ON THE AGRESSIVITY OF THE OCTOPUS, IN LITERATURE AND IN THE OCEANS

to have preferred another, rather surprising source: Jules Michelet. This famous author occasionally wandered from the safe highways of French history to the meandering paths of natural history. We owe to this eminent professor at the College de France some rather curious works. It is obviously one of those, *La Mer* (1861), that Victor Hugo used as his source of documentation on the octopus.

Just in case there might be any doubt, here are some excerpts from *La Mer*, where one will clearly recognize the seeds of the poet's inspiration.

One first discovers the origin of Hugo's learned antitheses in a striking contrast between cephalopods and other soft-bodied animals, represented as angels of pure innocence:

"Jellyfish and molluscs are generally innocent creatures - one might compare them with children - and I have lived with them in amiable harmony. No flesh eaters so far. Those who had to eat others did it without destroying more than they needed; most lived from life hardly begun: atoms and animal jelly, unorganized life. Thus, there was no pain. Neither cruelty, nor anger. Their small souls, so gentle, nevertheless cherished a spark, a tendency towards the light that comes from heaven, and towards love, revealed at night in those flickering lights which are the joy of the sea.

"Now I must enter in a much more sombre world: war, murder."

This world, so different from that of enlightened mussels and loving snails, is, you may have guessed already, that of public ennemy number one of the sea, the octopus, which Michelet prefers simply to call `the sucker':

"The sucker of the soft, gelatinous world is also of that same nature. In its war on molluscs, he remains a mollusc, and thus an embryo. He offers the strange spectacle, which would be so ridiculous and caricatural if it weren't so terrible, of the embryo going to war, of a cruel foetus, soft, transparent, but tense, breathing a murderous breadth. For it is not just to feed itself, that he wages war. He needs to destroy. Even when sated, bloated, he still destroys."

Only such a reputation as a killer born prematurely, worthy of one of James Hadley Chase's black novels, could lead Hugo to claim that the octopus "proves the existence of Satan", that it is a "terrible Sphinx advancing the terrible 'enigma' of Evil", and that the perfection of this malice has "induced even great minds to believe in the two-faced God, the fearsome two-faced deity of the Manicheans."

It is undoubdtedly in that passage, where the historian describes the attack of a swimmer by an octopus and the resulting struggle, that Hugo drew his information on the anatomy of cephalopods. Read for yourself:

"A man so struck while swimming must not lose his nerve in his struggle with its miserable ennemy. He must overcome his disgust, grab it, and quite simply invert it like a glove. It then collapses and falls.

"One is shocked, angry from having felt a moment of fear, or at least surprise. One must say to this warrior who comes, blowing, snorting, swearing:

" `Flawed brave, there is nothing in you. You are a mask more than a being. Without basis, without firmness. Your personality is but pride. You snore,

THE KRAKEN AND THE COLOSSAL OCTOPUS

steam engine, you snore and you are aught but a bag - then inverted, a soft and flacid punctured bladder, and tomorrow a nothing without name, washed out sea water'."

One might wonder, with some degree of alarm, whether when writing his historical novels Hugo consulted Lamarck's *Histoire Naturelle des Animaux sans vertèbres* or Cuvier's *Discours sur les ossements fossiles*.

Not satisfied with having endowed the octopus with a fanciful anatomy, the great poet of the *Contemplations* takes pleasure, following Michelet, in showering it with the most gratuitous insults. He calls it evil, treacherous; then, refining his scorn, "hate-filled glue", "spoil-sport of the meditator", "blasphemy of creation", and "sickness disguised in monstrosity".

One could hardly be less friendly. It was only fair that in 1866, soon after the publication of *The Toilers*, an eminent specialist of the day, Henri Crosse, came to the defence of the poor libeled cephalopods, and criticized Hugo, although in rather moderate terms.

After having denounced with some wit a few of the silliest passages found in that "strange chapter" the learned specialist gives vent to his bitterness:

"All that was missing for these poor animals to completely lose their reputation was for them to be exploited by contemporary literature, which is just what has happened to them." Whereupon Mr. Crosse firmly protests against *"the regretful invasion of science by literary types who are completely out of their element and who cannot do any better than the blind discoursing on colours.*

"There are some literary philistines who will praise this part as much as the nicer parts of the work. It is then the duty of those like us who have modestly, but conscientiously devoted their life to the cult of science to point out and to correct such coarse errors, most pernicious in that they come from an author of the quality of Mr. Victor Hugo and not from the pen of some unknown and unrespected author.

"We also note that, while this unfortunate chapter is loaded with glaring mistakes of all kinds and of supposed facts which are scientifically completely false, the most popular, if not the most intelligent, newspaper in Paris has hastened, with its characteristic flair, to chose it as an example of the book and to quote it in its entirety with thunderous praise.

"We see that the instruction of the public is in good hands, at least for the natural sciences. Mr Michelet, another literary type lost in science, had already drawn of the octopus in a portrait as amusing as it was fanciful. After Mr Hugo's there is no further contest; it will be easy to be more accurate, but there's no surpassing his gall!"

The years have passed and Crosse's bitterness is more justified than ever. Except for a handful of specialists, everyone has forgotten the learned malacologist, as well as his most illustrious predecessors in the study of cephalopods: Rang, d'Orbigny, Ferussac, Vérany, Blainville. Does anyone today read Cuvier or Owen's works? However, school benches are still filled with young people made to read fervently the immortal pages describing the unforgettable struggle between Gilliatt and a rubber octopus, possessed by the devil.

ON THE AGRESSIVITY OF THE OCTOPUS,
IN LITERATURE AND IN THE OCEANS

In a style which paraphrases Hugo's formulas, we could say that "stupidity steeped in pretension" and "rambling disguised as a masterpiece" have triumphed on all fronts.

Of course, for over a century now, hordes of novelists, journalists and film-makers have sought inspiration in *The Toilers of the Sea* and finished ruining the octopus's reputation. True, even before Hugo, the octopus, was hardly thought of as a harmless animal. In antiquity, Pliny said that there was no more atrocious death for a swimmer or a diver than to be bound and squeezed in the arms of an octopus and be carried away by it. During the Renaissance, Rondelet spoke in the same way, emphasizing the strength of the octopus's arms: "That is where the greatest and most marvelous strength of the octopus resides, more cruel and dangerous to kill a man than any other aquatic animal." In the nineteenth century, Cuvier, himself a great debunker of legends, took this view for granted and repeated that the eight arms of the octopus were a most dangerous weapon, not just for animals, but also for swimmers that it might draw under.

Man was thus considered to be just one of the many victims of the voracious mollusc. Aelianus, the moralist-chronicler from Praeneste, so fond of marvelous incidents, related how an octopus attacked and drowned a surprised eagle. The prolific Bolognese encyclopedist Ulisse Aldrovandi had gone as far as to state that an octopus could hold its own against a lion (if given a chance). In contrast, the reputedly `extravagant' Denys de Montfort appears rather moderate in having an octopus nearly strangle his favorite mastiff.

Hugo was the first who thought of representing the octopus as a creature harbouring a special hatred for humanity and entirely preoccupied by this evil instinct. It became a personal matter between us and this beast, promoted to the rank of hereditary ennemy. From then on, it became impossible to descend into Neptune's realm without having to face its official Cerberus.

The reaction to such excess was to be no less exaggerated. More recently, professor Stephen Riggs Williams from Miami University (Oxford, Ohio), an American zoologist who has worked extensively on the octopus, proclaimed that "a farmer in a cornfield is in more danger of being attacked by a pumpkin than a swimmer is of being attacked by an octopus". In his *Natural History of Nonsense*, professor Bergen Evans, a reputed debunker, makes special reference to the reputation of agressivity of the octopus: in his view, "it is for man one of the most inoffensive creatures".

In fact, upon examination, the octopus does appear as a rather nervous, even timid creature. As its soft and naked body makes it very vulnerable, it is understandable that when an ennemy approaches he should most often seek refuge in a crack in the rocks.

Everything in the octopus betrays a wish for dissimulation. It has carried the art of camouflage to a highly advanced level. The ease and speed with which he takes on the hue and pattern of its substrate, thanks to the pigmented cells distributed in its skin, is much superior to that of the proverbial chameleon (whose ability is also much overestimated). As if this wasn't enough, the octopus can, in case of extreme danger, lose its colour until it becomes transparent and then eject through its siphon a cloud of ink which completely confuses its ennemy[7] .

It was long thought that the only purpose of this maneuver was to hide for a moment the octopus from the sight of its adversary so as to allow it to disappear in some unknown direction; or rather, as it has also been suggested, to offer as bait to its ennemy a dark object on which it is tempted to pounce, leaving the prey for its shadow. Isn't it revealing that in abyssal depths, where a jet of dark fluid would be useless, some species of cephalopods eject a phosphorescent cloud whose luminosity is equally effective at blinding an ennemy.

The ink ejected by the octopus and its relatives seems to have some other, perhaps even more important, role. This is what Dr.G.E.MacGinitie, director of the Kerchoff marine station, at Corona del Mar, California, has been able to determine experimentally. This specialist had placed in the same aquarium a Californian octopus, *Octopus bimaculatus* and a wolf eel, a fish with formidable teeth and well known ferocity. The wolf eel immediately set upon the octopus, eager to come to terms with it; the latter had already gone through the whole colour spectrum. Finally, when the two opponents were but half a metre apart, the panicking octopus released its ink cloud, creating instant darkness. So far, nothing unusual. However, when the cloud gradually dissipated and the water had recovered its transparency, one noted that the wolf-eel seemed no longer to notice the presence of the octopus: it did not recognize it, even when it approached it closely. Its sense of smell seemed paralyzed, and remained so for an hour or two.

The obvious conclusion is that the famous ink jet is more than just a visual screen; it also works as a kind of stink bomb, similar to the angry skunk's smelly spray.

So, given its various means of camouflage, the octopus would seem mainly prepared for defence, which hardly seems compatible with a reputation of agressivity.

Is our bogey-monster a coward? In his classical opus *Mollusques Méditerranéens (Mediterranean Molluscs)*, published in Nice in 1851, Jean Baptiste Vérany bluntly expressed his scorn for the redoubtable monster.

"The effect of the suckers when applied to the skin, the snake-like motion of its arms, its strength, its hideous appearance, have, I believe, led to a gross exaggeration of the mischief of which this Cephalopod is capable; it is, in truth, stupid and hardly capable of hurting anyone."

The wind had turned. As early as 1879, Prof. Addison E.Verrill of Yale University, in New Haven, Connecticut, summed up current scientific opinion on the agressivity of the octopus:

"There is no satisfactory evidence that any of these species of octopus ever intentionally attacks man, or that anyone has ever been seriously injured by them. They are rather sluggish and timid creatures, seeking shelter in holes and crevices

---

[7] *It is the ink of the cuttle-fish that water colourists use under the name of* sepia, *which is simply the Latin name of the animal*

ON THE AGRESSIVITY OF THE OCTOPUS,
IN LITERATURE AND IN THE OCEANS

among rocks. They feed mainly upon bivalve molluscs, but will also eat fish, and might, perhaps, like lobsters and crabs, devour dead bodies. Their power and ferocity, as well as their size, have often been excessively exaggerated."

More recently, E.G. Boulenger concedes that there may exist some risk to swimmers when encountering large specimens, but, as he emphasizes, the danger is `more psychic than physical' due mostly by the unpleasant appearance of the animal and its viscous feeling.

Nowadays, many skindiving hunters, truly well acquainted with marine animals, will go as far as playing with octopus. According to them, it is a gentle and sociable animal. When they meet one, they follow it underwater, pet it, play with it, so much so that according to one of them "the octopus seems to greatly enjoy these games".

So, this is our "devil-fish", this "blob of glue steeped with hatred", finally transformed into an angel!

Does this mean that all stories of octopus attacks on people are the fruit of some fertile imaginations? That would be a rather bold claim. In a masterful book entitled *Animal Legends,* Dr Maurice Burton, of the British Museum, shows that many popular legends about the behaviour of animals, which scientists refuse to accept because they seem so unlikely, are really based on fact, but depend on exceptional or difficult to observe behaviour. It will thus be easy for me to show that there is some truth to the legend of octopus attacking people since it is actually based on some natural inclination of the animal.

Whoever has lived by the seashore in a region inhabited by many octopus could quote countless examples of apparently unprovoked attacks by these molluscs. Thus, Rev. William Wyatt Gill collected many such stories during his twenty years in Polynesia in the last century. "It has been disputed in Europe, he wrote, whether octopus ever attacks human beings. No native of Polynesia doubts the fact. A fisherman rarely goes alone in quest of octopi; he has usually a trusty friend in the canoe to take care of it." More recently, Sir Arthur Grimble saw a native from the Gilbert Islands suddenly immobilized in shallow water by an octopus, just at the moment when he was going to push his canoe onto the beach. The Gilbert Islander managed to free himself by a desperate pull before Sir Arthur reached him, but the suckers had pulled so hard that he left some skin behind. Even Thor Heyerdahl was grabbed by the ankle by a small, 90 centimetre long, octopus near the Kon-Tiki's final landing point.

One doesn't have to go to the other end of the world to find such examples. The Mediterranean is constantly the theatre of deliberate octopus attacks. The submarine hunter Guy Gilpatric describes many of these in his book *The Compleat Goggler.* The last time that I personally witnessed this kind of agression was in August 1955, on the Pierre-Plates beach, at the Ile du Levant. In front of an audience variously horrified, stupefied or merely interested, for the event is rather trivial, a Swiss swimmer walked out of the water with a 1 m 50 octopus attached to his leg.

This kind of incident doesn't usually leave traces, even in newspaper columns, because it is normally without serious consequences. What happens is almost always the same thing. The swimmer, or even a person quietly sitting on the sea shore with its feet dangling in the water, suddenly feels his ankle gripped by the tentacle of an octopus. People will occasionally shudder, many years after the incident, at the memory of this viscous and sneaky grip[8]. But the drama is usually a short one and it is enough to pull one's foot quickly out of the water for the monster to let go. Let's not forget that these marine molluscs cannot endure very long outside their natural element and always prefer to flee than to suffocate.

Those who think that octopus are just charming playmates will pretend that these beasties are merely trying to play with peoples' toes. Nothing could be more false.

Octopus are, without doubt, exclusive carnivores, beasts of prey, forever hunting. One who knew them well, the late professor G.C.Robson, of the British Museum, described them as follows in his authoritative monograph:

"Unlike the living Tetrabranchia and some of the Decapoda which are probably gregarious, the Octopodinae seem to be solitary, and, if it is not reading too much into our evidence, also irritable and ferocious. From their very obvious preoccupation with parental responsibility, seen in nesting, and their wary aggresiveness and furtiveness we receive the impression of a distinctive and peculiar disposition."

This opinion is both qualified and unequivocal. The octopus's hunting strategy is based on surprise. Although they will occasionally move to catch stationary or slow creatures, such as bivalves or gasteropods, they usually lie in wait in their lair for their more rapid preys, fish and crustaceans. Their tactic consists of grabbing them rapidly as they pass by, or of pouncing, all tentacles spread, and to envelop them in a net of sucker-studded bonds.

We now understand that if the octopus, and indeed all cephalopods, are masters of camouflage, it is not so much to hide from their enemies as to catch them more easily. Everything that parades before them is a potential prey. I have often performed the experiment during submarine explorations: all one has to do is to let one's fingers dangle in front of the hole where the octopus is hiding to observe its tentacles reach out and try to seize it.

One should not conclude that octopus preferentially attack people. Any whitish object would do; as if moved by a reflex, the octopus reacts automatically to the stimulus. This is the nature of the provocation of which the innocent bather is unconsciously guilty .

Fishermen from the Hyères Islands have told me that to catch octopus, they draw them out of their lair by waving in from of them an olive bough, whose leaves have a whitish underside, as do many fish. To my knowledge, octopus are not vegetarian!

---

[8] *I must admit that this was not the case for my Swiss bather. Experienced submarine hunter that he was, he had already captured many octopus, and he seemed delighted to have brought one back without having had to use his gun. As soon as he got out of the water, he flipped over the animal's mantle while holding it firmly to prevent it to struggle back to its element. In spite of my pleading, he did not even bother to kill the octopus before starting to tenderize its flesh by flinging it against the rocks. I hoped that next time the Great Swiss Hunter would meet with a more worthy opponent.*

ON THE AGRESSIVITY OF THE OCTOPUS, IN LITERATURE AND IN THE OCEANS

Their apparent taste for olive branches had already been mentioned by the grammarian Athenaeus in his *Deipnosophistai (Authorities on Banquets)*, the most erudite of culinary guides:

"Here is the evidence that octopus love the olive tree. Immerse a branch of this tree in the sea in an area where there are many octopus; hold it in the water for a short time and you will pull out as many of them as you wish, all embracing the branch."

In his Halieutica, a rhyming treatise on fishing, Oppian spoke with emphasis of the passion which the octopus felt for `Minerva's tree, the whitish olive tree'. "The attraction which draws him to this plant, the joy which inspires in him this oily branch, seems to be miraculous." Drawn by its fragrance, the octopus *"comes out of the sea, happily crawls out on land and approaches its darling tree"* in order to caress and embrace it. Oppian concludes by relating that *"fishermen, who have noticed this liking that the octopus has for the olive tree, use it to trap it; they tie together many leafy branches, attach a lead weight to them and troll with them from their boat; the octopus cannot resists this lure."*

Zoologists have always been puzzled by these texts; some have simply laughed them off. Now we can understand. They simply show that in the 2nd and 3rd century of our era, Mediterranean fishermen used the same kind of lure as they do today. It is of course not the smell of the olive tree that attracts the octopus. With its well-developed eye, comparable to that of mammals, the octopus must have excellent eye sight: for it, as for us, it is probably the dominant sense. It is then not surprising that it should be possible to fool him through an optical illusion.

Octopus are no more murderously menacing towards us than they are toward a simple olive bough. No less either. I would wager that, when it happens by chance that a small octopus of our shores grabs a human leg, they must be as frightened when they notice the nature of their catch, as would be a fisherman hooking a great white shark off a Paris wharf.

There remains, for those who worry about details, a significant objection. What happens when the octopus, naive agressor of appetizing toes, turns out to be quite large, and firmly anchored to the bottom? Will the victim not have some difficulty pulling his foot out of the water? Does he not risk having his skin slashed by the sharp beak of the voracious mollusc?

Fundamentally, yes. This should not cause alarm to those who like to soak their feet in the sea. Those octopus which venture near the sea shore are normally not much bigger than those sold in the Marseilles fish-stalls: hand-, or at best arm-sized. Those are easy to shake off at first contact. Given the central position of the mouth, the octopus cannot begin to bite its prey until he has completely embraced it with all eight of its tentacles.

Even if, in extraordinary circumstances, the animal turned out to be a large one, it would be rather unlikely that he should risk coming out of the water to pin down and bleed a human prey, all the while struggling and hitting it violently. Such a behaviour is unthinkable for this animal, which would have to come out of its natural element to carry it through.

The only real danger for a person is for a sufficiently large octopus to grab the arm or the ankle of a passing diver and hold him underwater. It wouldn't have to be enormous to succeed in doing this, since the adhesive strength of a multiplicity of suckers - there are nearly 2,000 overall - is quite large. Well anchored to the rocks, an octopus with a body no larger than a baby's head could easily immobilize a diver and perhaps drown him. Unless of course, the latter decided to defend himself, using one of the techniques described earlier.

The American biologist G.H.Parker measured, using a spring balance, the adhesive strength of the suckers of the californian octopus *Octopus bimaculatus*. He noted, for example, that to detach a sucker with a diameter of 2.5 mm, a force of 60 grams is required; for a 6 mm sucker, a force of 170 grams. Calculations show that an octopus with a span of 1.50 m adheres with all its suckers with a strength corresponding to a traction of 250 kilograms. Such an octopus, holding to a rock with only one fifth of all its suckers, would be equivalent to a 50 kilo ball attached to your ankle.

What are the chances to be grabbed in this way? For this to happen, one must wiggle one's limb for a while in front of the lair of a watching octopus; the animal is careful and wary, and rather slow on the draw. It is however rather rare that a diver remains for a long time at any one place on the bottom. Only pearl fishermen have to do this, and that is why they are more often than others subject to octopus attacks. They are of course aware of the danger and armed appropriately.

Of course, exceptional stupid accidents will always happen. Sir Grenville Temple relates such an event in his book *Excursions in the Mediterranean, Algiers and Tunis. (1835)* It is perhaps the only documented story, by a reputable author, of an octopus attack which led to a fatal drowning.

"A Sardinian captain bathing at Jerbeh, felt one of his feet in the grasp of one of these animals: on this, with his other foot he tried to disengage himself, but this limb was immediately seized by another of the monster's arms; he then with his hands endeavoured to free himself, but these also in succession were firmly grasped by the polypus, and the poor man was shortly after found drowned, with all his limbs strongly bound together by the arms and legs of the fish; and it is extraordinary, that where this happened, the water was scarcely four feet (1.20 m) in depth."

What is most extraordinary is just how the exact sequence of events ever came to be known, given that the only witness of the tragic accident was not in a position to describe them. One might perhaps suppose that the details were inferred from the position in which the victim was discovered. The accident is most plausible. Let's not forget that the victim is a sailor, unable to swim, as most of them are, who would have rapidly panicked. To avoid putting his head under water, he will have struggled blindly. With one's feet tied up, it is easy to fall and stumble, and to drown, even in shallow water. People have even drowned in their bathtub. Such accidents are no more likely than being struck by lightning or hit by a meteorite!

To conclude, to be grabbed in this way by an octopus, much effort is required: *one must have looked for it.* Indeed, this is how fishermen proceed in many countries: they count on the natural agressivity of the octopus to catch

Figure 20. A fight between two common octopus (Photo Jean Dragesco-Rapho)

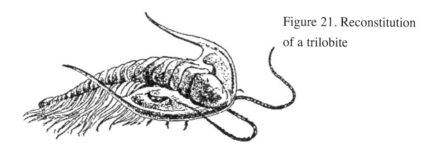

Figure 21. Reconstitution of a trilobite

Figure 22. Leonine sea monster, after Ambroise Paré, 1562

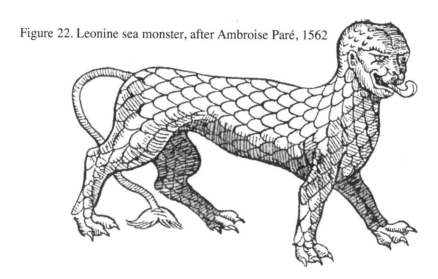

Figure 23. Sea-devil (*Manta birostris*)

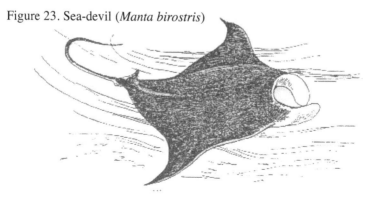

Figure 24. Fight between a hard-hat diver and an octopus, after the *Petit Journal Illustré*, 24 Jan 1909

Figure 25. Victor Hugo (1802-1885) at the time when he was writing *The Toilers of the Sea*

Figure 26. Jules Verne (1828-1905), at the time when he was writing *Twenty Thousand Leagues Under the Sea* (Photo Nadar)

Figure 27. Captain Nemo, staring at a giant squid through the port-hole of the Nautilus (Illustration from the original Hetzel edition of *Twenty Thousand Leagues Under the Sea*

Figure 28. A giant squid attacks the *Nautilus* (ibid.)

Figure 29. The crew of the *Nautilus* fight back

Figure 30. A Manta, as depicted by the illustrator of the French edition of B.H. Révoil's book, in which is related the event involving viscount Onffroy de Thoron (1870)

Figure 31. The *Plongeur*, of Bourgeois and Brun, the inspiration for Verne's *Nautilus*

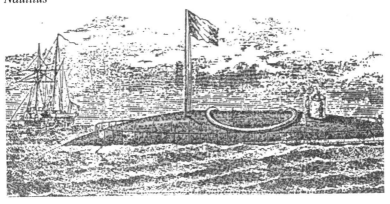

Figure 32. Sailor pinned down by a small octopus, after P.H. Grosse (1860)

Figure 33. The fight between Gilliatt and the octopus (illustration of Gustave Doré for *The Toilers of the Sea*, by Victor Hugo)

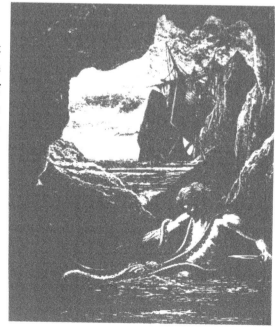

ON THE AGRESSIVITY OF THE OCTOPUS,
IN LITERATURE AND IN THE OCEANS

them bare-handed. Needless to say, this a sport which requires particular skill and courage. Our Mediterranean fishermen are experts at it.

As soon as an octopus has been spotted from the boat, using an underwater viewing frame, the fisherman dips his arm in the water and gently wiggles his fingers. The octopus soon comes up to check out this live bait with the tip of its tentacles; one has to keep still and wait until it has embraced your whole forearm. If one tries to pull the beast out of the water before it has gained a sufficient grasp, it is likely to slither back into the water like an eel. One must endeavour to be caught by it, rather than try to catch it. Once it has a solid grasp, and has by the same token let go of most of its anchoring points, it may be abruptly jerked out of the water; one then uses the other hand to grab its mantle and turn it inside out.

This is not an easy job with a large specimen!

My friend Pierrot Cestino, a fisherman of the Ile du Levant, told me how in 1932, off the fort of Antibes, he grappled with a 9 kilogram octopus, a rather large animal, but nevertheless of a size rather common on the French Mediterranean shores. At the point when he had just quickly jerked the animal out of the water, it had thrown its nearly 1.50 m long tentacles around its neck and arms before he had managed to invert its mantle. Partly immobilized, he could do nothing else than try to bite its head in an effort to crush its brain. The animal responded by gradually enfolding his face, even penetrating its nostrils with the extremity of its tentacles. Fortunately for him, his brother was nearby and could intervene to pull him off, suffocating, from the animal's grasp. For six weeks, poor Cestino carried prints of sucker marks on his shoulders and back; he was convinced that without his brother's help he would not have mastered this "damned octopus".

I would not have bet much on Pierrot's life if this battle had taken place under water. He, of course, admits that he never would attempt to bring out an octopus if he did not have his head above water.

For my part, I never try to catch an octopus with bare hands when diving without equipment, unless it should be on a sandy bottom, or one covered with eel grass, on which it cannot get a solid hold.

One must truly respect the courage of some Pacific Islands native divers. Some of them are so daring as to allow themselves, in order to capture them, to be completely embraced by octopus with a 3 metre span, in water of a depth of two to four metres. This information comes to us from Sir Arthur Grimble, a senior British civil servant, who started his career in the Gilbert Islands, in the Pacific, later becoming governor of the Seychelles. Pointing out the "invincible obstination" with which the octopus holds on to its prey after having embraced it, he shows that "it is this very trait which dooms the animal when he comes to face man, the ultimate predator."

"The Gilbertese happen to value certain parts of these as food, and their method of fighting it is coolly based upon the one fact that its arms never change their grip. They hunt for it in pairs. One man acts as the bait, his partner as the killer. First they swim eyes-under at low tide just off the reef, and search the crannies of the submarine cliff for sight of any tentacle that may flicker out for a catch. When they have placed their quarry, they land on the reef for the next stage. The human

47

bait starts the real game. He dives and tempts the lurking brute by swimming a few strokes in front of its cranny, at first a little beyond striking range. Then he turns and makes straight for the cranny, to give himself into the embrace of those waiting arms. Sometimes nothing happens. The beast will not always respond to the lure. But usually it strikes.

"The partner on the reef above stares down through the pellucid water, waiting for his moment. His teeth are his only weapon. His killing efficiency depends on his avoiding every one of those strangling arms. He must wait until his partner's body has been drawn right up to the entrance of the cleft. The monster inside is groping then with its horny mouth against the victim's flesh and sees nothing beyond it. That point is reached in a matter of no more than thirty seconds after the decoy has plunged. The killer dives, lays hold of his pinioned friend at arms' length, and jerks him away from the cleft; the octopus is torn adrift from the anchorage of its proximal suckers, and clamps itself the more fiercely to its prey. In the same second, the human bait gives a kick which brings him, with quarry annexed, to the surface. He turns on his back, still holding his breadth for better buoyancy, and this exposes the body of the beast for the kill. The killer closes in, grasps the evil head from behind, and wrenches it away from its meal. Turning the face upwards towards itself, he plunges his teeth between the bulging eyes, and bites down and in with all its strength. That is the end of it. It dies in the instant; the suckers release their hold; the arms fall away; the two fishers paddle with whoops of delighted laughter to the reef, where they string the catch to a pole before going to rout out the next one.

The most amusing point of this story is that Sir Arthur, having expressed such a strong interest in this fishing technique, had to agree, for fear of losing all face with the natives, to play himself the role of human bait.

"There is only one trick that the decoy-man must never forget, they said; he must cover his eyes with a hand as he comes close to the kika (octopus) or the suckers might blind him.

"I learned afterwards, explained Sir Arthur, that it is not so much the danger of losing one's sight when a sucker pulls on the eye-ball, as that of accidentally, under the pain experienced, drinking some water, which would make one heavier and the work of the rescuing companion more difficult."

What really sticks to the memory of this remarkable experience of Sir Arthur's is the horrible sensation of helplessness, of paralysis, which takes hold of a man embraced by an octopus, even a medium-size one.

"...I felt my whole right arm pinioned to my ribs....A mouth began to nuzzle below my throat, at the junction of the collar bones ...My mouth was smothered by some flabby moving horror..."

A lone diver would not be able to get away from such a situation.

It would seem that the danger of being attacked by a large octopus would be the greatest for a hard-hat diver, holding still in one place and attentive to his

ON THE AGRESSIVITY OF THE OCTOPUS,
IN LITERATURE AND IN THE OCEANS

work. Even then, he would have to be standing right across the mouth of the lair of one of these devils of the depths. One is usually more careful in unfamiliar surroundings. However it is indeed in shipwrecks, where divers are often called to work, that large octopus find a comfortable lair: there are many instances of this fact.

There have been innumerable tales of deliberate attacks of divers by octopus. Although it is not always easy to verify their authenticity, this doesn't mean that they are complete fabrications. There are two options: either the diver returns safely to the ship after his adventure, whereupon one may suspect that he has imagined the whole thing, or at least exaggerated the incident; or he remains at the bottom, prisoner of its attacker, and no one will ever know the exact cause of his death. So, the only completely believable case is that where the victim manages to bring back some concrete proof of the aggression, either a piece of tentacle, or even better, the whole beast.

Here is an example, with so many guarantees of authenticity that it has been cited in the monumental, 20 volume, *Manual of Conchology* of the usually very skeptical American malacologist George Washington Tryon. Tryon merely quoted the tale of the incident, published at the time by the Australian newspaper *Warrnambool Standard*. In *Cassell's Natural History*, which also mentions this story, there is a report from the principal witness (although the incident is incorrectly located in Melbourne). From these two sources, which agree in all essential details, we can reconstruct the event.

On Nov 4, 1879, a government appointed diver, M.J.Smale, was working on clearing reefs from the Moyne River in Belfast (= Port Fairy), in the province of Victoria. He had exploded some dynamite at the bottom of the river and scattered a large number of rocks. In spite of the very strong current, he dived again to fasten the boulders with chains so as to hoist them onto a punt. As he couldn't move one of the larger rocks, he leaned forward over it and felt with his hand to see if he might not be able put a small dynamite charge underneath it.

"My arm, he said, was scarcely down, however, before I found it was held by something, and the action of the water was stirring up the loose clay, and therefore I could not see distinctly for a few minutes, but when it did clear away I saw, to my horror, the arm of a large octopus entwined round mine like a boa constrictor, and just then he fixed some of his suckers on the back of my hand, and the pain was intense. I felt as if my hand was being pulled to pieces, and the more I tried to take it away the greater the pain became, and, from past experience, I knew this method would be useless. But what was I to do, lying in this position? I had the greatest difficulty in keeping my feet down, as the air rushed along the interior of my dress and inflated it, and if my feet had got uppermost I should soon have become insensible, held in such a position, and if I had given the signal to be pulled up, the brute would have held on, and the chances would have been that I should have had a broken arm. I had a hammer down by me, but could not reach it to use on the brute. There was a small iron bar not far from me, and with my feet I dragged this along until I could reach it with my left hand. And now the fight commenced; the more I struck him, the tighter he squeezed, until my arm got quite benumbed, but after a while I found that the grip began to relax a little, but he held on until I had almost cut him to pieces, and then he relaxed *his hold from the rock, and I pulled him up. I can assure you I was completely exhausted, having been in that position for over twenty minutes."*

If one is to believe the description which appeared in the Warrnambool (the nearest town) newspaper, the diver must have had a rather curious appearance as he finally emerged and dragged himself onto the punt:

"This huge ugly looking thing appeared to be entangled all over him, holding him in a firm embrace. However, Mr. Smale's fellow workmen were not long in freeing him from the unfriendly hug of his submarine companion. The body portion of the octopus was only about the size of a soup-plate, with eyes in its head like those of a sheep, but it possessed nine [?] arms, each about four feet [1.20 m] in length, at the butt as thick as a man's wrist, tapering off at the end to as fine a point as that of a penknife; thus it could spread over an area of 9 feet [2.75 m] in diameter. All the way along the underpart of each feeler are suckers every quarter of an inch [6 mm], giving it immense power. Mr. Smale declares that it was powerful enough to keep three men under water."

Besides a reference to nine arms, perhaps an error on the part of the reporter, or more likely of the printer, who took an 8 for a 9, all the details of this event have the ring of truth. In any case, one can certainly not label the description of the event as another example of Hugonian pathos or Riesebergian suspense.

The size of the "monster" is rather modest. Comparing with struggles with smaller octopus, its strength does not appear exaggerated.

Among more recent reliable testimonies, one must include that of a Royal Navy diver, Henry J. Bruce; that of one of his American colleagues specialized in underwater footage for Hollywood movies, John D.Craig; and that of an experienced pearl diver, the Swede Victor Berge.

Bruce was attaqued at a depth of 12 metres off Gibraltar by an octopus which had made its nest in an old pair of jeans. When his curiosity drove him to lift the old garment, it performed a macabre and mesmerizing dance before leaping upon him. To free himself from the octopus hiding in the pants, the diver had to stab him repeatedly with his knife.

We shall return later to Craig's story. Berge was attacked, without the least provocation, by a large octopus at a depth of 36 metres in Macassar Strait, between Borneo and the Celebes. While he was defending himself with his knife, the beast shook him with such force that his head struck repeatedly the side of his helmet, almost hard enough to knock him out. In despair, he sent out an SOS - four hard jerks on the pull-up cable - "Pull till the line breaks!" The combined strength of three men did not succeed to free him from the grasp of the octopus, solidly anchored to the rocks. Only by tying the rope as tight as possible to the ship in a wave trough could they manage to free him as the swell lifted the ship up.

There may perhaps not exist a proof sufficient enough to convince certain specialists that "any species of octopus has ever deliberately attacked a

ON THE AGRESSIVITY OF THE OCTOPUS,
IN LITERATURE AND IN THE OCEANS

human being"; however, the personal experience of professional divers, or the stories related by their colleagues has been enough to convince them, for they take elementary precautions against these so-called "indolent and timid" creatures:

"To guard themselves when working in octopus infested waters, divers carry a nitric gun, reports Frank Lane. When an octopus draws near, the gun is 'fired', and on inhaling the acid-impregnated water the monster is killed instantly."

While it is rather easy to cast doubt on the opinion of scientists who, locked up in their ivory tower, have little opportunity to measure up with the octopus, it is another thing to doubt the testimony of underwater hunters. How can one reconcile the sympathy expressed by scientists towards the octopus with the much more cautious approach of divers?

In my view, this difference of opinions is readily explained by the fact that the octopus and the hard-hat diver share the same habitat and are thus natural ennemies.

The octopus is essentially a benthic cephalopod, living at the bottom, that very same bottom where the diver walks with his lead-filled shoes. The snorkeling hunter and the scuba diver on the other hand, are pelagic creatures; they almost always stay at intermediate depths, approaching the bottom or rocky walls only fleetingly. Incapable of staying long in one place, they offer little opportunity for the octopus to grab them as they pass. A snorkeler, in particular, usually swims close to the surface, while large octopus stay below a certain depth, precisely where the hard-hat diver descends to work, but where the snorkeler ventures only during the short time during which he can hold his breadth.

Note that when a snorkeler wishes to kill an octopus, he does not tarry in front of its lair to give it a chance to embrace him so that he will then have to tear it off its home: he harpoons it from a safe distance. He takes the chance of playing with an octopus only when he encounters it swimming at mid-depth, or when he has flushed it and forced it to flee in the open.

In such a case, the shy octopus, so taken to hiding, cannot feel but insecure. Pursuing it at that time and preventing it from reaching rocks and their interstices, one notices that it tires rapidly and can be overtaken easily. It is not built for swift and lengthy swimming, but for slow crawling. Let's also not forget that the octopus is after all only an invertebrate, and that with its primitive circulatory system, capable only of a slow rate of re-oxygenation of its tissues, it is incapable of sustained muscular effort. In open water, deprived of its best means of defence, far from any rock to hold on to, or any crevice to crawl into, it can be approached without danger: it is thinking only of fleeing. Soon, it hardly reacts at all to teasing and petting. That is not to be taken as a sign of its beginning to like our company: the poor thing is merely exhausted!

The strange lack of agressivity of the swimming octopus may also perhaps be attributed to its ability under those circumstances of assessing the actual size of its opponent, something which he cannot do when he sees only part of it in front of its lair. Why the devil would it consider tackling such an oversized prey? Animals never kill for pleasure, like man, but for food, and one cannot imagine eating what one cannot master[9]. If we often need the help of smaller creatures, it is usually to eat them!

Finally then, the aggressiveness or passivity of an animal is a rather relative thing, most often linked to the size or the defensive capacity of the potential prey. Who would think of the cute shrew as aggressive? Nevertheless there are few more voracious or ferocious animals: it eats each day more than its own weight in meat, twice as much even for a pregnant female. This is a continuous hecatomb of insects, worms, molluscs, even small frogs, lizards, baby birds and small mammals. Our assessment of the beast's behaviour is purely anthropocentric. The naturalists of antiquity were better advised, recognizing it as cruel and rapacious.

Let's return to the octopus. Should a large specimen come face to face with a man, in mid-water, would it necessarily flee? It is doubtful. He might well attack, especially if it felt threatened.

In the film *The Sea Around Us*, inspired by Rachel Carson's marvelous book, one views a duel between an octopus and a much larger shark. At first, the ever prudent cephalopod attempts to escape behind a cloud of ink. However, once the battle starts, he fights with tenacity and ends up strangling his opponent, having closed all its gill slits with an eight-fold gag. I will accept only with great difficulty that at equal size, in the water, a shark is not a more dangerous opponent for an octopus than a man.

I also doubt that placed nose-to-nose with a really impressive octopus the undersea hunter would be much inclined to play with it. I could easily relate a few cases of such encounters: one in particular in a crevice of the rocks between the island of Port Cros and the Rock of the Gabiniere, where the diver readily admitted having nearly died of fright at the sight of a pair of sinister eyes, as large as ping-pong balls, suddenly fixed upon him. He left as fast as he could, without further ado!

Even if, in some circumstances, an octopus may be thought to be inoffensive, one must recall that the kindest and most inoffensive animal might sometimes become dangerous when it frightens and panics. Is it not said that nothing is more dangerous than an enraged sheep? In the case of the octopus, its aggressivity is irrational and automatic when it lies hiding in its lair. Even if Victor Hugo painted a rather fanciful portrait of the octopus and opened himself to ridicule by moralizing over its behaviour, it remains that the struggle between Gilliatt and one of these cephalopods merely one metre long is not unrealistic.

This is why I would advise divers to be very prudent and not to rely blindly on sundry statements which zoologists still fiercely argue about. To emphasize the point, I will relate one last incident, whose authenticity can readily be verified.

Some years ago, a young woman from the island of Port-Cros, Miss Buffet, was exploring an underwater cave when she was seized upon by a rather small octopus.

---

[9] *The hunter who harpoons an octopus without the least intention of eating it should never try to justify himself by claiming that he has rid the world of a `noxious creature', of a `killer'. That would also excuse shooting him on the spot! And this time, rightly so!*

ON THE AGRESSIVITY OF THE OCTOPUS, IN LITERATURE AND IN THE OCEANS

Concerned above all with reaching the surface, she had the greatest difficulty in getting rid of the animal, which had wrapped itself tightly around her. When she finally ripped it off, it had already severely gauged her flesh, from the shoulder to the breast.

One can only shudder about what would have happened to the young lady if the octopus had been any larger...

Here again, as in the case of the Australian hard-hat diver, the incident did not have serious consequences, mainly because of the rather small size of the aggressor. Thus, the really dangerous cephalopods, from our point of view, would obviously be those whose size exceeds ours. Only those would indeed be interested in attacking us to chop us up into small pieces and eat us by small bitefull, as they do for abalone, crabs and lobsters.

But do such monsters exist? Are they not perhaps only the figment of the over-fertile imagination of poets and thriller writers? Are they merely the exaggerations of sailors thirsting fantasy, or of boastful divers?

So far, we have spoken only of octopus of a length not exceeding that of a man, weighing at most twenty kilos. Are there any bigger ones? Are there not other cephalopods of a larger size? This is what we will discover in the following chapters, going from surprise to surprise.

First, however, it will be useful to cast a glance at that other novel which is responsible for popular exaggerations about cephalopods and, by reaction, for equally outrageous denials on the part of scientists; Jules Verne's *Twenty Thousand Leagues Under The Sea*. What Hugo did in *The Toilers of the Sea* to the anatomy and behaviour of the octopus, Verne did in his novel for the identity of giant cephalopods.

Chapter 3

A FAMILY ALBUM OF CEPHALOPODS

There is a tendency today to think of Jules Verne as a scientific prophet[1]. In fact, he was a superb writer of novels for young people, an outstanding achievement which should suffice to his glory. Although one may think of him as the father of modern science-fiction, it is not him but rather Herbert-George Wells, a real scientist, who really launched this genre.

When one takes the trouble to consult the scientific literature available at the time when Jules Verne was writing his "Voyages Extraordinaires", one notices that the novelist never really invented anything, and that his futuristic views are rather naive, and often impractical extrapolations.

To stick to *Twenty Thousand Leagues Under The Sea*, written around 1867-68, one notes that Captain Nemo's famous *Nautilus* is simply a copy of the submarine *Le Plongeur* designed by a naval captain, Simon Bourgeois and an engineer, Charles Brun. This vessel was launched in Rochefort in 1863 and performed well over a number of dives. Although it was propelled by compressed air, not long before, a professor at the University of Montpellier had built a small scale model of a submarine propelled by an electric motor, just like Jules Verne's *Nautilus*. Even the name *Nautilus* was not new for a submarine: the American Fulton had used it for his prototype, designed in 1798 and constructed in 1800. Fulton was an

---

[1] *Philippe De Baleine wrote in* Paris-Match *(2 Feb 1952) that "The miracle of Jules Verne is his geographical and scientific precision. The marvels which he describes do not contradict reality but transform and enhance it. Starting with scientific premises current in his time, he built a fantasy world half-a-century in the future. He invented scuba diving and hunting, the atomic bomb, television, V2's, spelunking, electricity as a motive force, low pressure gas lamps, liquid-air bombs, the use of the ocean's thermal energy, microphone and loud speakers, tanks, skyscrapers, advertising by projecting light on clouds, astronautics, helicopters, cruise missiles, etc..." Realising at that point that he might perhaps have been carried away by his enthusiasm, Mr. de Baleine reconsiders: "It is perhaps not exactly correct to say that he invented all these thing. More precisely, his genius lies in that he foresaw their applications before all the scientists of his time." Everybody knows of course that scientists, typically absent minded, invent only for fun and never give any thought to the applications of their discoveries.*

idealist, full of utopic proposals. He thought that he could end all wars by making them intolerably horrible through the invention of his fearsome subsurface vessel. Jules Verne obviously modelled Captain Nemo after Fulton. When one realizes that this unsung genius, whom Napoleon labelled as a charlatan and a crook, and who met similar incomprehension in England, had for his motto *Libertas maris, terrarum felicitas*, it is difficult not to recognize him when thinking of the proud and wary figure of the commander of the *Nautilus*.

Regarding the improvement brought by Captain Nemo to the Rouquayrol-Denayrouse diving apparatus used by his crew members, it would surely have killed them through brain hemorrhage at their very first dive[2]. It is rather easy, at a time when divers descend to a depth of 40 metres, to imagine them promenading at 150 metres; however, one has to provide them with some rather more realistic apparatus, satisfying the laws of physics as well as those of physiology.

Alas, everything in Jules Verne's work is merely haphazard. He certainly read extensively but, lacking scientific training, he did not always choose his texts with much discernment, and often regurgitated the information in a distorted way for lack of having properly assimilated it. Everything is approximation when not coarse error. Do not re-read Jules Verne in later years if you have in the mean time acquired some solid background: you would be disappointed and perhaps, in some places, horrified. You might even experience some bitterness in noticing that this marvelous story teller led you astray in your youth. With his pedantic pseudo-science, he stunned you under an avalanche of technical terms and erudite words which actually hide the emptiness of his knowledge. It doesn't take much sleuthing to trace back the references where he drew his documentation, and from which he copied or rewrote long passages without much apparent understanding. Listen, for example, how he pontificates in a pseudo-doctoral tone, through the mouth of Mr. Aronnax, adjunct professor at the Museum of Natural History of Paris. Convinced at first that the *Nautilus* is some kind of marine monster, he defines its zoological nature in these words: "I believe in the existence of a powerfully organized mammal, belonging to the branch of the Vertebrates, like the baleen and toothed whales, and the dolphins."

Might perhaps Mr. Aronnax know of any mammals which do *not* belong to the Vertebrates? Does he think by any chance that among mammals only cetaceans possess vertebrae? Poor Museum that would be afflicted with a zoology professor who would express himself in such a bizarre manner! Not to mention that this gentleman massacres the names of experts and of animals, calls seals "magnificent cetaceans", believes that whales live for a thousand years, speaks of museums where they preserve the *skeleta* of octopus and thinks that infusoria are the builders of coral reefs. Nothing astounds him: he sees without surprise a giant salamander, freshwater animal

---

[2] *This point has been the subject of a brilliant demonstration by P.de Latil and Jean Rivoire in* A la recherche du monde marin, *pp 148-151*

par excellence, swimming offshore in the Sea of Japan; he finds quite normal the presence near the South Pole of arctic animals, such as auks and walruses; he is not even surprised to observe living trilobites, even though all his colleagues believe that their group became extinct three hundred million years earlier!

What surprises me, personally, is that during his journey at the bottom of the oceans, Mr Aronnax *never* ever met with a creature which was not familiar to him, and to which his ass of a lackey could not immediately stick a Latin label. Nowadays, every oceanographic expedition brings up something new from the abyss. However, for Mr. Aronnax, the sea no longer had any zoological mysteries. All that Jules Verne's scientists have going for them is their knowledge; they don't have the curiosity and wonder that distinguishes the true scientist.

Since our novelist drew indiscriminately from every source, he ended up confusing them all. Thus, he rashly blends together the dugong, a lazy and peaceful 3-metre long Sirenian, with, on the one hand Steller's sea-cow, its now extinct cousin from Bering Strait, and on the other a pinniped armed with powerful tusks, the walrus, into a ferocious 7-metre long dugong which attacks his heroes somewhere in the Red Sea...

There is also the famous episode of the struggle pitting Captain Nemo and his crew against a horde of giant squids, six to eight meters in length. That such titans do exist had already been proved a few years earlier, in 1861, by a documented report from the captain of the destroyer *Alecton*, commander Bouyer. Monsters of that species had subsequently received the scientific name *Loligo bouyeri*, i.e. Bouyer's squid. We shall return to it below.

Of course, with his usual sloppiness, Jules Verne doesn't fail to speak of a commander Bouguer (!), but worse, he insists on calling giant cephalopods sometimes squids, sometimes octopus. What a zoological mish-mash![3]

Following Jules Verne, hundreds of writers and reporters have repeated this mistake. The long-term result has been a deep misunderstanding whose importance will become clearer below.

No! Do not by any means go back to your childhood library if you really loved Jules Verne, that magician who undoubtedly inspired many a research career. Do not tarnish your memories of his top-hatted adventurers. Do not destroy the charm of your juvenile enthusiasm. It feels so good to think of an erudite Jules Verne, the universal and prophetic sage who guided the imaginary travels of our younger days.

---

[3] *I have quoted here only a few zoological mistakes among many, but Verne's ingorance was equally monumental in other fields: physics, chemistry, geology, astronomy, and so on...Just to quote a striking example, he places the divers from the Nautilus in water at a temperature between 6°C and 7°C below zero, i.e. in a block of ice!*

THE KRAKEN AND THE COLOSSAL OCTOPUS

Cephalopods, which play such an important role in fiction, to the point that they are almost an indispensable prop of any underwater adventure, occupy a much more modest and discrete place in scientific literature. Specialists who devote their research to cephalopods are so rare as to be thought somewhat eccentric by their colleagues. Moreover, through some strange quirk of fate, many of them, and not the least prominent, ended up losing their mind! All of which contributes to the aura of mystery and weirdness which surrounds them. One can easily imagine how a romantic biographer could exploit such circumstances... madness taking hold of the brain of a scientist after repeated encounters with nightmare creatures...

Reality is rather more pedestrian. Nothing looks less like an infernal workshop than the laboratory of a teuthologist (as are called zoologists specializing in the study of cephalopods). They are neither the stage for desperate struggles, reminiscent of the agony of Laocoon and his sons, nor for any other morbid confrontation. As any other zoological laboratory, they are filled mostly with books, journals and filing cabinets. The only horrifying glimpse might be afforded by enamel dissection pans, used for examining long-dead gelatinous flesh: actually a rather less frightening spectacle than the display of a fish store. There is nothing particularly scary about the labeled jars lined up on the shelves or piled up on the work benches. In loving contemplation of the shriveled and discolored specimens which they contain, specialists are most likely to think of them as "splendid" or "magnificent". At worst, they might remind the layman of bowls of marinated herrings or salted tunicates, to be perceived either with mild disgust or pangs of hunger, depending on taste.

Should anything have any effect on the mind of the cephalopod specialist, it would not be the fearsome aspect of the specimens at his disposal, but their rarity.

Cephalopods as a group, are rather poorly known. Even professional naturalists sometimes make serious mistakes about them. The information available on the anatomy of the majority of species is extremely meager, obtained from the examination of only a few rare specimens, sometimes only of fragments found on a beach, or in the stomach of a shark, or of a whale. Many cephalopods live deep in the ocean; those within range of ordinary trawls are generally too quick or too smart to be caught. Chance remains the collector's most effective collaborator. Besides some nearshore species, most of what is known about the biology of these creatures remains conjectural.

However, to ensure that all readers shall be able to find their way in the diversified world of cephalopods, and participate with a minimum of stress in the thrill of the research which has led to the discovery of the giants among them, I will first trace an outline of their systematics.

The first subdivision of the class Cephalopoda is based on a feature which is not externally recognizable: the number of gills. On that basis, we distinguish the sub-class Tetrabranchia, which have two pairs of gills, from than of Dibranchia, which have a single pair. Because this trait cannot be distinguished among fossils, since gills leave no traces, these names have been rejected by modern teuthologists in favour of Protocephalopoda and

A FAMILY ALBUM OF CEPHALOPODS

Metacephalopoda. The distinction is, however, irrelevant, since Protocephalopoda include those which have, or are suspected to have had 4 gills, while Metacephalopoda include those which have, or are suspected to have had 2 gills.

Not to worry! It is quite easy to recognize at a glance those cephalopods which belong to one or the other group. Actually, the first group is currently represented only by the genus *Nautilus*, the only living cephalopod which inhabits a beautiful spiral shell[4]. The second group includes all other living species. Among these, many still do have a shell, but it is always much reduced in size, and carried internally, or rather hidden under the mantle. The only cephalopod which might superficially be confused with a nautilus is the argonaut (*Argonauta*), whose female seems to possess a shell. However the fragile, whitish structure inhabited by this creature is not a shell, but a nest, secreted by two of the arms of the mother-to-be, broadened in the form of paddles. In English, the same generic name is used for the nautilus and the argonaut, the first one being called the pearly nautilus, because of the nacreous interior of its shell, and the second the paper nautilus, because its pseudo-shell is white, thin and translucid. In fact, members of these two genera are only distant cousins and have very little similarity. The male argonaut does not have any arms broadened into a shell-like structure and looks a lot like its relative the ordinary octopus: like it, it has eight arms studded with suckers. The nautilus, on the other hand, has a very large number of arms, about a hundred, which are short and smooth, more like those of a sea anemona than the powerful arms of the other cephalopods. This accounts for the distinction which some zoologists prefer to draw between Tetrabranchia and Dibranchia, labelling them as Tentaculifera (tentacle-bearing) and Acetabulifera (sucker-bearing) respectively.

There is thus no possible confusion between the nautilus, the South Pacific shelled cephalopod, and the other representatives of the group. Since this animal never exceeds 20 centimetres in length, we shall not be concerned with it in our study of giant cephalopods and may thus at the same time forget about the whole sub-class Tetrabranchia or Tentaculifera.

We shall have no trouble finding our way through the other group, that of the Dibranchia, or Acetabulifera: its principal members, the octopus, the cuttle-fish, or sepia, and the squid or calamary, are well known. All three are often found on the menu, at least in France and other Mediterannean countries. The stomach provides ready access to the mind!

---

[4] *Protocephalopods used to be very abundant. In the Palaeozoic, a wide variety of kinds of nautilus dominated the marine fauna. In the Mesozoic, they were replaced by representatives of a neighbouring group, whose shell looked like a ram's horn, and which were called for that reason Ammonites, Ammon being the name taken by Jupiter when he changed himself into a ram to escape from the Titans. From the Tertiary on, no one group continued to dominate the ocean's fauna and only five or six species of* Nautilus *somewhat miraculously survived.*

THE KRAKEN AND THE COLOSSAL OCTOPUS

Among octopus, the body looks more or less like a spherical bag, of which the head is an integral part. As its name implies, these molluscs have eight legs[5], also called arms because they are used for grasping as well as for locomotion. These eight legs are all of about the same length and are all armed on their underside with a double row of suckers. Only in the musk octopus (*Eledone moschata*) is there a single row of suckers.

Squids and cuttle-fish have a rather different body structure. Their head is clearly differentiated from the rest of the body: there is a distinct narrowing at the neck. The body itself is elongated and fringed with a horizontal fin. In the cuttle-fish, this fin surrounds the whole body; in the squid, it consists of a pair of posterior lobes, similar to an airplane's rudders. Actually, these lateral fins of the squid are not merely steering surfaces; they are capable of twisting motions and act mostly as propellers.

In addition to the eight arms which they have, like octopus, squids and cuttle-fish also have two very long tentacles, swollen at their distal end into a kind of flattened club studded with serried ranks of suckers. In the cuttle-fish, these two tentacles are retractile; at rest, they are drawn back in appropriate furrows. To catch its prey, the scuttle-fish, or sepia, harpoons it, whipping out its tentacles. Squids prefer to wiggle these tentacles in front and around them; they function as antennas. A number of researchers like to distinguish these tentacles, calling them *pedonculated* arms, from the eight shorter or *sessile* arms.

In the literature, even in the scientific literature, the name tentacle is bestowed, rather legitimately, upon the arms of the octopus, but also, and somewhat confusingly, upon the sessile arms of the squid. Here, just to keep things simple, I will often speak without further distinction of arms, or tentacles. Let it be clearly understood that, whenever there might arise any ambiguity, arms will always refer to the eight "sessile arms" and tentacles to the two "pedonculated" arms.

In summary, there should be no difficulty in distinguishing the octopus group, from that of squids and cuttle-fish, by their total number of arms; the former being the order Octapoda (with eight arms), the other that of Decapoda (with ten arms).

The three types of cephalopods, besides their general appearance and their number of arms, differ in a number of fundamental anatomical characteristics. Originally, all cephalopods used to have an external shell, as nautilus still does, made of a series of successive compartments. However, this shell was not always rolled up in a spiral; it could be linear, in the form of a cone; or curved in the form of a horn-of-plenty, or of a bishops' crosier, depending on the degree of folding. Among one of the ancestors of all of the modern Dibranchia, the belemnites of the Mesozoic, the shell had already begun to degenerate. Having become too small too house its occupant, it had

---

[5] *It used to be thought that these eight legs originated in a division of the basic fleshy foot found in all primitive molluscs. Nowadays they are thought to have grown from the buccal region: a kind of specialized set of lips. The siphon would be the only vestige of the ancestral foot.*

become gradually enfolded into its flesh. However, in those molluscs, it was still composed of three distinct parts: a strong calcareous rostrum. similar to the tip of a javelin (in Greek *belos,* hence the name belemnites): within the rostrum, the tip of which stuck out of the body, there was a conical cavity, divided up in small compartments, the phragmocone, last trace of the internal partitions of the shell; finally, a horny blade completed the structure on the dorsal side. Among today's Dibranchia, these various parts have evolved into a wide variety of forms which can guide us in establishing an unambiguous classification.

Among cuttle-fish, the rostrum has shrunk and virtually disappeared. In contrast, the phragmocone has completely invaded the dorsal region and its thick edges have become fused: this is the well known cuttle-bone often found on beaches. Bird fanciers often suspend it in the cage of canaries so that they may use it to sharpen their beak. When crushed, this "bone" is used in pharmaceutical products, for example, tooth powder.

I should mention here one curious cephalopod, related to the cuttle-fish, in which the phragmocone has degenerated less than in other dibranchia, including belemnites. This is spirula, a thumb-size cephalopod living at depths below 3,000 metres. Spirula is still housed in a spiral shell made up of distinct compartments; however this shell is completely hidden under the mantle. Quite a trick, to live in an external shell which nevertheless appears internal since it is hidden from sight!

Among squids, in contrast to cuttle-fish, the phragmocone as well as the rostrum have completely disappeared; there remains of the complex belemnite shell only the horny dorsal blade, which has become a kind of sharp stick called the sword (*gladius*), or pen. This is what Themistocles had in mind when he told the Eretrians: "You are like squids, with a sword but without a heart, capable only of fleeing!"

To accuse squids of being heartless is pure calumny, based entirely on ignorance; the second insult is perhaps more justified and refers to the peculiar mode of propulsion of cephalopods. Among squids, as in octopus, the locomotory funnel faces in the same direction as the tentacles. As a result, when the animal swims, its propulsion system pushes it hind-side first. There is thus some ground to say that they move backwards. To say that they know only how to flee is however a gross exaggeration. They can, also like the octopus, bend their siphon towards the back and move forwards, tentacles first. Their backward jet propulsion system is used only in emergencies, when a rapid escape is necessary. Using only the wavy motion of their hind fins, or of their lateral arms, often covered with a membrane, squids can readily move in any direction. They can even lazily row along using their paddle-shaped longer tentacles.

The squid actually owes its actual name, calamary, to its pen and ink sac. Already in 1558 Guillaume Rondelet, the father of ichtyology, wrote that "...in Languedoc, this mollusc is called calamar because one finds in it everything necessary for writing: ink and a small knife, which looks at one end like a penknife, and at the other like a quill."

Scribes formerly used a sort of portable writing kit, in latin *calamarium* (from *calamus*, a reed which Romans used to write with.). The name calamary, is also appropriate because the two longer tentacles of the animal are reminiscent of quills dipped into an ink pot.

The point to remember is that squids have under the skin of their back a "pen" or "sword" which is the last vestige of the complex shell of their ancestors. The latter reaches its ultimate degree of atrophy in the octopus, where it consists of two minute stylets, which are used for muscle attachment.

Thus, no shell, or hardly at all among Octopods. And among Decapods, only a chalky bone in cuttle-fish and a horny pen in squids.

The difference in the anatomical structure of the octopus and their ten-armed cousins has a direct impact on their choice of habitat and their behaviour. The octopus is built basically for crawling and is essentially a bottom, or *benthic* cepahlopod. The squid, thinner and equipped with fins, lives in mid-water and is a *pelagic* animal. These differences readily explain their characteristic hunting techniques. Hidden in its lair, the octopus awaits the passage of its prey, and pounces on it unexpectedly. The streamlined squid catches its prey by overtaking them. To use a comparison from the world of land predators, the octopus attacks its prey by surprise, like a leopard, a forest predator, while the squid runs it down like a lion, a hunter of the open savannah.

Squids, sharp-ended like bullets, have such propulsive power that they may, either in play, or carried away by their momentum, fly in great leaps above the water surface. In this manner, they occasionally land on the beach or more commonly at sea on the deck of ships. This ability to glide in the air was already well known in antiquity. Pliny, Oppian and Aelianus all mention it, and the erudite latin poet Varro (116-27 B.C.) even claimed on that basis that the Latin name of the squid (*loligo* or *lolligo*) must have originally been written *volligo*. Although this is very speculative, no alternate etymology has been suggested.

It is amusing to note that, until recently, flying squids were thought to be legendary, dismissed and forgotten as yet another old-wives' tale. Thus, as late as 1833, flying squids were announced as a new phenomenon, never observed before. That is certainly how a British traveller, colonel W.H.Sykes, saw it when he presented to the members of the Zoological Society three specimens of a species of slender squids[6] which had leaped onto the bridge of the *Lady Feversham* during her return trip to England in 1831. During that meeting, Prof. Richard Owen claimed to know of another example of this remarkable facility: one Dr. Henderson had offered to the Museum of the Royal College of Surgeons two specimens picked up on the deck of a ship in

---

[6] *The particular species was at that time called* Loligo sagittata. *However, soon afterwards, all slender squids were to be split apart from ordinary or pen-squids by Alcide d'Orbigny, who created for them in 1835 the new genus* Ommastrephes, *meaning those who turn their eyes (from the Greek* ommata, *eyes and* strephein, *to turn).*

the Mediterranean. No one in that learned assembly thought of exhuming the precious wisdom of antiquity.

In the diary of his *Voyage dans l'Amérique méridionale*, published from 1835-1843, d'Orbigny described such flights in more detail:

"We saw at night leap onto the bridge of a ship, 15 to 20 feet [4 to 6 m] above the sea surface, an Ommastrephes bertrami, undoubtedly trying to escape from some fish."

He added that this skill was also found in other squids, particularly *Sepioteuthis* which has well developed fins, likely to make it a successful glider. It has recently been discovered that a clawed squid, *Onychoteuthis banksi* is also capable of gliding flight. This fact was discovered in 1949 by a British specialist, Dr. W.J.Rees, who, while examining the collections brought back by various oceanographic expeditions, noticed that specimens of that species had never been caught in nets but had always been picked up from the deck of ships.

Thor Heyerdhal relates that during his famous Kon-Tiki expedition, there was one night during which their balsa raft was literally bombarded with squids. From that to claim, as did Trebius Niger, a lieutenant of Lucullus quoted by Pliny, that squids may rain in such great numbers upon a ship as to submerge it, there seems to be considerable exaggeration. In any case, it is clear that among flying squids we are very far from the octopus, which are truly the crawlers of the cephalopod world.

The cuttle-fish, or sepia, less streamlined and thus not as good a swimmer as the squid, has a status intermediate between that of the squid and that of the octopus. It does not live alone in a hole, but it also does not swim very far above the bottom. It often lies on sand or mud, where it digs in like a flat fish and hides from its preys using its camouflaging abilities. It can then harpoon them by surprise by uncoiling its tentacles. During the summer, the cuttle-fish approaches the shore, where it lays its eggs. These are attached in clusters on sea grass or algae, and are often called sea-grapes. What cuttle-fish do during the rest of the year is not well known; they have been caught in trawls more than 100 kilometres offshore, but it is probable that they stay on the continental shelf and do not venture into abyssal depths.

Thus, it is in principle impossible to confuse the octopus, the cuttle-fish and the squid, representative types of the three living great sub-orders of Dibranch cephalopods: Octopoids (within the order Octopoda), Sepioids and Teuthoids (in the order Decapoda).

Things get more complicated when we stray away from these representative specimens! Even the number of arms is not absolutely fixed: there is a Decapod which, through secondary loss of some tentacles, has only eight arms (*Octopodoteuthis*).

Further, one should not imagine that all Octopoids are benthic cephalopods, like the common octopus. In 1838, a Danish specialist, professor Eschricht, discovered an octopus equipped with two small lateral fins. He confused it for a squid and called it *Cirroteuthis*, meaning ciliated squid. It was actually a swimming octopus, for which a new sub-order had to be created: the Cirroteuthoidea. It did not take very long to discover other ciliated octopus, but it was found that they were not all pelagic: some lived on the bottom, especially at great depths. So, Octopoidae (ordinary octopus) and Cirroteuthoidae (ciliated octopus) could not be distinguished merely by the nature of their habitat.

While speaking of ciliated octopus, I shall refrain from commenting at length about a most peculiar representative of these monsters, living deep in the abyss, whose name *Vampyroteuthis infernalis* (infernal vampire-squid) sounds like the title of a horror story. This is a "living fossil", described by Chun in 1903, to the study of which an American specialist, Miss Grace Evelyn Pickford has devoted her career - quite an unusual occupation for a young woman. According to Miss Pickford, the presence of an extra pair of atrophied tentacles makes this octopus a close relative of squids, which justifies the creation of a new order just to accomodate it.

Finally, all Sepioidea do not, like the cuttle-fish sport a fin which completely surrounds their body. Spirulidae, which we mentioned above, and Sepiolidae, a kind of miniature cuttle-fish, only have a pair of tiny hind fins, even smaller than those of squids. As a counterpart, there is a Teuthoid, *Sepioteuthis*, which has lateral fins stretching all the way around its body.

Of course, those are exceptions, but they must not be forgotten when comes the time to identify an animal known only through a quick glimpse or a small material fragment. Actually, the only decisive criterion in classifying Dibranch Cephalopods into one of the four living sub-orders (Octopoids, Cirroteuthoids, Teuthoids and Sepioids) is the shape of its shell, or what remains of it. However, knowledge of the habitat and an external examination are usually enough to quickly distinguish between octopus (ciliated or not), squids and cuttle-fish.

However, a terrible confusion still reigns in the popular, and sometimes even in the scientific, literature devoted to these marine molluscs.

Before Victor Hugo, the word "pieuvre" did not even appear in French dictionaries. It is the great poet who, through his prestige, popularized the word that Norman fishermen used in their dialect for what was then known as "poulpe" (from *polypus*: many-footed). An entirely legitimate linguistic licence, since the action of the novel *The Toilers of the Sea* takes place in the Channel Islands. However, the appearance of this new word sowed a cloud of confusion among the non-initiated. If "pieuvre" was an acceptable synonym for a "poulpe", one might use the same word to describe squids, which also have many feet. From then on, various French authors started confusing octopus and squids. What a bouillabaisse!

Jules Verne had no qualms about calling the giant squids which attacked the *Nautilus* "poulpes", i.e. octopus[7]. *Twenty Thousand Leagues*

---

[7] *In the movie which he made on the basis of Jules Verne's novel, Walt Disney did not make this same mistake. His rubber squid looks rather lumpy, but it is nevertheless a squid, with entirely correct anatomy.*

A FAMILY ALBUM OF CEPHALOPODS

Under the Sea has become such a classic of juvenile literature that this error [at least in French] takes an early hold in the mind of young people. It is finally not even surprising to find that squids are called "pieuvres" in *Faune des Océans*, the French translation of *A Natural History of the Seas* of the great zoologist Ernest G. Boulenger, former director of the aquarium of the Zoological Society of London. It is obviously an error made in translation[8].

To make matters worse, the dimensions of octopus and squids are generally measured quite differently. When stressing the large size of a cephalopod, it is logical to quote its extreme dimensions. Thus, for an octopus, which spreads out radially, it is normal to quote the span: the greatest distance between the tips of laterally opposite arms. A squid, on the other hand, is much more elongated and the longest tentacles stretch out two or three times longer than the body; the logical dimension is then the total length, from the tip of the body to the end of the longest tentacles.

In my opinion, the latter manner of describing the size of a squid is rather unfortunate. The length of the rather elastic whip-like tentacles is quite variable, depending on their degree of stretching, and in some species, they become visible in a live animal only at the moment when it throws them at its prey. Nobody would ever think of including in the length of a chameleon that of its extraordinarily extensible tongue! I will thus often speak below of the length of the body-mass of the squid, which will refer to that of the body proper plus the head and the eight short arms.

Even when we restrict ourselves to the dimensions of the bulkier part of these animals it is rather difficult to obtain precise measurements. To quote a reputed Norwegian scientist, Dr. Erling Sivertsen:

"Cephalopods are molluscs, a fact which becomes quite evident when one tries to measure a specimen hauled up on the beach. It spreads out more or less flatly depending on the nature of the substrate, and after a few days, or even only a few hours on a dry beach, its body loses so much water that it visibly shrinks. At the same time, the arms become so elastic in their consistency that varying measurements can be obtained by stretching them. That is why comparison of different specimens is difficult and yields uncertain results."

Please remember then, that throughout this book the dimensions quoted are given only to provide some general order of magnitude.

To make things worse, the anatomy of cephalopods is so unusual for human beings, that many people seem unable to imagine it correctly, and make the worst mistakes when it comes to infer some measurements from others. It is of course rather strange to have to find the length of an animal's body by subtracting from its total length that of the sum of the lengths of its

---

[8] *This also happened in the French and the English translation of Thor Heyerdhal's* Kon Tiki Expedition, *where translators made the same mistake...excusable perhaps because there exists in Norwegian only a single word Blekksprut to denote all types of cephalopods.*

head and its arms.

One mistake, which even excellent naturalists sometimes make, is to calculate the span of an octopus by multiplying its total length by two. It is just the length of the arms which one should double! For the common octopus (*Octopus vulgaris*) the head and the body are together about one sixth of the total length. Thus, a three meter long specimen would not have a span of six meters, as I have seen one scientist calculate it. The total length of 3.o m is made up of a central body bulk of 0.50 m plus 2.50 m long tentacles: the span is then 5.0 metres.

The difference is significant. However, when squid and octopus are mixed up, the error in calculation becomes catastrophic!

In the fall of 1954, I wrote for a magazine devoted to underwater spear-hunting an article on giant cephalopods in which - by some quirk of fate - I emphasized the errors arising from a confusion between squids and octopus. To make my presentation more topical, the editor prefaced it with a report released shortly before by a press agency (France Presse)

IN NORWAY

"A giant octopus has recently been found, dead, in the waters of Trondheim fjord.

"According to Mr. Erling Silversten, director of the Academy of Sciences, the tentacles of the monster, which was found drifting lifeless in the fjord, measure 17 metres in length, and its body has a diameter of more than 2 metres."

The clumsy style of the release and the misspelling of the name of my esteemed colleague Dr. Erling Sivertsen, director of the *Kongelige Norske Videnskabers Selskab* might seem irrelevant from a zoological point of view. They nevertheless revealed a poor translator and suggested an unfaithful rendition of the original text which excited my suspicion.

As it is quite unusual to find dead or dying octopus adrift at sea, while on the other hand this is quite common for squids, I could suppose that the press-release most probably referred to the latter. Given the length of the tentacles and especially the diameter of its body, this seemed to be a giant specimen, whose central body-mass might reach 18 metres, if one was to believe the measurements given.

Fully convinced from the error in translation that the animal was an octopus, the editor calculated the span of the beast by multiplying by two the length of the tentacles, which yielded 34 metres. Then, thinking that an octopus is built like a starfish, with a central body from which the arms spread out, he added an extra two meters for the width of the body, and wrote up this sensational headline:

AN OCTOPUS WITH A 36-METRE SPAN

Finally, to cap it all, he had the illustrator draw a "sketch of the giant octopus with a human scale". The result was truly impressive!

Actually, the dimensions mentioned in the dispatch were completely incorrect. As Dr. Sivertsen was kind enough to inform me, the length of the tentacles of the squid - for it was indeed a squid - was not 17 metres, but 7.10 metres; the body was not 2 metres in diameter, but 2.14 m in length. This makes a great difference when one is dealing with an animal which is as thin as a cigar! By tracing back the source of the information, I had significantly shrunk the "giant octopus" of the press agency.

As surprising as this may seem, it was not so much the exaggeration of the dimensions of the mollusc which made it so incredible as its mistaken identity. What was originally imagined, after undue stretching of its limbs, as a gigantic creature, weighing around 60 tons, had changed into quite a different animal, weighing at most ten tons, but, given the state of our knowledge, even more incredible! And this, in spite of the testimony of Dr. Sivertsen and my own apparent assentiment...all because of a "small" error in translation and the eagerness of a copy writer.

This monster, as have many others, was created in the mold of ignorance and loose reporting, to which writers like Hugo and Verne have mightily contributed, but to which also some scientists have put their pinch of salt.

Part Three

THE FAMILY TREE OF THE FEARSOME KRAKEN

"The possible is a formidable matrix. Mystery forms itself into monsters. Fragments of shadow issue from the inherent mass, rend themselves, break loose, roll, float, condense themselves, and borrow from the surrounding darkness; they are subject to unknown polarizations, become alive, and compose for themselves an indescribable body out of the darkness, and an indescribable soul out of noxious vapors, and depart - these larvae - to dwell among living things. It is somewhat like shadows converted into beasts."

VICTOR HUGO, *The Toilers of the Sea*

Chapter 4

GIANT POLYPS OF ANTIQUITY

―――――

"What song did the sirens sing?" asked Sir Thomas Browne, "Under what name did Achilles hide among the women?" One need not be a XVIIth century philosopher to worry about such apparently futile questions. Today, the meticulous zoologist seeking traces of tentacled monsters in the writings of antiquity is quickly brought to an equally peculiar question, for example:

"To which zoological species belonged the monster Skylla which Ulysses had to face on his return journey?"

It is encouraging for a researcher to learn that such an erudite man as Sir Thomas was concerned with this kind of question, applying to such problems the taste for enquiry which triggered the renaissance of the scientific spirit in England.

"Bothersome questions, perhaps," he continued, *"but not beyond the scope of reasonable conjecture."*

Actually, it is quite possible, as already did Eusebe de Salverte at the beginning of the last century, to show that in describing the monstrous Skylla of the Odyssey, Homer must have been inspired by some giant octopus.

Circe had warned Ulysses: only two marine routes would lead him back to his kingdom, but he had to avoid at all costs one of them, where his ship could be smashed on a dangerous reef, the *Planktai*. Although the second route was much preferable, it was not without peril: at one point it passed between two rocks, each one guarded by a monster. The first rock reached to the sky and was pierced by a tunnel: this was the lair of Skylla. At the foot of the second, much lower rock, the mighty Charybdis swallowed and vomited, three times a day, the deep and dark waters of the sea.

Charybdis is obviously a whirlpool. The fantastic portrait that the witch traced of Skylla, "which not even gods would be glad to encounter", is rather confusing; however, behind the distorting lens of epic imagination, one will readily recognize an octopus, although clearly one of impressive dimensions[1].

> *"In that cavern Skylla lives, whose howling is terror.*
> *Her voice indeed is only as loud as a new-born puppy*
> *Could make, but she herself is an evil monster. No one,*
> *Not even a god encountering her, could be glad at that sight.*
> *She has twelve feet, and all of them wave in the air. She has six*
> *Necks upon her, grown to great length, and upon each neck*
> *There is a horrible head, with teeth in it, set in three rows*
> *Close together and stiff, full of black death. Her body*
> *From the waist down is holed up inside the hollow cavern,*
> *But she holds her heads poked out and away from the terrible hollow,*
> *And there she fishes, peering all over the cliffside, looking*
> *For dolphins and dogfish to catch or anything bigger,*
> *Some sea monster of whom Amphitrite keeps so many;*
> *Never can sailors boast aloud that their ship has passed her*
> *Without any loss of men, for with each of her hands she snatches*
> *One man away and carries him off from the dark-prowed vessel."*

Quite wisely, Ulysses chose to sacrifice six sailors to the Skylla's voracity rather than to see his whole ship sink into Charybdis' gluttonous maw. Of two evils, he knew how to choose the least. Whoever invented the expression "to fall from Charybdis into Skylla" obviously did not share this wisdom.

That is however not the point. What we are concerned with here is the identity of Homer's barking, polycephalic monster.

One may argue that cephalopods do not bark and have only one head, like everyone else; one may also wonder at the number of necks and feet that

---

[1] *According to one of the most reputed commentators of Homer, Victor Bérard, the two rocks are to be found in the Tyrrhenian Sea, north of Sicily, more precisely between Lipari and the island of Vulcano. In fact, one finds in current nautical instructions, a description which is strikingly reminiscent of the advice of Circe. There are still two routes available to a coastal steamer leaving Monte Circeo, on the coast of Latium, for a southern port: they go on either side of Sicily. The eastern route goes through the Strait of Messina, where there is no dangerous reef, but a small whirlpool, the Garofalo. Along the western route, going past Trapani, the guide describes two remarkable rocks: "The northernmost, the Pietra Lunga, 47 m in height, is a volcanic promontory pierced at its base by a hole through which small boats may bass; the other, the Pietra Menalta, is much lower and usually covered with sea gulls." Shift the whirlpool to the base of the Pietra Menalta and you will find in its details Circe's description. Homer wasn't too particular about geographical precision, as is demonstrated by his putting on the "road not to be taken" the Planktai, a feature borrowed from the older epic of the Argonauts, and which actually refers to the Bosphorus. It would seem that the poet has merely tried to point out the various kinds of hazards menacing his heroes in Sicilian waters.*

GIANT POLYPS OF ANTIQUITY

Skylla is supposed to have. It remains nevertheless that this passage is strongly reminiscent of an octopus lurking in its lair. Anchored to the cracks in the rock by two of its arms, it spreads out the other six to catch its prey. The end of the tentacles, often rolled up in balls, are readily perceived as heads. One may even suppose that the "three rows of teeth, close together and stiff, full of black death" refers to the rows of suckers lined up on the arms. Although there are only two rows of suckers in most octopus, they are packed in such a tight mosaic that only by examining a dead specimen is it possible to distinguish their arrangement. Finally, even though the octopus has no voice, it nevertheless does make a curious sound with its siphon when it is taken out of its element and expulses water in convulsive spasms out of its mantle. In a pinch, this small noise might remind one of the yappings of a puppy[2].

Skylla was thus an octopus, but an octopus on the scale of the heroes and demi-gods of Homeric Greece[3].

After all, it is neither more, nor less, legitimate for a poet from the seventh century before our era to endow the octopus with six jaws than for old man Hugo to announce that it has "a thousand infamous mouths". At least, the former did not, like his modern imitator, adopt a pompous didactic tone in presenting such an impossible anatomy.

All this said, it would be completely inappropriate to take the homeric epic as the first report of a giant species of cephalopods. Might as well go even further back, to the story of the Hydra of Lerna, with its nine ever-regenerating heads, which haunted a swamp in Argolid before Hercules disposed of it.

Otto Koerner, a prominent German homeric specialist, did not hesitate to interpret the Skylla episode as proof of the existence of gigantic octopus in the Mediterranean. Let's merely conclude from the text of the Odyssey that stories of octopus capable of seizing sailors have circulated since the deepest antiquity and that it is difficult to attribute such deed to a small octopus. Let's nevertheless remain circumspect. If the great Hellenic poet had no qualms about transforming the Garofalo, an inoffensive whirlpool, into the deepest chasm, could the prototype of the barking monster not have been, in a similar vein, a small octopus, just large enough to fit in a bouillabaisse?

Nevertheless, stories of monstrous cephalopods circulated widely in the Middle Ages, and especially during the Renaissance, even among the most cultivated people. This did not last. Soon enough, the critical spirit of Positivist Science rebelled against such "ridiculous fables". Where were the

---

[2] *It is interesting to note that a number of translators of the Odyssey have simply omitted verses 86-88 of the twelfth book, with the excuse that Skylla could NOT have had a puppy's voice. Others have corrected Homer by stating that the beast had a voice "like a nursing lioness".*

[3] *Perhaps the name Skylla itself might provide a slight cue as to the monster's identity. According to some philologists, it would derive from the sanskrit* skad *(to tear) and would mean "the crusher". One might wonder whether the name 'squid" might not have a similar origin.*

remains of such titans? It was pointed out that on Mediterranean shores an octopus rarely exceeds two metres in span, a cuttlefish is only 30 cm long, and a 80 cm squid, tentacles included, is already a large specimen. For many scientists of the last century such dimensions were effectively considered as records.

In a curious about-face, the most incredulous have had to change their views, and modern zoologists have had to acknowledge that the poets of antiquity and the authors of medieval bestiaries were perhaps closer to the truth that their learned detractors. The story of the progressive discovery of gigantic cephalopods is an embarrassing one for the Doubting Thomases of Science.

After Homer, there is a gap of at least three centuries before reliable references on unusually large cephalopods are found again, in Aristotle's *History of Animals*. This time, we are dealing with a conscientious and meticulous zoologist - the first worthy of the name - rarely to be found reporting lies or exaggerations. After having been the tutor of Alexander the Great, he had become his friend, and if one is to believe Pliny, the great king had put at his disposal vast resources to assist with his private investigations, including the acquisition of specimens of rare and exotic animals. The great philosopher had at his disposal sufficient means not to have to rely on suspicious tales or hearsay.

Besides the ordinary squid (*teuthis*), known not to exceed a few decimetres in length, the great natural philosopher mentions that there also lives in the Mediterranean a giant species (*teuthos*). "There are some, he wrote, which are up to five cubits long." Estimates of the length of the Greek cubit, which varied from region to region, leads to a length of 2.31 to 2.64 m[4].

"The great squids, continues Aristotle, are rare. Their shape differs from that of the smaller squids in that their pointed part is wider. A circular fin, not found in the smaller species, surrounds the entire pouch. Both the great and the small squid inhabit the high sea." In his Deipnosophistai, the Greek physician Athenaeus specified, a few centuries later, that the large teuthos differed from the smaller teuthis by its reddish colour.

Aristotle's meticulous studies of the anatomy and biology of cephalopods show that he usually based his knowledge on personal observations, even sometimes on dissections. His allusion to the "high-sea" habitat of squids suggests however that he may not have had the opportunity to examine himself the larger specimens, which he said were rare. He must

---

[4] *As Aristotle probably used the Athenian cubit, 47.5 cm long, the first estimate is probably the most reliable. That is also the value used by Paul Gervais in his study of the Mediterranean squids (1863). The length of 3.10 m quoted by his colleague Moquin-Tandon (1865) and many other authors following him is unjustifiable. To this day, we find traces of the Greek name for the squid in the Provencal* touteno *and the Italian* totano.

Above: Figure 34. The pearly nautilus (*Nautilus*)

Below: Figure 35. *Spirula*: left, ventral view; right, lateral view showing a section of the internal shell

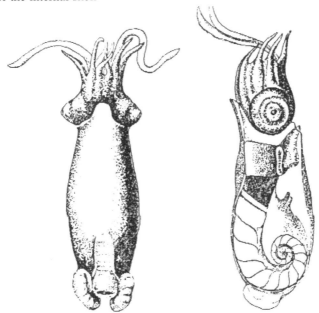

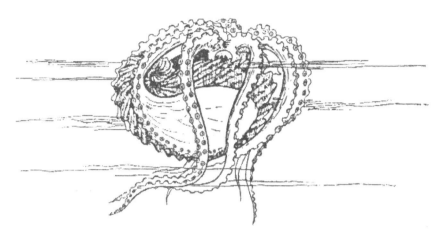

Figure 36. A female paper nautilus (*Argonauta*) in her shell

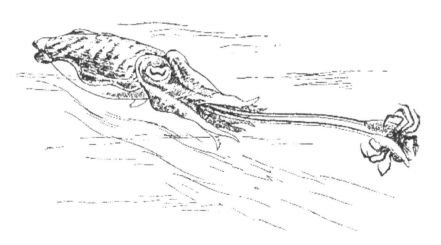

Figure 37. Cuttle-fish, or sepia, catching a crab

Figure 38. Sepia or cuttle-fish in mating colours (Photo Le Cuziat-Rapho)

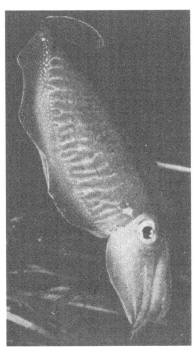

Figure 39. The musk octopus, with a single row of suckers. (Photo from Maurice Burton's *The Romance of Natural History* 1953)

Figure 40. Cuttle-fish "bone" and squid's "pen" or "sword"

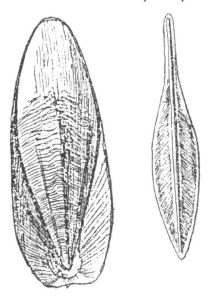

Figure 41. The suckers of a squid (left) and of an octopus (right)

 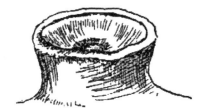

Figure 42. The suckers of the pedunculated arm of a squid (Photo Ronald Le Sueur

Figure 43. The suckers of the common octopus (Photo Otto Croy)

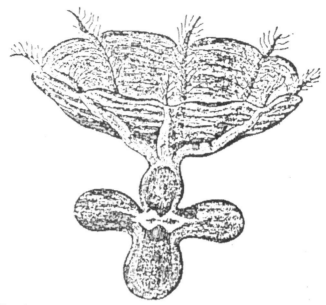

Fig 44 Ciliated octopus, showing the extreme extension of the umbrella, the membranous web that unites the arms

Figure 45. Basket star, or Gorgon's head (*Asterophyton arborescens*), after Rondelet

Figure 46. The (imaginary) octopus with a span of 36 metres

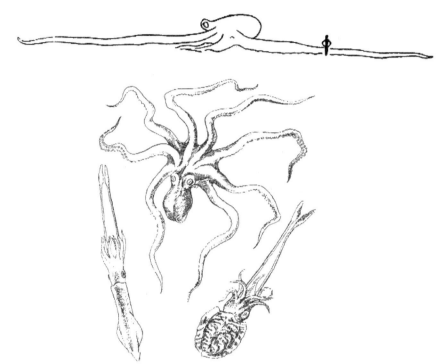

Figure 47. The three principal kinds of cephalopods: above, the octopus;
below left, the calamary or squid; below, right, the sepia or cuttle-fish

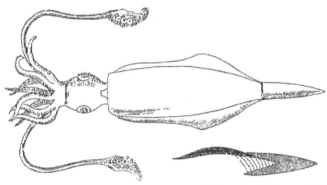

Figure 48. Reconstitution of a belemnite, with details of the shell (shown in
section)

Figure 49. The hydra, after a bestiary of the XIIth century

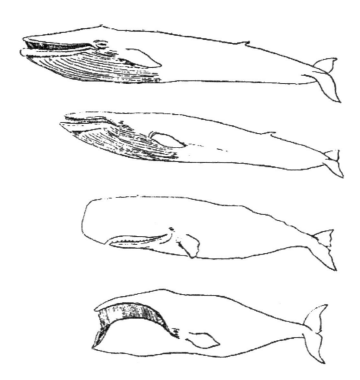

Figure 50. The largest cetaceans: from top to bottom, blue whale, fin whale, sperm whale, right whale

certainly have received the information from fishermen, which he was in the habit of interrogating, and who had probably caught some of these giant squids.

Pliny the Elder, sometimes called "The Naturalist", although he aimed more at offering a synthesis of the knowledge of his time than at recording his observations of nature, was not quite as particular about the quality of his information. On the topic of giant cephalopods, the Roman encyclopedist quotes General Lucius Lucullus, whose memoirs had been published by Trebius Niger, one of his lieutenants in the kingdom of Grenada. As Pliny had himself been colonel of cavalry under Tiberius, he could hardly be suspicious of the military, nor doubt their judgment.

Trebius related that in Carteia[5], in the Baetica, an octopus (*polypus*) crawled each night out of the sea to eat up fish from the salting vats. To end this larceny, a tall palisade had been erected around the brine vats. In vain, however, for the octopus, climbing a nearby tree, managed to cross the obstacle. One night however, the dogs heard it and alerted the fishermen by their barking. It was then discovered that the beast was of a monstrous size and managed to hold the dogs at bay by its terrible smell. It would also whip them with its tentacles or hammer them with its more robust arms. It could only be overwhelmed after it had been pierced by numerous tridents[6].

To establish the authenticity of what Pliny rightly calls a "prodigy" the animal was carved up and brought in pieces to the highest authority of the territory: the proconsul of Baetica.

"Its head was shown to Lucullus: it was of the size of a barrel and of the capacity of 15 amphorae [around 400 litres]. Its barbs [i.e.arms or tentacles] were also presented to him. Their size was such that a man could hardly embrace them, as knotty as clubs and 30 feet long [about 10 metres]. The cavities with which they were covered looked like basins and could hold the volume of an urn. Its teeth were of a size commensurate to that of the animal. The rest of the beast, weighing over 300 kilograms, was conserved as a marvelous specimen."

This information is undoubtedly based on real events, otherwise Trebius would not have had the gall in his relation to quote his own direct superior, General Lucullus. However, his report contains some rather equivocal elements.

First of all, a number of details suggest that the monster was not an

---

[5] *Carteia, or Carteja, a town in Baetica (Andalusia), near Calpe (Gibraltar). Today: Rocadillo.*

[6] *It would appear necessary to quote here the original text, often poorly translated, which made the identification of the animal impossible*: Namque et afflatu terribili canes agebat, nunc extremis crinibus flagellatos, nunc robustioribus brachiis clavarum modo incussos, oegreque multis tridentibus conficii potuit. (Naturalis Historia, *lib. IX, cap.XXX*).

octopus, but rather a squid. The narrator makes a clear distinction in the manner in which it whips the dogs with its *crinibus extremis* and how it clubs them with its *robustioribus brachiis*. This is a clear allusion, on the one hand to the long tentacles which the squid uses as whips to harpoon its victims, and on the other to the more massive arms which it uses to enfold and devour them. Later, it is said that the cavities on the arms, that is the suckers, looked like basins. This is another characteristic trait of squids. While the suckers of octopus are rather flat and deep only in their centre, so that they look like coins with a large hole in the middle, those of squids look more like drinking cups and even have stems, which makes the resemblance more striking.

It is however rather unlikely that a pelagic cephalopod, built for rapid swimming rather than for crawling, should have come on land to steal fish from brine vats. I am also convinced that this kind of animal would have been absolutely incapable, with its long streamlined body, of climbing over a palisade: stranded on the beach, a squid is not even capable of crawling back into its element.

Octopus on the other hand are quite capable of such prowesses. One can watch them crawling on the rocks along the shoreline. For those who might doubt the ability that octopus have of climbing over obstacles, I will quote the following observation, made in the last century by a renowned specialist.

At the Naples Aquarium, the Swiss zoologist Kollman had organized a duel between an octopus and a large lobster, so as to study their fighting tactics. After noticing that it was always the octopus which attacked and overwhelmed its opponent, Kollman had interrupted the unequal struggle and saved the lobster's life by placing it in a nearby basin. The following day, he discovered that there remained of the unfortunate crustacean only the carapace: during the night, the octopus had climbed out of the water, crawled over the barrier, many centimetres high, which separated it from its ennemy and had quietly eaten him.

Octopus are a constant problem for their wardens because of their frequent escapades. They have been found on top of bookshelves, in the stairs, the street, or even - ultimate horror - in the teapot of an English gentleman.

In the case of the "polypus" of Carteia, it is possible that two independent incidents may have accidentally been lumped together. One might reconstruct the events as follows. The theft of salt fish was perhaps due to an octopus, or perhaps simply to two-legged wanderers of our own species. However, when a giant squid happened, probably much against its will, to strand on the beach - and such incidents shall be repeated again and again in History - this horrible monster was immediately blamed for all previous mishaps occurring in the region.

It is also not impossible that, in spite of appearances, a titanic squid might have been responsible for the thefts. After all, no one had seen it climb over the fence with the help of a tree: that was only a supposition. It is not impossible that a giant squid, caught in shallow waters and incapable of escaping to the deep sea might have spent a few days near the shore at

Carteia. Becoming hungry, the nocturnal animal might have by night crawled to the very edge of the water and undertaken to explore the area with its ten-metre-long tentacles. We may reasonably assume that the brine vessels were near the shore. The squid might have been able to catch a number of fish over the fence without having to climb over it.

The story of the Carteia "polypus", which will be retold with minor variations in 1509 by the ex-duke of Genoa Battista Fregoso (known in literature as Fulgosius), is perhaps not the only reference in Pliny to some monstrous cephalopod. In another part of his work, devoted to the aquatic fauna (book IX of his *Natural History*), one finds an ambiguous passage which some have interpreted as an allusion to such monsters:

"In the Sea of Cadix [the Atlantic] is the tree (arbor) which carries such gigantic branches that it has never been able to enter the Strait [of Gibraltar]."

That Pliny is indeed talking about an animal (and not a tree) is confirmed by the fact that in the book (XXXII, 53, 2) which sums up his zoological work, he begins with the words *arbores, physeteres, ballaenae...*(trees, sperm whales, balleen whales...) his enumeration of the largest marine creatures[7]. With its long narrow tentacles, an octopus may remind one of a tree, although such an analogy would be even more appropriate for a star fish with branching arms of the group Ophiurids: *Asterophyton arborescens*, for example the basket star, or Gorgon's Head. However, the latter is only a few decimetres in diametre. In any case, an animal with arms so long that it couldn't make it through the Strait of Gibraltar can only be the product of a runaway imagination, or, more likely, a mistake in interpretation by some copier, confused by the term *arbor*.

In his penetrating commentaries on book IX of Pliny, J.Cotte suggests a direction in which the key to the enigma might be found:

"I note that Pagellus erythrinus[8] is commonly called arboro in Adria, alboro in Venice and arbun in Spalato. Couldn't such a name, applied to an animal which does not enter our sea because of its chemical composition, perhaps be responsible for Pliny's story."

As for me, I wonder if there isn't, behind this story which has puzzled generations of naturalists, merely some typographical error, or a misinterpreted word taken by Pliny from some foreign language. In Morocco at that time, one of the largest sea monsters was called *anbar*. That same word was used in the coptic version of the Bible to describe the giant fish which swallowed Jonah. Once latinized, it was spelled *ambar*. As we shall see later,

---

[7] *The word physeter is usually translated as sperm whale. I have reasons to believe that this is a mistake and that it referred to a certain type of sea serpent (cf my book on this subject).*

[8] *A sea-bream commonly found in French fish markets under the name pageau.*

this word refers to the sperm whale, which is common around the Azores, but which only rarely ventures into the Mediterranean. An ancient naturalist might well have written of the existence in the Atlantic of a gigantic animal, called *anbar*, which never crossed the sill of the Pillars of Hercules. From there to conclude that it couldn't do so because of its size was but a small step for an imaginative mind. The word *anbar* could easily have been mistakenly copied as *arbar* and then transformed into the more familiar *arbor*. Once this metamorphosis carried through, some new commentator could naturally attribute the difficulty that this strange tree-like animal would have in passing through shallow water to the length of its branches.

Thus, to Pliny's knowledge, the greatest marine animals would have been the sperm whales, the baleen whales and the sea serpents: a rather reasonable list.

The passage referring to the *arbor* is followed by another one, almost as obscure, but which is more likely to refer to an octopus. There is mention of some other inhabitants of the Atlantic called "wheels" (*rotae*), because of their configuration. They are distinguished by four rays, spinning like the wheels of a windmill around an axis centered on their two eyes[9]. An incorrect translation has led some to believe that these were animals with the kind of our-fold symmetry found in corals. This description actually refers to animals with four arms on each side of their head, where the eyes are striking because of their size. One will easily recognize a cephalopod with its eight continuously moving arms. As there is no indication given of the size of this animal, I would not have bothered mentioning it if this particular passage wasn't at the origin of an error of interpretation to which I shall return later.

Three centuries after the account of the pseudo-octopus of Carteia, one finds in the *Treaty of the Nature of Animals* of Claudius Aelianus a surprisingly reminiscent story.

The likable Greek sophist was mainly motivated by a moralizing purpose, laudable in itself, in writing his opus of seventeen books, mostly a compilation of anecdotes. His goal was to show, through the description of the marvels of nature, that animals are better than men: that most of them are endowed with "beauty, intelligence, industriousness, justice, temperance, affection, love and even piety", all qualities which are so often lacking in human beings. This moralising purpose leads us to suspect however that the author might not hesitate in exaggerating to reach his goals.

On the subject of octopus, he asserts that they sometimes grow so large that, in old age, they even attain the bulk of whales and other cetaceans. To back this up, he relates that in the territory of Pozuolli in Italy, a monstrous octopus, apparently scornful of what the sea had to offer for prey, was in the habit of emerging from its element and to come aground to exert its ravages. To reach the store where merchants stored their salt fish in large

---

[9] Apparent et ROTAE apellatae a similitudine, quaternis distinctae radiis, modiolos eorum oculis duobus utrinque claudentibus Ionis *(lib IX, 3,1)*.

barrels, it passed through a large underground cave which the city used as sewer. After seizing the first handy barrel, it would squeeze it in its arms until it burst and then eat its fishy contents.

Surprised to find all these broken barrels and eager to discover the source of these continuing depradations, the marchants had finally posted an armed sentinel. One night, the octopus returned, burst another barrel and feasted on the scattered salt fish. The hidden guard, witnessing this moon-lit banquet, was so horrified by the size and the hideous appearance of the poacher that he did not dare attack it. When, on the next day, he told of what he had seen, he was greeted by general skepticism. He was suspected of hallucinating, called a dreamer and a spinner of tall tales. However, as the damage was far from imaginary, it was resolved to post a number of guards so as to seize the impudent thief.

As soon as darkness fell, the monster again sneaked in through his usual subterranean passage. This time, the astounded guards could have no doubt as to its identity. Some blocked its escape, others leaped at it with axes and cutlasses, cutting its arms, carving its flesh, finally killing it by chopping it in small pieces.

Is this perhaps a version of the incident of Carteia, distorted and embellished through the centuries, relocated in Italy through some error in the telling? One might well think so. But it might also be some new event, upon which may have been grafted details of the earlier story, by then classical for having been repeated by many well-known authors. This is often the fate of an extraordinary event: people are such creatures of routine that they love to give a normal appearance even to abnormal incidents. Something exceptional occurs again, even after a lapse of a few hundred years? It is reassuring to cast it into the mold of its first occurrence. Thus are legends born. Soon it will be told that giant octopus usually come onto land at night to pillage stores of salt fish.

Thus is also born incredulity, of a quite unjustifiable kind. After all, if the incident of Pozuolli is so similar to that of Carteia, it is perhaps because octopus are particularly fond of salt fish. Who could ever ignore the windfall that the discovery of a mine of food represents? What is certain is that if the "polypus" of Pozuolli ever existed, it must have been an octopus, not a squid such as the remains of the beast shown to Lucullus.

What is of interest for us at this point is that Aelianus did not hesitate to report the existence of cephalopods as massive as the largest cetaceans. Let us recall that the largest cetacean known today is a baleen whale, the blue whale (*Balaenoptera musculus*). Two large specimens were weighed, piece by piece: one 20 metres long (51 tonnes); another, 27 metres long (119 tonnes). As whales of that species often reach 30 metres, an even sometimes 34.5 m, the record length, it is reasonable to estimate that they may reach a weight of 150 tonnes.

In the Old World, blue whales are to be found only in the North Sea and in the Atlantic, and never seem to enter the Mediterranean. It is thus doubtful that the Hellenic world would have known about this species. However the two runners-up in size among cetaceans are sometimes seen in the waters that surround Greece. These are the sperm whale (*Physeter macrocephalus*), where the male ranges from 13 m to 25.5 m in length, and the fin whale (*Balaenoptera physalus*) with a length that varies between 18.5 and 25 metres (in that species, the female is generally slightly larger than the male).

We shall retain from all this that even in antiquity some cephalopods (octopus or squids?) were already ranked besides the largest cetaceans as marine monsters, i.e. animals of abnormal dimensions. Following a number of other philologists, J.Cotte has pointed out in this respect that formerly the word *cetus* (*ketos* in Greek) from which the word cetacean derives did not denote only these large marine mammals: it used to include ALL large marine animals including sharks and other large fish, such as tuna, as well as seals[10]. It could then be used for cephalopods as well, provided they were of sufficiently large dimensions.

The reader will then not be surprised to find in the next chapter a long digression on the mistaken identification of the whale with an animal called *cetus*. Besides their entertaining qualities, these anecdotes will help us to clarify some of the most obscure chapters of the history of giant squids and octopus. In the Middle Ages, this history is closely tangled in a messy knot of legends about whales, sperm whales, turtles, and the Great Sea- Serpent itself. They all revolve around the legend of a beast so vast that it is mistaken for an island.

As there is no sword sharp enough to cut this Gordian knot, we shall have to laboriously attempt to unravel it.

---

[10] *In the Iliad (XX, 147) Homer speaks in these words of the walls of the divine Hercules: "Athene and the Trojans had in former days erected this battlement to guard him from the ketos which pursued him from the shore to the plain." If one candidly translates ketos by whale, as did Leconte de Lisle, one creates a rather amusing situation, quite inappropriate in such a tragic epic.*

Chapter 5

THE MEDIEVAL FABLE OF
THE ISLAND-BEAST

To mistake a marine monster for an island, to cast anchor on its back and to feel what one imagined to be dry land sink into the sea after lighting a fire on it, is truly, shiver my timbers!!! the height of vexation for a navigator. Judging from the number of natural history writers who mention it, this strange incident seems to have been commonplace in the Middle Ages.

If in those times mariners seemed to have been particularly absent-minded, naturalists were no less inattentive, since from their descriptions of the events it is impossible to decide whether the misidentified beast was a cetacean, a reptile, a crustacean or a cephalopod. At least, Pliny was more precise.

In those days, it is true, an incredible chaos reigned in the understanding of the animal world, particularly with respect to marine creatures. Among the learned, it was thought at that time that the highest achievement was to translate, with more or less success, their Greek or Roman predecessors, sometimes via Arab intermediaries, and then to copy from each other. What was of utmost interest was to disclose the "properties of things" as well as their "differences", and especially to point out "marvels", with the sole purpose of finding allegorical meaning to these "properties", of drawing some moral lessons from these "marvels", and of glorifying God through its most surprising creatures. All this left little time or incentive for medieval zoologists to engage in original observation, or indeed to check on the veracity of their most elementary claims. Thus, the smallest mistake of a copyist could give rise to a new "marvel". Many names of animals were also translated into other languages in completely incorrect terms.

The bizarre search for allegory and moral significance in the natural world stemmed back to the Hellenic decadence. However, the immoderate taste for exaggeration and extravagance was certainly an oriental influence, as one may judge from the dimensions assigned to the whale by people of the orient.

When, in the fourth century B.C., Nearchos, Alexander the Great's admiral, had sailed on his orders to explore India, from the Indus to the Euphrates, he had related that before his departure from Cyisa, a number of whales had been seen, many of which were about 100 cubits long. Nearly 50 metres? This slight exaggeration might be understandable since it is difficult to estimate the length of a cetacean at sea. Later, Nearchos was to find a whale stranded on the Persian shore. "Some sailors went ashore and measured its length at 90 cubits". A somewhat more reasonable length. However, if the Macedonian voyager had sought information among the natives, he would have reported much more impressive dimensions.

The most debatable part of Pliny's work is that in which he reported information from the Orient, gathered among people more concerned with the marvelous than with precision. Thus, he mentioned the existence in the Indian Ocean of whales of more than an hectare in area! To have such a broad back, they must have been more than 200 metres long!

Among Arab authors, who took up the flag of science after the Greeks and the Romans had dropped it in favour of that of virtue, one long found similar exaggerations. Thus, in his famous *Book of the Golden Prairies and Mines of Precious Stones*, written in 914, the Baghdad geographer, naturalist and historian Abou'l Hasan 'Ali ibn al-Hosain al-Mas' Oudi pointed out that it is in the midst of the sea of Zendj, meaning Oman, that one finds a fish called *al-oual* (the whale):

" It sometimes reaches, wrote Mas'Oudi, a length of 400 to 500 omari cubits [between 200 and 300 m], which are measures used in the country, but its ordinary length is about 100 cubits [around 50 m]. Often, during calm periods, it shows the tips of its fins above the sea surface; they are comparable to the mainsail of a ship. Occasionally, it raises its head and ejects from its gills a column of water which rises above it as high as an arrow would be shot. Sailors fear its approach and rattle sticks and beat drums so as to keep it at a distance."

In the XIIIth century, another Arab scientist, Zaharigga ibn-Mohammed ibn-Mahmud al-Kaswini, who was long held in the Muslim world as a high authority in matters zoological, spoke in nearly identical words of the same fish in his book *The Marvels of the Animal Creation*. However, he placed it somewhere "off China", and made it more than 300 cubits long!

In China itself, in old treatises such as the *Tsi-hiai,* there was a report of an even more extraordinary *pheg* whale, which shook the sea over three sea miles.

The prize for most outrageous story clearly goes to Hebrew reports. For certain rabbis, the whale was more than 1500 stadia in length, that is over 250 kilometres! In one of the books of the Talmud, the *Bara-Bathra*, it was said that a ship had to sail for three days above one of these cetaceans to go from its tail to its head...

On the coat-tails of Christianity, the mystical *Weltanschauung* of Asia had gradually invaded the West and submerged its more rational sense. Moreover, it is through the Arabs that western monks were to renew contact with Hellenic science. Thus flourished through the Middle Age a whole series of increasingly extravagant fables about the *cetus,* soon identified exclusively with the whale.

The most popular zoological works in Europe in those times were

THE MEDIEVAL FABLE OF THE ISLAND BEAST

Bestiaries, strange catalogues in which a series of animals were reviewed mainly in order to reveal the symbolism hidden behind their structure or their character,

The first bestiary published in French seems to have been that of Philippe of Thaon (or Thaun), a small village near Caen. It is easy to date, because the poet-cleric had the charming idea to dedicate his work to the beautiful Aelis of Louvain, whom he calls "Queen of England". As she had married Henri Beauclerc, king of England and duke of Normandy in 1121, the work must have been written between that date and 1135, the year of the death of that king, after which Aelis lost her royal title through her subsequent marriage to William of Albini, count of Arundel.

The bestiary of the Thaon cleric is no more than a kind of translation, in rather awkward and pedestrian verses, of a work then widely distributed in the western world: the *Physiologus.* It is thought that this antique best-seller had originally been written in Greek, probably in the second century A.D. The prominent Alexandrian exegete and theologian Origen already quoted it in the third century. It is clearly in Egypt, the true intellectual centre of the western world at the beginning of our era, that was created the first collection known under the name *Physiologos.* The apostles who had taught in the christian communities of the Orient, particularly in Alexandria, had recognized the effectiveness of examples taken from nature. When there was mention of some animal in the passages which they commented from the scriptures, they would report what was said about it in the works of natural history, usually Greek, and arranged to draw from their presentation some moral lesson in agreement with christian doctrine. Overall, the *Physiologos* was originally a collection of homilies concerning the various animals mentioned in the Bible.

This edifying collection had at an early date been translated into Syrian, Armenian, Arab, Slavic and Latin (in the latter, at the latest in the VIIth century). Although the curiosities of the animal world mentioned therein were originally intended, in the spirit of its author(s), to attract the attention of the reader, so as to make him more receptive to some moral lesson, the *Physiologos* could not avoid becoming mainly an initiation to zoology.

In the words of Charles-Victor Langlois in his book on the *Connaissance de la Nature et du Monde au Moyen Age (Knowledge of Nature and the World in the Middle Ages),* Philippe de Thaon appears to have been "the first of those many writers whose ambition was to teach the elements of science in the language of lay people."

So, what does this ancestor of all French popular science writers tell us about the whale? (For clarity, this passage is written in modern language; no effort has been made to reproduce it in verse.)

"The cetus is a very large beast; at all times it lives in the sea. It takes sand from the bottom and puts it on its back; it comes to the surface to rest quietly. The navigator sees it, believes it to be an island and hastens to it to

prepare his meal. The whale feels the fire, the ship and the people, and it dives, drowning them if it can..."

In the mind of the times, the treacherous whale represented the devil; the sand which dissimulated the danger stood for the riches of the world. Attracted by the latter, a man of little faith would rely on the promises of enjoyment which they held. But that was only an illusion: Satan would soon drag the careless into the flames of hell[1],

This fabulous story is found in all later bestiaries, but each new author, or rather adaptor, has added his own details. In the *Divine Bestiary* of Guillaume de Normandie, written shortly after 1208, there is no further mention of *cetus*: only the word whale appears in the text. At this point, it is the very colour of the scales of the monster which gives the illusion of sand[2]. This mirage fools the sailors... and we know the rest:

> *Entrer cuident en bonne place*
> *Lor ancres gnietet, lor feu font,*
> *Lor mangier cuisent sor ce mont.*

That is:

---

[1] *This old allegory, found in all bestiaries, as well as in the homilies of the Fathers of the Church, is also the clue to Herman Melville's story of Moby Dick. Literary critics wander into the most nebulous speculations when they have to interpret what they do not understand. They bestow upon the white sperm whale which captain Ahab pursues with such fury the most various interpretations: from God, through the obsessive and hateful memory of the author's frigid mother, to all the tabus which bind the human spirit. One might believe that those who have correctly interpreted it as Satan, the very personification of Sin and Evil have only accidentally hit upon the correct solution. True, Melville's work is often confusing because of the multiplicity of symbols which it contains, but there is nothing particularly obscure for those who are familiar with christian mythology. The whaler who abandons his profession to lead his men to disaster in order to kill some invulnerable monster is obviously the image of the grandiose stupidity of religious fanatics, so insistent on extirpating from the world the evil which makes up its very fabric, that they forget to live their life. Perhaps it was best for Melville that the public should not understand his passionate criticism of the puritan world in which he had been raised.*

[2] *Well before then, Pliny claimed that the whales (ballaenae) had hair, which was somewhat closer to the truth, since whale fetuses still have a few hair around their mouth. It is only after the original studies of Pierre Belon (1551 and 1553) that the truth came to be known about the body of this animal "which has neither hair, nor scales, but is covered with a uniform, dark, hard and thick leather under which there is a layer of lard a long foot thick."*

THE MEDIEVAL FABLE OF THE ISLAND BEAST

> Thinking they are in a good place,
> They throw their anchors, light their fire,
> Cook their food on this mount.

Besides cooking on the back of the poor beast, the imprudent sailors drive stakes into it to tie up their boat, with the fatal consequences one can readily imagine. Thus are fooled "the sorry and weak miscreants which have faith in the devil". When they expect it the least, "the fellow, hurt by the fire" dives into hell and drags them there with it.

All bestiaires did not share the same religious inspiration. In the *Bestiaire d'Amour* of Richard de Fournival, each zoological description is a pretext for a gallant madrigal to some reluctant beauty. The author's thesis was to demonstrate to the lady of his dreams that she could not possibly afford not to yield to his pleading and to share the passion which he felt for her "his beautiful tender love". One might perhaps find such frivolity rather surprising coming from a cleric, son of Philip Augustus's physician and brother of the bishop of Amiens. However his attitude was quite in keeping with the mores of the time, and his youthful literary excesses did not keep him from finishing his career as chancellor of the Church of Amiens.

In his sweet bestiary, written around 1250, the young tonsured skirt-chaser did not hesitate to stretch the mariners' visit on the living island:

"They came to a kind of whale which was so large that when it holds its back above the water the sailors believe that it is an island, because its hide is everywhere like the sand of the sea. And the mariners tie up to it as if it were really and island and stay there 8 days or 15 and cook their meats on the whale's back. And when it feels the fire, it dives with all aboard to the bottom of the sea."

In the eyes of the amorous Richard, this means that "one must rely least on what is most resemblant". There is the kind "who says that he is dying of love feels neither evil nor pain" and deceives honest people. However, he is not of that ilk and wishes to reassure his lady that his suffering is real!

Said lady was not convinced at all. With much spirited wit, she replied to the bestiary of Messire de Fournival by another, wherein she skillfully turned against her pretender all his arguments. To her mind, ladies and maiden who have the imprudence of relying on seductive clerics, so skilled in curtesy and handsome words, are just like the sailors which take a whale for an island. If they take too much pleasure in listening to their words, it may be for the misfortune of all. Dissipated clerics risk loosing considerable benefits, which they could receive by becoming cannons or bishops. As for the maiden, if they don't behave wisely they compromise their chances of finding a gentleman knight who might bring them honour. So... Who knows whether Richard was convinced by such arguments. It is a fact, however, that he chose the rich benefits.

THE KRAKEN AND THE COLOSSAL OCTOPUS

Enough of amourous jousts and of such profane use of the characteristics of beasts. Thank Heavens, Richard de Fournival was an exceptionally rare original among authors of edifying bestiaries.

Thus embroidered upon from work to work, the story of the island-whale reached its apogee in a religiously inspired version, where direct symbolism, probably thought to be too weak in its evangelical efficacity, was replaced by a outright miracle.

First, here are the facts in their classical form, as they are found in various transcriptions of the legend of Saint Brendan (Brandan or also Brandaines), a benedictine monk born around the year 484, who founded the abbey of Clonfert, in Ireland[3]. Tradition has it that, in the company of 17 monks of his congregation, Brendan crossed the seas in a small boat of reeds covered with oiled skins, in search of a Delicious Island, supposed to be the Lost Paradise. One day, when the time came to celebrate Easter, he saw a strange, round little island, without rock, or sand or grass, but only naked ground. After having tied up their boat with ropes, Brendan's companions set foot on the islet, and as appropriate for that day, each celebrated mass. So far so good. However, they then lit up a fire to cook a cauldron full of meats and prepared to have their meal. The meat was not even cooked when the island began to shake in an alarming way. The monks barely had time to jump back into their boat before seeing the ground vanish below the sea surface.

On the Lord's advice (or perhaps because he had read the *Physiologus*?), Brendan had prudently remained on board. When his brethren, surprised and terrified joined him, the holy man found the right words to comfort them:

"It is not on an island that you have celebrated Easter, but on a beast, the first and greatest of those that live in the sea. So wished it our Lord to enhance our faith, for the more marvels we witness, the more we shall believe in Him. And know that this great beast is called Jasconius. From the origin of times, it has being trying to put its tail in its mouth, but it is so enormous that it never succeeds. In this, it symbolizes for us eternity..."

While the illustrators of Saint Brendan's voyages have always described the monster as a whale, the original text quite clearly refers to a rather more flexible animal. One is strongly reminded of the Snake Ananta of ancient Hindu cosmogony, of the Sea-Serpent Jörmungandr of Scandinavian mythology, or of the alchemists' Worm Ouroboros, all of which bite their tail to symbolize infinity or eternity. However, as sea-serpents were probably even less familiar than whales to people of the Middle Age, the Jasconius quickly took on the shape of a whale.

---

[3] *The name Clonfert comes from the gaelic Cluain Fearta Bhreanainn. As cluain means "plain", or "quiet area", and fearta "miracle", "prodigy", the expression means "the plain of the miracle of Bhreanainn", or "the retreat of the miraculous Bhreanainn", the latter being the original celtic name of our holy abbot.*

THE MEDIEVAL FABLE OF THE ISLAND BEAST

The adventure was clearly laden with moral lessons. But it remained a minor news item, based on an encounter with some marine monster until some clever commentators figured that there could be much greater advantage derived from such an incident.

We thus find that at the end of the XIth century, or at the very beginning of the XIIth, the most erudite Belgian hagiographer Sigebert of Gembloux attributed to St. Maclou an even more edifying adventure.

This Maclou (also Mahout, or Malo) was born at the end of the Vth century in the valley of Lian-Carvon, Wales. His name was probably McLeod, similar to that of many Irishmen of today. After studies in Ireland, precisely under the direction of abbot Brendan, he had come to settle in Brittany in 538 and had created an ermitage near Aleth. A few years later he had become bishop of that ancient city, next to which was later built another which was to bear his name: Saint-Malo.

According to some romance language versions of the *Voyage of Saint Brendan,* Maclou was one of the 17 monks participating in the memorable expedition. According to other commentators, it is only later that the Welch monk had left in search of the island discovered by his former superior. In any case, it so happened that after a long voyage at sea he found himself, according to Sigebert de Gembloux, in a rather delicate situation.

It is indeed painful for a pious clergyman not to be able, on Easter Sunday, to celebrate Holy Mass on land and with all the required pump. Maclou was fervently praying to the Almighty when all of a sudden a new island emerged from the deep under his very eyes. That was a first miracle, which only half astounded him, so preoccupied was he with the idea of consecrating to God fortwith this virgin territory. Immediately taking action, Maclou had the altar and all other necessary accessories carried to the island and began to officiate in front of the kneeling crew. The entire ceremony was carried out without incident and all 180 sailors received Holy Communion. However as soon as the faithful, the priest and the altar had returned on board, the island sank in a mighty whirlpool down into the abyss! It was a whale, which had remained immobile for the whole duration of the Holy Office[4].

This pious cetacean had really nothing left in common with the sinister devil of earlier bestiaries! It appeared rather as a remote descendent of that which had so conveniently emerged to save the prophet Jonah from drowning. One thing for sure: it was difficult to top this story in evangelical efficacy.

It is doubtful whether our good abbot Brendan would have created by himself the tall tale of the monster mistaken for an island, a tale much older

---

[4] *This adventure has also been attributed to other pious personalities, including of course St. Brendan himself. In his Enchanted World , Ferdinand DENIS has even attributed it to Erich Falchendorff, bishop of Trondhjeim at the beginning of the XVIth century, a person who still has a role to play in this whole affair.*

than he was. However, during his long travels through the world, the holy voyager had undoubtedly met with some large marine animals, and had probably waxed enthusiastic about their being so large that at first sight one could mistake them for sand banks or rocky islets. This is far from a silly metaphor. Even in these days of reason, the illusion persists. We find it in the writing of the writer Georges Blond, a mariner by trade, who in his *The Great Adventure of the Whales,* recalls his first encounter with one of them:

"As for me, I once saw one at sea, in the Atlantic, at the latitude of Gibraltar. At first, I couldn't believe that it was an animal. The sea broke on its brown back as on a small floating island. I felt a completely new emotion, of a cosmic order, take a hold of me, comparable perhaps in quality to what people feel when they experience an earthquake or witness the emergence of a volcanic island."

Take this story, imagine that it goes from mouth to ear for a number of years, and then have a fan of Georges Blond write it down on paper, translate it, and translate it again, and then put it, for a final rewrite, in the hands of journalists eager for sensational news and you will be surprised to hear that "Captain Blond" and his crew were almost swallowed by the sea once for having lit a fire on the back of a whale, which they had mistaken for an island.

When a sailor sees what he thinks is a new island, mustn't his first thought be to set foot on it, so as to relax in conditions of comfort and stability not found at sea? Consider also that, in some other part of his book, the author may have noted, quite correctly, that a variety of organisms, mainly lice and barnacles, but sometimes also algae, attach themselves to the skin of whales, as they do to coastal rocks or to drifting wrecks, and you will understand how some remote commentator may be brought to speak of mosses, which may later grow into bushes and finally be embellished into trees. After that, a very large marine creature will really look like an island.

In this way, through a curious detour, another ancient legend joins that of the island-beast: that of floating islands.

In former years, navigators often wondered why they could not find again the lands inaccurately located on the rough charts of their predecessors. Movements of the earth's crust which had submerged islands, or raised new ones where none had been seen before, gradually gave rise to a bizarre belief: some islands, probably made of ultra-light materials such as pumice, must be floating. Herodotus claimed that the Egadi Islands of the Thracian Bosphorus were formerly floating isles. Pliny also assured us that the island of Deloa, one of the Cyclads, had earlier sailed on the seas.

From there to believe that some of these fleeting lands were alive, was but a small step. What proved, after all, that these undiscoverable islets taken by ancient navigators for new lands were not the back of some marine monster roasting in the sun?

This is why St. Brendan was perhaps a contributor, in spite of himself, to the popularity of the fable of the diving island. He was one of the first westerners to bring back from his voyages a description of remote islands, which later geographers were to place either to the west of the Cape Verde islands, or off Ireland, or even in Brazil or in the Indian Ocean. Faced with the fleeting character of these lands, some assumed that they might be floating. One finds in geographical treatises, even up to the Renaissance, notes such as this one, taken from the work of Pedro de Medina. It is precisely about Brendan's isle:

" Not far from the island of Madeira, there used to be another island, named Antilia, which is not seen today; this island has settled itself in the Antilles, to which it has given its name."

Erudite poets, who had heard at that time of the existence of marine monsters similar to islands, could not fail to perceive a resemblance between these and lands which were thought to be wanderers. One illusion fed another, which soon gave the island-beast such credentials that Renaissance writers endorsed it with unshakable confidence. For example, we see Olaüs Magnus, archbishop of Upsala and Primate of Sweden and Gothland relate in 1556 in his *History of the Northern People* that there exists in the northern seas an enormous cetacean on which mariners have occasionally thrown their grapnels and set foot.

The western world received such a respectable testimony on the marvels of the North as striking confirmation of its beliefs. Actually, good Olaf was by no means an observer of nature; it was under the bright sun of Rome, where he spent a large part of his life, that he wrote his documentary work on Scandinavia, based in great part on rumours. He never actually performed his functions as bishop of Upsala. His older brother John had actually filled that office, while he was only an arch-deacon at the cathedral of Strängnäs. When Gustav Wasa, after liberating Sweden from the Danish yoke introduced the reformation in his new kingdom, John tried to oppose the king, who forced him into exile to Rome. His brother Olaf thought it more prudent to resign his own office and to follow him there. When John died, the pope formally transferred his title to his younger brother, but Olaf remained in Italy for the rest of his days and never took possession of his episcopal see. He continued to live in Rome, in the convent of St Brigit, founded by Swedes. What else could he do but to reminisce about his lost country with those other refugees who had remained faithful to the Papacy, and to try, nostalgically, to inform his hosts about his homeland?

Thus came into being the *Historia de gentibus septentrionalibus*, more or less accurate memories of a small group of emigres. Not surprisingly, its contributions included the most outrageous fables, as well as a collection of often distorted anecdotes. Much of the documentation probably came from the Vatican library. On zoological matters, it is certain that the *Physiologus* exerted a strong influence, either through Olaf's Scandinavian informants (there existed an Icelandic version of that book as early as the

THE KRAKEN AND THE COLOSSAL OCTOPUS

XIIIth century), or through his Mediterranean colleagues.

In support of his story of the cetacean mistaken for an island, Olaüs Magnus draws, for example, on the testimony of one Jorath[5], as well as on that of two Fathers of the Church, both from the IVth century: St Jerome, who translated the Bible into latin, and St Ambrosius. The latter, archbishop of Milan, claimed in the fifth book of his *Hexaëmeron*, a collection of homilies about the six days of creation, that when a *cetus* appears at the surface of the water a whole island seems to emerge, with high mountains reaching to the sky. Jorath on the other hand claimed that the monster allows roots, fruits and bushes to grow on its back. It is thus quite understandable, for us as it was for Ambrosius, that these ephemeral apparitions should have led to many errors, with confused mariners inscribing these islands on their maps.

There is no dount that the *Physiologus* was the great inspiration behind all these commentaries. This famous best-seller permeates the writings of Ambrosius to the degree that His Eminence was one of those to whom its authorship was attributed, until more ancient traces of it were discovered.

Thus the fable launched by the ancestor of all *Bestiaries* received over time such a broad diffusion that the various versions, more or less distorted, which it spawned finished by supporting and corroborating each other, until they gave it an aura of indisputable truth.

The first act of this comedy of errors was to be played when the concert of anecdotes regarding the island-beast came to the ears of a contemporary of Olaüs Magnus, the Swiss Conrad Gesner. The latter's prestige was to be enormous for centuries, and he is justifiably held as one of the most erudite and best organized minds of his time. Physician, zoologist, botanist, and philologist, he is also recognized as the father of bibliography.

When he was only eleven years old, this young Zurich boy had been highly impressed by a particularly odious act of war: the sack and incineration by the Turcs of the library of the King of Hungary, in Bude. This destruction of the witnesses of human thought, and the awful feeling of irreparable loss that he felt, were to inspire him to devote his life to titanic works of compilation and synthesis, so as to limit future damage. His magnificent *Historia animalium* (1551-1587), which is undoubtedly his master piece, has deserved him the nickname of "Germanic Pliny".

This Helvetic (rather than Germanic) Pliny was also the first to

---

[5] *This author of a work called De animalibus is quoted by many encyclopedists of the XIIIth century, among others in the De Proprietatibus of Barthelemy l'Anglais, the Speculum Naturale of Vincent de Beauvais and the De natura animalium of Albert the Great. He must have been very well known at that time, since these writers only referred to important sources. I have not tried to find more about him since a consciencious historian like Charles Victor Langlois confesses not to know about him and adds that: "...the orientalists which I have consulted have not been able to identify him."*

90

undertake the elaboration of a *Bibliotheca universalis* (1545-1555): what we would call today a general bibliography. Limited to dead languages (Latin, Greek, Hebrew) it included no less that 15,000 titles. For each one, Gesner usually provided an abstract, or quoted some excerpts, for he had read, or at least perused, practically everything.

One should not be surprised that he mentioned in his *History of the Animals* that in northern seas navigators had often perished for having cast their anchor onto the back of a whale mistaken for an island. He must have read that often enough to think that he should mention it. Gesner specified that in their language nordic sailors called this confusing cetacean *Trolwal* (or rather *Troldhvalr*, meaning ogre-whale, or evil whale in Norvegian), while in German it was called *Teufelwal* (devil whale).

The island-whale had now entered zoological literature through the main gate.

At this point, the reader will have gathered the impression that during the Middle Ages stories of island-beasts gradually crystallized exclusively around the whale. That is not the case at all. We have already seen that the Jasconius on which St. Brendan and his companions trod was some kind of sea-serpent. When one consults Arab authors, one also finds that in the Orient, these same fables relate to a completely different animal.

"With respect to the marine turtle, [wrote Kaswini in the XIIIth century, in his Marvels of the Animal Creation], it is enormous to the point that mariners mistake it for an island. A merchant relates what follows: "We found in the sea a high island, on which there were green plants. We climbed onto it immediately and dug holes so as to make a fire for cooking. Upon which the island started to move and the sailors said: Climb back on board, this is a turtle and our cooking fire has burnt it; come quickly if you don't want it to carry you away! Because of the immensity of its body, said the merchant, it looked like an island, and with time, dirt had accumulated on its back, so that it had become like a field where plants grew."

Another Arab author, Ibn al-Wardi, from Aleppo, confirmed this story in very similar terms in his *Pearl of Marvels*: he often drew from Kaswini's work. It is clearly upon the work of these two authors that is based the story of the island-beast found in the first voyage of Sindbad the Sailor in the *Thousand and one Nights*. No point telling it here in detail: it is the usual story of "the low little island which looks like a prairie because of the grass that grows on it", and on which the sailors alight to relax, eat and drink until it shakes violently and it is noticed that it is alive. Strangely, in Sindbad, the animal is a whale, not a turtle. However it is certainly in the texts of Kaswini and Wardi that the author of this passage has found his information since he also quotes the story of fish 100 to 200 cubits long which are chased away by striking sticks, and that of the fish, one cubit long, with the head of an owl, both of which are reproduced in this same voyage of Sindbad.

The reappearance, in Arab folklore, of the island-whale is readily explained. It is difficult to attribute a precise date to the composition of the *Thousand and one Nights*. It is a collection of popular tales, possibly of Egyptian origin, which gradually grew over the centuries and seems to have gathered with equal enthusiasm marvels told in western works as well as those circulating in the Orient. Until recently, the earliest known manuscript of the famous Arab tales dated from 1536; the texts available to European translators were of much later origin. But the stories were much older. Mas' Oudi, who died around 956, already laughed at the absurd fables included in "books transmitted to us and translated for us from the Persian, the Hindu and the Greek... such as the *Book of a Thousand Khurafas* (Tall Tales), which people also call the *Book of A Thousand and one Nights*." In 1948, miss Nabia Abbott, an Arabic expert at the University of Chicago, discovered fragments of a manuscript dating at the latest from the IXth century[6].

It is thus clear that the details of Sindbad's first voyage, drawn from Kaswini and Wardi, who lived in the XIIIth and XIVth century respectively, were added at the earliest in those times. One may supppose that it is under the later influence of western writings that Kaswini's island-turtle was transformed into a whale.

The fables of the island-turtle, the island-whale and the sea-serpent island are so similar in their smallest details that they must have the same origin. But how is it possible that the beast of the story be so different in Oriental, Western or Nordic legend?

The first answer that comes to mind is that the shift from one marine animal to another is a simple transposition of a story meant to emphasize extraordinary size. Each independently, the Orientals and the Norse would have adapted the story of the island-whale to the largest animal of their region, or at least of their own mythology, respectively, the turtle, which holds up the world according to Brahmanic cosmology, and the fabulous sea-serpent of northern folklore.

This analysis leads us to ask ourselves whether the whale is actually the original protagonist of the story. We believe this to be so because this animal is *to our knowledge* the largest of all. But wasn't it perhaps by mistake that the island-like *cetus* was originally identified with a whale?

Let's return for the sake of thoroughness to Philippe de Thaon's *Bestiary*, where we find rather strange details on just how to hunt the supposed whale:

---

[6] *These fragments were part of a collection of Egyptian papyrus of the VIII, IX and Xth centuries. They consist of four ruffled brown pages covered with small Arabic characters similar to those used around the IXth century. Four dates written on the manuscript when it was consulted suggest that the copy was made around the year 800. The earliest date recorded is that of the month of October of the 266 of the Hegira, that is the year 879 of our calendar. The front page is entitled The Thousand Nights, which shows that the title Alf Lailah wa Lailah (The Thousand and One Nights) was given only subsequently. This change might be attributed to a superstitious attitude of the ancient Arabs regarding whole round numbers.*

THE MEDIEVAL FABLE OF THE ISLAND BEAST

"The cetus is of such a nature that when it wants to eat it yawns, and the opening of its mouth releases a smell so sweet and attractive that small fish, which like this odour, enter its mouth. Then, it closes it and kills them."

Guillaume of Normandy also refers to such a sweetness of breadth, this time specifically about the whale, which thus attracts fish to its mouth:

> Et cil les transglout toz ensemble
> En sa panse, qui est si lee
> Come ce fust une valee[7].

The translation of *cetus* by whale, as we have noted earlier, is entirely gratuitous: it is only one of the interpretations given to this ancient word. The famous dominican monk Albrecht von Bollstadt, also called Albert the Great, because he was held to be the most eminent scientist of the Middle-Ages, believed for example that the *cetus* was the male of the *ballaena*. As we already know, *cetus* was a global collective for all large marine animals.

Nothing in the physiology or the behaviour of whales proper, meaning baleen whales, suggests that they should be described as having a sweet breath. However, given what we know today, we might ask ourselves whether these legendary features of the *cetus* couldn't perhaps describe the sperm whale.

This large toothed cousin of the baleen whales feeds mainly on cephalopods, and in order to catch them he uses a trick reminiscent of the fishing technique of Mediterranean fishermen: he dangles at right angle his narrow lower jaw, studded with ivory knifes. It is presumed that in the darkness of the deep ocean their whiteness confuses squids in the same way that octopus are lured by an olive branch. It would then be true to say that when it wishes to eat the sperm whale yawns wide to attract its prey.

In addition, even if we don't really know about the sweetness of its breadth, some of its excrements, known under the name of ambergris are exquisitely perfumed once exposed to air...it would be easy to confuse, because of its smell, one extremity of the digestive tube for the other.

Ambergris is a vital clue of the puzzle with faces us and deserves careful attention: it will, in a way, provide the link between cetaceans and cephalopods. This enigmatic product is of interest for a number of reasons, in part because its market value surpasses that of gold. Who wouldn't be interested in a treasure which may be found by anyone on the beach?

Ambergris has probably been appreciated for thousands of years for its various properties, the principal of which is to absorb odours and to

---

[7] *"And it swallows them all together
In its belly, which is so wide
That it is like a valley."*

release them in a finer and more permanent form. This explains the extensive use which Orientals made of it in the art of perfumery, a practice later followed by Westerners, no later than towards the end of Antiquity. From North Africa to the Sunda Islands, all around the Indian Ocean, ambergris was also used for culinary purposes, to enhance the aroma of spices and the bouquet of wines. Finally, ambergris was used in the fabrication of Indian pastilles which refined Parisians of the last century sucked on to improve their breath.

The authors of Antiquity whose texts have reached us do not mention ambergris. This product must however have been known by the ancients if one is to believe the words of writers such as Simeon Seth and Aetios (it is thanks to the latter, a Greek physician who lived in the VIth century at the court of Justinian I in Byzantium, that we have access to numerous works which have since been lost). In addition, ambergris has been found in Celtic tombs in Brittany. The Phoenicians, great travelers that they were, must surely have traded in it.

It is thought that the ancient Egyptians burned ambergris in their temples, in the same way as incense is used. Muslims have long used it in this manner, particularly when they went to Mecca on pilgrimage. There is some in most of these smelly cones called pastilles of the seraglio.

In a more practical vein, Arab scientists of the Middle Ages -ar-Razi, Avicenna, Serapion, Averröes, Mesues - included the precious product in the preparation of some drugs, attributing to it calming and antispasmodic properties. Western apothecaries unquestioningly followed in their footsteps, and the therapeutic popularity of ambergris continued until the last century.

In 1691 the *Pharmacopoeia Londonensis* characterized ambergris in these flattering words: "Excellent corroborative; it is discutient, resolutive, alexipharmic and analeptic." In plain words, the jargon of Mr. Purgon meant that the product was excellent in restoring strength through its stimulating action; that it reduced swelling, provoked the expulsion of noxious substances and accelerated recovery. The London pharmacopoeia added that: "it strengthens the heart and the brain, revives spirits: the natural, the vital and the animal.Its delicately sulfurous nature makes it an excellent perfume; it a good prophylactic against pestilence and protects the humours against infection." The author of this "*Art of Healing and Practice of Chemistry*" also explained that the product, when fully ripe, perfumed everything that it touched, soothed migraines, reduced the swelling of ingrown nails, warmed up frigid people, prevented apoplexy and epilepsy, firmed up all parts of the body and cured sterility. A true panacea!

Only a century and a half ago, the most learned Dr. Hippolyte Cloquet was still talking in his *Faune des Médecins* (1822-1825) about the excellent results which he had obtained by administering ambergris in treating nervous dyspepsia and chronic pulmonary irritations.At that same time, the *New Codex* of the Faculty of Medecine still contained the formula for a number of tinctures based on ambergris, for external as well as internal use.

Arabs and Turcs, sensual but born tired, appreciated ambergris mostly for its supposed aphrodisiac properties. In Morocco, it is still used as an ingredient of *madjoum,* a kind of jam used in harems. It will thus not be surprising to discover the "divine" marquis de Sade recommending ambergris compote to crown a meal planned to stimulate amorous tendencies.

Today, in Europe, ambergris is used almost exclusively in perfumery, to enhance the natural aroma of flowers or of musk, and to increase its lasting power. The great Guerlain said of it that: "Ambergris plays the role in perfumery that cream plays in cuisine; if I don't include it, my perfumes don't sell."

Ambergris is usually found in the shape of waxy conglomerates floating at the sea surface or stranded on the beach.

"It is believed, wrote in 1675 the chemist Nicolas Lemery, that is originates exclusively from Oriental seas, although it has sometimes been found on the coast of England and in many other places in Europe. The greatest quantity is found on the coast of Melinda [East Africa] and near the mouth of Rio de Sena River [the Zambezi]."

When really fresh, ambergris generally has an unpleasant odour of stagnant manure, but after being washed by the sea and slightly oxydized in air, it gradually loses all unpleasant odour and begins to smell like humus, or like freshly plowed ground. Put in a cellar to age, it completes its purification and there only remains a musky residue, delicate and penetrating, reminiscent of benzoin.

The name ambergris (grey amber) is not always justified. Originally, the product is always a blackish brown and resembles tar. After a long time in sea water, it clarifies a bit. Once hardened, its colour varies from dark brown to mottled green or to a dirty white. It is quite possible that the term ambergris may have originated from a popular distortion of some quite different term. At the end of the XVIIth century, the German scientist Justus Klobius claimed that while this substance was commonly called *ambra grysea* in Latin, it should really be called *ambra chrysea*, meaning golden amber, because it was worth its weight in gold.

Dr. Ellis Troughton, of the Australian Museum, who has often been called upon to manipulate samples of ambergris has described it as follows:

"Its texture resembles somewhat that of dry cow dung which has been broken up. It does not crumble like beef fat under pressure and does not stick to the fingers, and it does not become brittle unless it has been for many years in storage. When slightly heated, it may be modeled like plasticine and it looks like shoemaker's glue."

Because of its rarity and of high market demand, ambergris has always fetched a high price. Its value varies of course depending on the purity of the product as well as on the supply of the current production. Ambergris of the highest quality has been sold at the equivalent of 10 US dollars per gram. A few years ago, it was still worth its price in gold: 15 to 30 US dollars per ounce. Following the introduction in perfumery of various synthetic products, its price has fallen, but it is still commands nearly one US dollar per five grams.

This fabulous substance, which has justifiably been nicknamed "floating gold", is found in the shape of blocks sometimes weighing a few kilos and, quite rarely, up to a few hundred kilos. The discovery of such a block is as profitable as that of a gold mine. People who have heard of the sudden fortunes made by finding ambergris keep an eye open when walking along the sea shore. An eye, and especially a nostril.

It is not easy to recognize ambergris when one has never seen it, or never smelled it. Neither the colour, which is so variable, nor the smell can be used as a sure guide. This is why curators of museums of natural history are so often harassed by people, their eyes shining with expectation, who submit to their expertise pieces of soap, lumps of lard, or of tar, sponges, pieces of decomposed meat, clinker, rotten wood, and even shrunken shoes and all kind of other flotsam.

To determine unambiguously whether one is dealing with ambergris, one must recall first that it always floats on water, that it softens at a moderate temperature so as to become malleable, and that it dissolves cold in pure alcohol or in ether. When one sticks into it a red hot needle or wire, a resinous blackish-brown liquid boils out, which does not stick to the metal nor to the fingers. Brought to the flame of a candle, a piece of ambergris immediately volatilises into a white vapour. Finally, when set on the desk of a busy museum curator, false ambergris is readily recognized by the more or less explosive reaction which it provokes.

In the West, the origin of this prestigious product remained unknown until rather recently. There was of course no lack of opinions expressed, from the Middle Ages to the end of the XVIIIth century. In 1667, the learned Dr. Klobius could already quote eighteen. It would actually be rather difficult to invent one which had not already been proposed.

Some have attributed to ambergris a mineral origin; others claimed a vegetal source; others still, and neither the most numerous nor the most respected argued for an animal origin.

The most original hypothesis is without contest that of the famous German botanist Leonhard Fuchs, who, in the XVIth century, simply denied the existence of the natural product. According to him, it was manufactured from a variety of aromatic ingredients. One has to admit that Fuchs was often right. Given the great value of the product, master artificers vied in ingenuity in defrauding rich amateurs. At the Renaissance, our old friend Olaus Magnus noted that: "It is falsified with aloes wood, with muse and storax, and with other drugs. But this is easily uncovered: falsified ambergris softens like wax, while the real product never does."

From today's perspective, the most absurd opinions are those which claimed that ambergris was a mineral product. The idea that it was a concretion similar to meerschaum was of course not so silly. However, the great Italian botanist of the XVIth century, Andrea Cesalpini, thought that ambergris was some kind of precious stone; others thought that it was native sulfur. Finally, many people thought that it might be some sort of tar vomited by submarine springs, some bituminous substances from the guts of the earth. This is the explanation which was generally accepted in learned circles in the XVIIth and at the beginning of the XVIIIth century. It also found favour with the most erudite jesuit Gaspard Schott from Antwerp, in his *Physica Curiosa* (1667), as well as with most chemists, led by Nicolas Lémery and Etienne-François Geoffroy. The *London Pharmacopoeia* pompously endorsed it. Buffon was indecisive, unsure whether ambergris had a mineral or animal origin.

The geological explanation gave birth to a new race of adventurers: amber seekers. Since there are islets made up of lava, why aren't there also some made up of odoriferous bituminous blocks? The world of piracy was thrown into a flurry of activity, especially among the Dutch, when a French traveller by the name of Isaac Vigny came back with the news that there indeed existed such an island in the immensity of the sea...

The opinions which claimed a vegetable origin for ambergris seem to us somewhat more reasonable today, since we know after all that most aromatic substances come from plants. Thus, the botanist Serapion the Younger declared around the year 1000 that ambergris was a mushroom, a kind of submarine truffle; his contemporary the Arab physician Avicenna also subscribed to this idea, as did, five centuries later, the brilliant philologist and physician Jules-César Scaliger. In the mean time, Averröes, an Arab physician shunned in the Middle Age because of his incredulity, had expressed the opinion that the floating gold was a kind of camphor. There were also those who took ambergris for a marine plant, analogous to the sponge (there was a long dispute about the nature of this "zoophyte", so much so that the peaceful Erasmus said that one "should throw the sponge over the story of the sponge". And finally there were those, even more off track, who claimed that it was the fruit of coral, also then considered as an underwater tree.

A more reasonable hypothesis was that of the French botanist Aublet, who saw in ambergris a vegetable gum accidentally fallen into the ocean. A similar opinion, but based on a somewhat more fanciful story, was put forward by one of the glories of the XVIIth century, the English chemist Robert Boyle, on the basis of a manuscript which his compatriots had found on a Dutch vessel which they had captured. This document, dated Batavia, 1 March 1672, stated that ambergris came from a tree which, whatever the distance that it grew from the shore, always pushed its roots towards the sea. In the tepid tropical waters, the roots secreted a thick gum which was gradually eroded by currents and floated to the surface. It would not be too surprising to imagine that the letter was a hoax intending to divert the efforts of English researchers. One sure thing is that the "Skeptical Chemist", as he called himself, completely fell for it.

The identification of the perfumed ambergris with a solidified resin become so firm that it finally became confused with the *succinum* of the

THE KRAKEN AND THE COLOSSAL OCTOPUS

Latins (the *elektron* of the Greeks), and the term "amber" was extended to the latter as well. French merchants proposed to distinguish the two kinds of amber by their colour: the true amber henceforth to be called ambergris and the other "ambre jaune"[8].

Among the supporters of animal origins, we shall mention only for the record, those who thought that ambergris was the liver of a fish, seal vomit, or crocodile excrement, opinions which were actually closer to the truth than all others seen so far. We must also admire the calm assurance of Jean-Baptiste Denys, who cooly wrote in 1672 in the *Journal des Savants* that:

"Ambergris is a mixture of wax and honey gathered on the sea shore by bees, which after being cooked and melted by the heat of the sun falls into the sea where it undergoes another transformation... through the movement of waves and the admixtion of saline particles from the sea, it changes into this precious substance."

This opinion gained many supporters.

The most poetic explanation was without contest that which explained the marvellous ingredient as the feces of a bird fed with aromatic plants. In Madagascar, this extraordinary bird was called *Ashibobuch,* while in the Maldive Islands it was known as *Anacangrispasqui.* This explanation accounted both for the original stench of the substance and for its exquisite perfume after aeration: not a completely stupid story! It also explained another feature of ambergris. The Portuguese Garcia del Huerto had pointed out in 1563 in his book *Simples, Drugs and Products of India* that one often found the beaks of birds in lumps of ambregris. To catch other birds, one also had to be a bird, and the *Ashibobuch* must then have been a raptor which, besides aromatic plants, also ate other, smaller birds.

Charles de l'Ecluse, a professor at the University of Leyde pointed out in vain in 1605 that the beaks found in ambregris came not from birds, but from cuttle-fish. He had no influence on the supporters of the perfumed bird dropping theory. They had another weighty argument to use against their opponents: one also found sometimes within ambergris bird claws. Of course, one could not suspect in the XVIIth century that there also existed clawed squids!

De l'Ecluse, a great botanist and traveller, to whom we owe the description of many new animals in the New and the Old Indies - both ends of the world at that time - was surprisingly close to the truth when he spoke of ambergris as a kind of bezoar created in the stomach of a cetacean, a kind of concretion or ball similar to that found in the stomach of other mammals. He had taken this opinion from a Burgundian called Servais Marel whom he had

---

[8] *Yellow amber is the fossilized resin of an extinct conifer, Pinites succinifer, similar to today's red balsam. It is also a gift of the sea, gathered exclusively on the shores of the Baltic Sea after violent storms. Hence the confusion.*

met in Frankfurt. Alas! Nobody was convinced!

Many people in the world had already known for a long time the true origin of ambergris. These were the natives of the countries surrounding the Indian Ocean. One must believe that westerners did not think much of the opinion of those who they probably considered as savages, or barbarians. Some world travellers however did not have their mind stupidly fixed on the idea of the presumed superiority of their own civilization. In the XIIIth century, Marco Polo quite directly wrote about the riches of the area of Madagascar: "...it is well know that whales produce amber." In 1705, in his *Cabinet de Curiosités d'Amboine,* the Dutchman Georg-Everard Rumph also revealed that the Malays called this substance *jjan tajj,* that is "fish excrement", where fish is taken as meaning an aquatic animal in the broadest sense. The German explorer and naturalist Engelbert Kämpfer, the "Humboldt of the XVIIth century", related that in Japan, ambergris was commonly called *kunfura no fuu,* meaning whale feces[9]. He added that the lower quality product, called *Mokos[10]*, came for the intestines of a cetacean 5 to 7 m long, while that of higher quality was found on the seashore.

In the Arab world, it had been public knowledge since the Middle Ages that ambergris was sometimes harvested from the guts of some kinds of whales. Avicenna and Serapion thought that it was because of the liking that some large cetaceans had for this "kind of marine truffle". They even claimed that these whales devoured them with such gluttony that they died of indigestion. There was some excellent observation at the basis of this belief: one rarely ever finds any ambergris in a healthy sperm whale, but usually in thin, exhausted, even moribund specimens, or even in floating or stranded corpses.

A precious new clue about the identity of the source of ambregris was obtained when the son of a Moorish prince, Al Hasan Ibn Mohammed al Wasan fell into the hands of christian pirates just after he had finished exploring a still unknown part of black Africa. Brought to Rome to the Pope, who had the good sense of treating him with great deference, the learned Arab converted to Christianity, took the name of John Leo the African and wrote, in 1520, a *Historical Description of Africa* which was a revelation to the western world. Under the item fish, one may read:

---

[9] *More correctly, kujira no fuu, the first word meaning whale.*

[10] *Mr Félix-Archimède Pouchet pointed out in 1843 that these are the very dimensions of the midget sperm whale (Kogia) and that Kämpfer's text undoubtedly refers to this small and poorly known cetacean. The accuracy of his diagnosis was confirmed by the following evidence, obtained from Mr. Dautremer, interpreter for the French legation in Japan: "The Koguio - rather than Kogia - is called in various provinces Ouki kouzira or also Makko kouzira." This latter designation being the Japanese scientific word, meaning "perfumed". It is clearly this significant word Makko that the German traveller had written Mokos.*

THE KRAKEN AND THE COLOSSAL OCTOPUS

"Ambar is a fish of terrifying shape and size; it is seen only when it is dead, when the sea casts it onto the beach. Its head is as hard as if it were made of stone, and there are some that are twenty-five toises in length and some even more; so much so that the name whale wouldn't fit them badly. Those who live on the shore of the ocean say that it is this fish that ejects ambregris, but they known not whether it is sperm or feces."

This extraordinarily interesting text undoubtedly referred to the sperm whale (*Physeter macrocephalus*), with its characteristic large and powerful head, capable of smashing a ship. It finally revealed where amber got its name, thus eliminating lots of speculation, and for the first time, described it as a product of the sperm whale itself, also called *ambar*.

Alas, the learned of the day, among whom Conrad Gesner, Olaus Magnus and Jerome Cardan, systematically disregarded the hypothesis that said that ambergris was the feces of this cetacean. They preferred to see in it the sperm of the male. Most probably, they found it simply indecent that the most delicious perfume in the world, the most comfortable drug, could be nothing but whale turds. Here is how Olaüs Magnus interpreted the facts, with a gusto rarely seen nowadays among members of the clergy:

The male of the whale reproduces in the same manner as a man does with a woman, but because he stays in only briefly, much sperm spills into the sea, changing in many ways, but keeping the colour of the sea. Mariners navigating through the sea gather it with great curiosity (as I have once seen it done at sea) and deliver it to apothecaries to refine it, who then call it Amber, using it to cure gout and paralysis as a most precious balm and unguent."

One should not imagine that the learned prelate was confusing ambergris with the oily substance called *spermaceti*, literally "cetus sperm", which is extracted from the head of the sperm whale. Later on, he actually speaks of the "grease which the whale has in its head, so abundant that one fill thirty or forty barrels of it."

The only problem with the sperm version was that occasionally one found amber in the stomach of a stranded sperm whale. The bishop of Geneva, Francois de Sales, best known for his *Introduction to the Devout Life*, related at the end of the XVIth or the beginning of the XVIIth century, that "in a port of Galicia, there was caught a whale in the belly of which was found a mass of amber similar to the trunk of a tree." This demonstrated in his eyes that in spite of what impious critics claimed, challenging biblical texts, a whale did not have such a small gullet that it could only swallow small fish. If one of them was capable of having swallowed a piece of amber as large as a tree trunk, another could easily have swallowed the prophet Jonah! The holy man's conclusions were quite correct, for, in contrast to baleen whales, the sperm whale, which this whale indeed was, can swallow enormous preys[11]. However, the premises of his logical argument were wrong, because of course the sperm whale had *not* swallowed the block of amber.

Popular opinion according to which ambregris was merely a favorite meal of the sperm whale left completely unsolved the mystery of its origin. What was then needed to solve the problem was to combine this opinion with a variety of ancient explanations. Thus, ambregris became successively the product of the digestion by the whale of a truffle, of marine bitumen, of honey, of a fragrant fruit, and even, as an ultimate complication, of the excrement of some land animal. All this was already a step forward beyond those explanations where the sperm whale played no role at all.

Unfortunately, some normally well-informed scientists thought they had grounds to reject any connection between ambergris and cetaceans. Father Eusebius Nieremberg, a Spanish jesuit of Tyrolese origin, stated in 1635, after a thorough enquiry on the techniques of whale hunting, that he had never heard of whalers finding any ambergris in the guts of their prey. The worse is that this was probably true, since *at that time the sperm whale was not yet hunted*! Even if a few American Indians dared to face it in their fragile canoes, the beast's terrifying teeth scared off the most courageous Europeans. If you think about it, it was only natural that no ambergris should be found in the guts of baleen whales. The information revealed by John Leo the African, and repeated by Jerome Cardan, according to which the only and true provider of the precious product was the *ambar*, "whose head was as hard as a stone", had been completely forgotten.

In the final instance, it is the ignorance or scorn of ancient wisdom and of the beliefs of foreign people which explains the persistent popularity, as late as the end of the XVIIIth century, of fanciful theories seeking an origin for ambergris in submarine bitumen, vegetable gum or denatured honey.

According to legend, it is in 1712 that an American whaler, captain Christopher Hussey, first had the audacity to harpoon a sperm whale in the open ocean. This prowess was the beginning of a new industry. The lure of profit finally succeeded where research and scientific speculation had failed, as a result of ignoring appropriate reports and the advice of people living on the shores of the Indian Ocean. When professionals systematically began to search for ambergris at its very source, the mystery that had baffled generations of scientists was finally dissipated. Even then, it was not without some difficulty, for people are usually reluctant to reveal the exact location of a treasure.

In 1724, an eminent Boston surgeon, Dr. Zabdiel Boylston[12] declared that the origin of ambergris was finally known, thanks to the revelations

---

[11] *The translation of the "big fish" which has swallowed Jonah by ambar in the coptic version of the Bible was clearly a judicious choice.*

[12] *Dr. Boylston was the first western physician to practice vaccination against small pox. He performed this vaccination on his own son on 27 June 1721.*

made by some Nantucket whalers[13]:

"In flensing a male sperm whale, they discovered inside it by accident around 50 pounds [22 kilos] of this drug; following which, these fishermen and their colleagues became more and more curious to search in all the whales of this species that they killed; and they found it in smaller quantities in many males of that species, and in no other, but only in about one per cent of them. The fishermen say that the product is held in a bag, or bladder without issue, which they have sometimes found intact but empty. This bag is found only near the genital parts of the "fish". Ambergris, when it is taken from the whale, is moist and has an extremely strong and repulsive smell."

Dr. Boylston prudently concluded that:

"Is ambergris, according to the above report, produced naturally or accidentally in this "fish"? I leave to others, better informed, the task of answering this question."

This information seemed to support the old belief that the precious substance was nothing other than the seed of the sperm whale. Or perhaps, it was also suggested, the bladder described might be some kind of musk gland. A year later, the matter took an unexpected turn, following details communicated to Mr Paul Dudley by a Boston whaler, Mr Atkins. The latter had been hunting sperm whales for ten or twelve years; he seemed an experienced and reliable witness. According to his meticulous anatomical description, the large oval bladder where ambergris marinated in the form of a lump 8 to 30 cm in diameter was at the root of the penis. It actually opened onto the latter and was connected at the other end to the kidneys. This bag was full of a slightly oily, orange coloured fluid which smelled even stronger than what floated in it.

On this matter, another friend of Mr Dudley, the Reverend Prince pointed out quite pertinently that this bag could only be the urinary bladder, and that the lumps of ambergris where therefore equivalent to bladder stones. Mr Dudley, who perhaps knew little anatomy, preferred not to express any opinion on the matter.

This bladder story shed little light on the question. Soon after these whalers' stories, there appeared in print the *Natural, civil and ecclesiastical History of Japan,* in which Engelbert Kämpfer called ambergris "whale feces". This opinion met with no support whatsoever; it was held as "a story related by orientals for lack of accurate observations". Mr Atkins "exact observations" carried more weight, although they were to turn out to be strangely false.

---

[13] *Together with the nearby small town of New Bedford, Nantucket became the capital of the American whaling industry.*

THE MEDIEVAL FABLE OF THE ISLAND BEAST

In 1741, a sperm whale stranded in Bayonne. In its intestine, there was found a lump of ambergris weighing more than 5 kilos. This amount was nothing extraordinary; in the days of Louis XIV, the Dutch East India Company had bought from the king of Tidor a block weighing 182 pounds [about 80 kilos], which it had sold in Lorient for 11,000 rixdales, the equivalent of more than 2,000 pounds sterling[14]. It was the first time in the west, since the rather vague statement by François de Sales, that there was an opportunity to verify *de visu* the true origin of ambergris. Alas, the opportunity was wasted: it was thought that the sperm whale had merely eaten the lump of ambergris found in its intestine.

It is only in 1783, after having interrogated many New England whalers, that an Austrian physician, Dr.Franz-Xaver Schwediawer[15] established without any doubt that ambergris is an intestinal product of the sperm whale, exclusively of the sperm whale, and of both sexes.

That ambergris came from sperm whales and from no other whales of any kind was of course confirmed by the fact that it was found only in the warm waters inhabited by that cetacean.

When a sperm whale is harpooned, the whalers had said, it usually vomits its last meal and releases its excrements, as do other animals at bay. The sudden release of sphincters is one of the best known consequences of deadly panic. After such a purge, there is not much of a chance, still according to whalers, of finding any ambergris in the whale. However this unfortunate evacuation occurs only in strong and vigorous specimens. Weak, sluggish and

---

[14] *This is hardly a record. There are at least four examples today of the discovery of masses of ambergris weighing more than 400 kilos, although in one of these cases it is not proven that the sample came from a single whale. The latter is a 447 kilo block which, according to Beneden and Gervais, used to belong to the Dutch East India Company. The other blocks were all found directly in the guts of sperm whales: one of 400 kilos in 1882 by fishermen on Pitt Island; another of 420 kilos on the coast of Australia by Norwegian whalers. The latter was estimated at 3 million golden francs in value, i.e. 2 billion old francs (US$ 4 M, at 1994 exchange rates). The unchallengeable latest case dates from 1953. During its annual whaling campaign in Antarctica, the crew of the Southern Harvester collected in the rectum of a 15 metre long male sperm whale an ambergris block weighing 422 kilos. It was an elongated form 1.65 m long and 75 cm across. Nice catch, which must have brought in nearly 40 million old francs ($US 80,000).*

[15] *This Dr. Franz Schwediawer, better known under the name Francis Swediaur for his research on syphilis, is a rather mysterious personality, whose nationality has been much debated; he was said to be Austrian, English, and even French. In fact, he was born in Steyer, in High Austria on 24 March 1748. Having graduated as a physician in 1772 from the University of Vienna, he practiced in that city for many years. He then moved to England, where he published a periodical, the Foreign Medical Review. In 1784, at the time of his presentation on ambergris to the Royal Society of London, he still had Austrian citizenship. Five years later, however, he came to settle in Paris, where he called himself Francis Swediaur and obtained French citizenship. He died in Paris in August 1824.*

sickly sperm whales rarely react so violently when they are harpooned to death, and ambergris is often found in their guts. This is also happens in stranded individuals, who did not die in the throes of violence.

Dr Schwediawer then wondered if it weren't perhaps the excessive accumulation of ambergris in the intestine which, by causing a dangerous obstruction, was responsible for the sickly state of some sperm whales.

On this matter, according to eye-witnesses, it is indeed in the digestive tract itself, "most often about 6 or 7 feet [about 2 m] from the anus, and never more", that the smelly concretion is always found. Schwediawer presumed that it was the intestinal coecum that some whalers had mistaken for a bladder specifically destined by nature to secrete and store the precious substance. Dudley was wrong too, he said, in thinking that ambergris is formed in a kind of musk bladder associated with the root of the penis: it is also found in females. It is clear that the beaks of cephalopods so often found in ambergris couldn't have passed from the intestine to the kind of bladder described by Dudley and Atkins (and even less so into the closed bag described by Dr Boylston).

Franz Schwediawer also did not think that Charles de l'Ecluse was right in thinking that ambergris was like a bezoar. However, it all depends what one means by a bezoar. To counter the opinion of the subtle Clusius - as de l'Ecluse was known - Schwediawer invoked a variety of obsolete arguments about the nature these kinds of concretions, thought at that time to be "a phlegmatic recrement", i.e. rejects arising from some internal irritation. In the eyes of the Austrian physician, the precious smelly material formed only when the swallowed squid beaks were not immediately eliminated and formed a conglomerate with the feces. "This is why, he concluded, we can define ambergris as abnormally hardened feces of *Physeter macrocephalus*, mixed with a few undigestible remains of its food.

The fact is that there almost always found within the smelly concretion a variety of cephalopod beaks - up to a thousand in one case - and other equally undigestible remains coming mostly from squids: horned pens, the toothed rinks of suckers, clawed from armed species, etc...[16]

The explanations provided by Dr. Schwediawer did not of course explain the mystery of how a material of such humble origin could posess such a delightfully unusual perfume. It was thought that ambergris could not possibly be made only of feces, solidified from their normally liquid state. They must have been denatured and subjected to some profound chemical transformation.

Some wondered whether the presence of undigestible remains might not irritate the walls of the intestine and stimulate some glands in secreting a

---

[16] *Sometimes the analysis of a lump of ambergris can bring surprises. In 1912 the American ornithologist Robert Cushman Murphy discovered, in a sample taken from a sperm whale killed off Haiti, mustache bristles from a pinniped so rare that it was thought to be extinct: the Antilles Seal (Monachus tropicalis).*

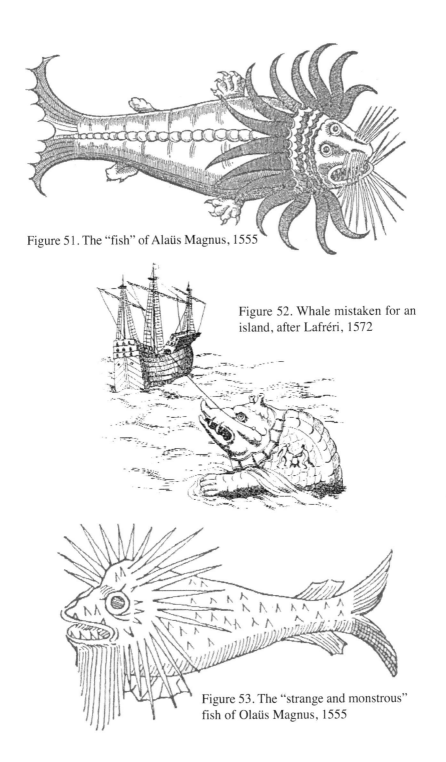

Figure 51. The "fish" of Alaüs Magnus, 1555

Figure 52. Whale mistaken for an island, after Lafréri, 1572

Figure 53. The "strange and monstrous" fish of Olaüs Magnus, 1555

Figure 54. Reconstitution of a pleiosaur

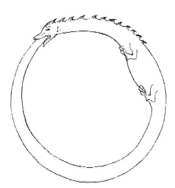

Figure 55. The legged snake Ouroboros, symbol of eternity

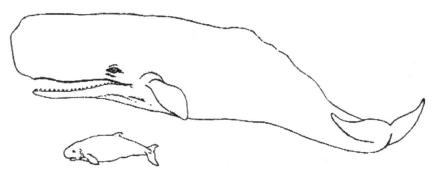

Figure 56. Relative dimensions of the giant sperm whale and of the pigmy sperm whale

Figure 57. A 422 kilo lump of
ambergris found in 1953 by the
Southern Harvester

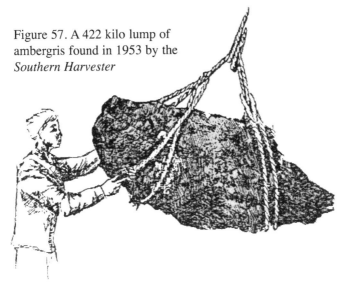

Figure 58. Giant sea-spider
(*Macrocheirus kämpferi*)

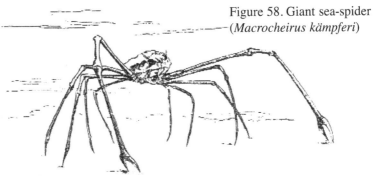

Figure 59. The leatherback, the largest known turtle

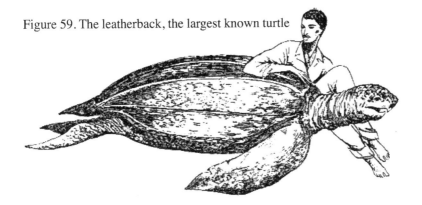

Above: Figure 60. Bishop Erik Ludwigsen Pontopiddan (1698-1764), the
first scientific commentator of the kraken (Universitetets Zoologiske
Museum, Copenhagen)

Below: Figure 61. Giant ammonite, (pachydiscus seppendradensis), next to its
discoverer, the zoologist and poet Hermann Lanois (1835-1905)

Figure 62. The "polypi" of Olaüs Magnus, 1555

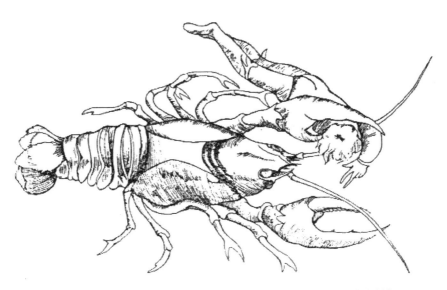

Figure 63. Sebastian Munster's "crab of wonderful size", 1556

Figure 64. Sea-grapes, after Rondelet, 1554

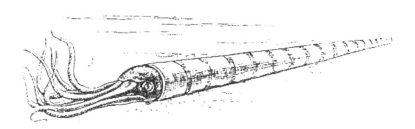

Figure 65. The marine microcosm, after Francesco Redi, 1684

Figure 66. Reconstitution of a fiant Ordovician nautiloid

Figure 67. St. Brendan's companions celebrate mass on the back of a whale (Engraving from the XVIth century)

Figure 68. Abyssal swuid (Abralia veranyi) Photo Popper-Atlas

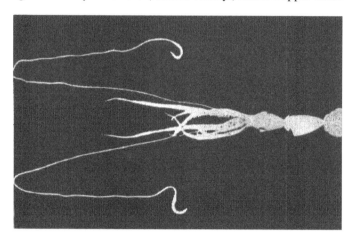

Figure 69. Brittle star, or ophiurid starfish (Ophiotrix, sp.)
(Photo Le Cuziat-Rapho)

substance which would enrobe these sharp objects in a compact, waxy layer.

Some light was soon to be thrown on this question by the analysis of the enigmatic substance, initiated as early as 1682 by the chemist H.-N.Grimm. He found, among other things, a specific organic compound which Pelletier and Caventou named *ambreine* at the beginning of the XIXth century. This fatty matter, which crystallizes as thin needles, is closely related to cholesterol. As gall stones often arise from the precipitation of this latter substance in bile, professor Pouchet suggested in 1843 that ambergris might be a pathological concretion of a similar nature. This opinion agreed with that of whalers, according to whom only sick sperm whales contained ambergris.

It turns out that ambreine alone, in spite of its delightful smell, cannot replace ambergris in the eyes of the perfume industry. Besides this substance, which makes up from 1/4 to 4/5 of its mass, ambergris also includes mineral salts, alcaloids and some acids. Chemists believe that its unique properties are due to the combined presence of ambreine and of a benzoic ester. Esters are formed by the combination of an alcohol and an acid radical. In this case, the alcohol is benzoic alcohol, which is also extracted from bezoin, and which explains in part the peculiar odor of ambergris.

It has also been asked whether the exquisite perfume of the embarrassing concretion might not simply arise from the squids and octopus which are the normal food of the sperm whale. The skin of cephalopods often contains glands which secrete an oily substance with a strong odor of musk. One of them is called the musk octopus (*Eledone moschata*). Pliny, referred to this animal as *Ozaena* (from the Greek word *ozê* to give out a smell, which has also given us ozone), and had already mentioned its strong smell, which was supposed to attract its mortal enemy, the wolf eel. He also related that the Romans dried and powdered these octopus to make perfume out of them.

So, it is not necessary to hypothesize the contribution of special secretions, nor even a radical chemical change in the ingested material to explain the delicate odor of the intestinal concretions of the sperm whale. It is also not necessary to call upon some pathological process to explain the formation of lumps of ambergris. A poor digestion is not really a sickness. Dr. Robert Clarke, of the Institute of Oceanography in Wormley, Surrey, UK, pointed out in this respect that the two sperm whales harpooned in 1947 by the *Southern Harvester*, from which blocks of ambergris weighing respectively 422 and 155 kilos were taken, did not appear ill in any way. Both of them had their stomach full of food and seemed in perfect health.

In my view, the explanation of the legend, which has been current for centuries among whalers, is that suggested by Dr. Schwediawer's informants. If ambergris is usually found only in the guts of sick individuals, it is because they are the only ones which no longer have the strength to empty their bowels under repeated harpooning. At least, this was formerly so. Nowadays, with modern hunting techniques, the final blow is often also the first, with an explosive harpoon shot in accurately by a powerful gun. The agony of the animal is much abbreviated and even strong and healthy individuals do not have the time to empty their bowels.

THE KRAKEN AND THE COLOSSAL OCTOPUS

Dr. Schwediawer had thus shown considerable flair in deducing that ambergris was the result of an abnormal hardening of the feces around undigestible rejects cluttering the digestive tract. Normally, the feces of the large cetacean are quite fluid in consistency; however, if it is held back by obstacles so that it spends a longer time in that part of the intestine where water is re-absorbed, it can gradually harden. In addition, under the action of a bacterium specific to the rectum of the Sperm Whale, *Spirillum recti physeteris,* it undergoes enhanced chemical transformation.

Thus the sperm whale is quite similar to the hero of the amusing farce of Fernand Crommelynck, *Golden Guts,* who transformed everything he ate into precious metal. The difference being that it is "floating gold" which comes out of the guts of the enormous cetacean.

The undigestible beaks which are at the origin of the ambergris producing process, also cause in the sperm whale serious stomach troubles, resulting in cavernous rumbles and gargantuan burps. When the enormous glutton comes to take a nap at the surface, these may be heard for miles around. Such monumental indigestions must occasionally end up with vomiting and diarrhea. Perhaps ambregris is then expelled as a happy side-effect.

The ancient Chinese author Li Shizhen was also not too far from the truth when, in his great pharmacopoeia of the XVIth century, the *Pan Tsao Kang Mouh (Bencao Gangmu)*, he called the precious product by the name *lung hsien hiang*, meaning "perfume of dragon saliva". He wrote that the aromatic substance was vomited by troups of sea dragons when they met in the southern seas at certain times of the year...

Once more, an unbelievable medieval legend turns out to have been based on some excellent observations.

After this long detour along the perfumed road that led us to the true source of ambergris, we now feel more confident to state that the *cetus* of medieval bestiaries is no other than the sperm whale. There should be no further hesitation in recognizing in it the marine monster which casually yawns to express its hunger and seems to attract its preys to its mouth by the sweetness of its breath.

Our journey has more particularly taught us that beyond ambergris, there is the musky flesh of some cephalopods, of these very cephalopods which also, according to legend, attract their prey by smell. Here is something to think about. Is it really necessary to go all the way through the sperm whale's gut to reach the source of the *cetus'* sweet breadth? One should eliminate go-betweens. Might the mysterious and ill-defined colossus not be some very large good-smelling cephalopod itself?

Reaching all the way back to the origin of all bestiaries, to the famous *Physiologus,* one notices that the hero of the story of the island-beast is a very special kind of *cetus,* called in Greek *aspido-chélonê,* meaning serpent-turtle. This particular animal is clearly distinguished from the whales, gigantic beasts said to owe their name to their ability to "blow water": in Greek, *ballein* means to throw, or to reject[17]. The *Physiologus* treats of whales and of the *aspido-chélonê* in two different articles.

This may finally explain why Arab natural philosophers spoke of an island-turtle rather than an island-whale; they had remained closer to the original legend that their western colleagues. What could then be that enormous serpent-turtle of which spoke the ancient *Physiologus*?

There exist a number of so-called "snake-necked" turtles, which pull their head into their shell by folding their neck back in the shape of an S. The Chelonian sub-order Cryptodera, to which they belong, includes most turtle species, including the giant of the group, the leatherback (*Dermochelys coriacea*). But even this champion among them, though living at sea, does not exceed 2.75 m in length, and is quite puny compared to a whale; it is difficult to see how it could have given rise to a legend intended to emphasize exceptional enormity. Even the largest known fossil turtle, *Archelon*, was only about 4 metres in length. Even its possible survival would not explain anything.

Of course, nothing forces us to believe that the *aspido-chelone* is a real turtle, any more than a cat-fish is a cat. A well known author of the XIth century, Hughes de Saint-Victor, wrote of the *aspido-chelone* that: "It is a marine bellue, which looks in part like a snake, in part like a turtle."

In this respect, one aquatic giant could well deserve the name "turtle-snake": the Plesiosaur, which combines the body and the paddle like flippers of a turtle with the extremely long neck and tail of a snake. It might even have exuded a strong musky odour, as do many reptiles. However, this Jurassic colossus disappeared, it would seem, millions of years ago. Unless, of course, it were the ever-living prototype of the Great Sea-Serpent... This would explain why the beast-island of Saint Brendan would have been described as snake-like. But, as Rudyard Kipling would say, that is quite another story.

One final explanation, that towards which I have been leading, suggests itself: might the name *aspido-chélonê* not be meant to be understood as "turtle with snakes", in the popular sense of turtle with snake-like hair, or snake-like arms? Such a nickname would be appropriate for a squid or cuttle-fish: the oval and hard body of a cuttle-fish is somewhat reminiscent of the shell of a marine turtle, while the arms of the cephalopod form a very realistic knot of vipers. Further, some of these molluscs reek of musc and also open

---

[17] *In fact, cetaceans do not reject water through their blow hole, but air from their lungs, from which water condenses by decompression.*

As to the popular etymology proposed by the Physiologus, it should be taken with some caution; it is more likely that as for the Germanic val, the English whale, the Arabic al-oual, the French word "baleine" stems from a primitive root quhal expressing a sense of enormity. Boisacq has also suggested an even more antique root bhlk, from which would have come the indo-european bhl-no, leading to old Swedish balin "blown up" and perhaps, the English bulk.

wide the crown of their arms to eat.

I advance this explanation only with every imaginable *caveat*, without even succeeding in convincing myself of its legitimacy. The fact is, however, that a number of XVIIth century writings argue in its favour. It is not impossible that the legend of the island-beast, which attracts its preys by its bewitching smell, might have originally been inspired by the observation of giant cephalopods. One thing will turn up: in the eyes of the Scandinavians, it is indeed a cephalopod which is the hero of the story.

Let's sum up, At the beginning of our era, the Greek *Physiologos*, written to draw moral lessons from the marvels of the world, spoke, on the one hand of a monster (*kêtos*) called *aspido-chélonê*, so enormous that it was likely to be mistaken for an island, and on the other of equally enormous whales which blew jets of water into the sky. Some careless translators or copyists soon dropped the name *aspido-chélonê*, which appeared obscure or strange[18] and rather than speaking of a particular *cetus*, or monster, they spoke of "the" cetus, or monster *par excellence*. Perhaps they had the sperm whale in mind when making this correction, but it seems indisputable that the *aspido-chélonê* was not a cetacean. Later, when it was undertaken to translate the Latin text into the vernacular, *cetus* was nevertheless translated by whale and the paragraphs originally concerned with whales were thought to be of minor interest and were either eliminated or incorporated in the *cetus* chapter.

From one corruption to the next, from one abbreviation to another, the metamorphosis of the enigmatic aspido-chelone was carried through.

To conclude this enquiry on the origin of the story of the island-beast, it would seem to me that it must be the turtle (or some large animal resembling it) which must be considered as the real prototype of the fabulous creature. As all our Indo-European languages derive from the Sanskrit, it is possible that our most widely spread legends might have their roots in old Hindu myths, especially those found in the *Rig-Veda*.

In his classical study on *Zoological Mythology*, professor Angelo de Gubernatis presented this illuminating revelation:

"In Sanskrit, the land, which rises from the waters in the form of an island (as the moon and the cloud are islands in the sky) receives the name of Kurma, meaning turtle (more literally the "convex" or "humped one"; "that which rises", "that which stands out".)"

---

[18] *The elimination of the word aspidochélonê is especially understandable since it had gradually been deformed by copiers to the point that it had become meaningless. First found under slightly altered forms such as aspidodelone (Latin manuscript from Cambridge) and aspidohelune, then truncated into aspis or espes (in Syrian) or aspedo (in Icelandic), it was completely distorted in fastitocalon (Anglo-Saxon) and fastilon (Leipzig Latin manuscript).*

THE MEDIEVAL FABLE OF THE ISLAND BEAST

There clearly lies the root of the fable which various people were to adapt, each in their own way, to emphasize the enormous size of the sea monsters with which they were familiar.

In some regions of Europe, people were not passively fooled by the corruption of ancient texts, usually distorted by translators and copyists; particularly in Scandinavia, where during the night of the Middle Ages, the scientific spirit was not to suffer the same decline as in the rest of the continent. While nowhere else on the continent anyone took the trouble of observing nature, since "everything was to be found in the works of Aristotle", the people of the North, less dominated by the influence of Mediterranean culture, kept their eyes wide open. They were thus the first to integrate new phenomena into the framework of existing science.

When the fable of the island-beast first appears in Scandinavia, it is at first in its classical form.

The first, not very explicit, allusion to the monster is found in the *Museum wormiamum* a catalogue of the Cabinet of Curiosities assembled in the XVIIth century by the Danish physician Olaf Worm, alias Olaüs Wormius. In order to present in his work a sort of synthesis of what was known at the time about cetaceans (in the widest sense), this professor at the Royal Academy of Copenhagen had relied on information drawn from a manuscript dating back to the end of the XIIIth or the beginning of the XIVth century, the *Speculum regale* (the Royal Mirror)[19]. The *Kongs-skuggsjó*, as this book was called in old Norse, was organized as a dialogue between a father and his son; it included information on the social and political life of the Scandinavian countries, as well as precious details on the natural history of the north, particularly about Greenland, Iceland and Ireland. There was, among other things, a nomenclature of twenty-four different cetaceans. The twenty second, which was said to be enormous, was rarely seen; it was known under the name of *hafguse*[20]:

"Those who have seen it say that its body looks more like an island than a beast. As no corpse has ever been found, there are some who claim that there are only two of its kind in the world."

This text is rather poor in details. Fortunately, further commentaries on this same *hafguse* are to be found in the work of a contemporary compatriot of Wormius, the *Anatomical History* by Thomas Bartolin, the man who first described the lymphatic system.

This famous Copenhagen anatomist, former student of Marcus Aurelius Severino of Calabria, tells us first of all that the *hafguse* is also

---

[19] *It was attributed to the founder of the Norwegian dynasty, Sverrir Sigurdson, because the author claimed that he was of the first rank in the court of Norway. Nothing is less certain.*

[20] *In Jutland, a gusen is an island of vegetation in an open area. Hafguse simply means "marine islet covered with vegetation".*

called "marine vapour" or also *lyngbak*, meaning "heather hill". He tells the story of the miraculous incident involving, according to him, bishop Brendan, who, having mistaken the monster for an island barely had time to celebrate mass on its back before it submerged. We know what to think of this story!

Finally, Bartholin bows to the wisdom of Providence, who, for fear that there would be neither food nor space enough in the world for such giants, resembling more a land than a whale, created only two of their race, sterile but immortal. Their feeding is also most economically planned: each one of them is hungry only once a year. After its lengthy digestions, it emits a horrible borborygm which, O great marvel! exudes such a sweet perfume that fish are drawn in from everywhere. The monster immediately opens a mouth "as large as a strait or a gulf" where the lured fish rush in...

The legend of the *cetus* (alias *aspido-chélonê*) had changed only little in Scandinavia over a millenium and a half, in spite of the adventures of its protagonist in other parts of the world. It seemed as immortal as the *hafguse* itself!

In the *Rare, Selected and Curious Medico-Physical Observations* of Christian Franz Paullinus, German naturalist of the end of the XVIIth century and author of various botanical and zoological treatises, we shall finally find some clues on the shape, and hence the nature of the mysterious island-beast[21]. The author gathered his information in Scandinavia among friends and relatives. Here is what Paullinus had to say:

"One of my most trustworthy friends, Ambroise Rhodius, physician and mathematician from Christiania, in Norway, told me that, when there are no winds to ruffle the water surface, it is not uncommon to see in the vicinity of Wardehus castle [Vardö] a marine monster common to the coasts of Lappland and Finmark [...] and which is called in those latitudes Seekrabbe [sea crab]. In appearance it is reminiscent of the heracleotic crab, or Maia, as described by Gessner. However, its size is enormous, and its body is so vast that a whole regiment could comfortably maneuver on its back."

"In the absence of winds, when the sea is flat, and the sun is shining, this monster gradually rises above the surface, with a nearly imperceptible motion, and exposes his back to the rays of the sun. It remains thus immobile until the sun descends towards the horizon and the air cools. Then, in the same imperceptible motion with which he had emerged, it slowly disappears under the water and hides in the depths."

"Those who have seen it say that it looks like a pile of moss-covered rocks when it shows up on the surface. It is quite true that if a row boat approaches it too closely, it stretches its arms, grabs it with its claws [hamis et uncis] and draws it to itself. The natives believe that it devours the people

---

[21] *The work of Paullinus appeared in 1706, but the observation quoted here had already appeared four years earlier in the Ephemerids of the Academy of the Curious of Nature, as an article under the author's signature, but dated 1677.*

that it catches in this fashion. When the sea is rough, it hides at depth and does not bother anyone."

Another one of my friends told me that sometimes some rather tall trees are seen on the back of this animal. Borelli confirms this story when he tells us that the back of whales is sometimes covered with lush greenery, to the point that sailors have thrown their anchor, set their tents and lit fires, deceived by the extent of the body, which compares with an island in size. When the monster feels the heat, it dives, to the enormous stupefaction of the mariners."

Here, there is no ambiguity. Paullinus draws a clear distinction between the beast which is likely to be mistaken for an island, but shaped like a crab, and the whale, which creates the same illusion.

Is the giant really a crab? That is quite doubtful. Truly gigantic crabs are found neither in zoological literature not in nature. At best, Aelianus relates that when Alexander was sailing through the Red Sea, he had seen crabs whose shell was 3.30 m around.

Actually, we know of some crabs which are not far removed in size from such monsters. The giant sea spider (*Macrocheirus kämpferi*) which lives off the coasts of Japan, has a shell which is at most 1.50 m around, but has such long legs that when laid flat its span exceeds 4 metres. That is however still a long way from the dimensions of true marine giants, like cetaceans. Even the king of story-tellers wouldn't imagine having a single soldier parade on the back of this crab, let alone a whole regiment.

Arab navigators, who in the Middle Ages plied the waters of the Indian Ocean, from Ormuz to China, had terrifying tales of even more monstrous crabs. They would bring their claws to the water surface, like reefs, and would close them on ships which were so imprudent as to sail through them.

In Siamese folklore there persists today a belief in giant crabs and scorpions capable of endangering sailors' lives by dragging ships to the bottom. Among Eskimos, it is enormous sea-spiders which are responsible for such evil deeds.

All these exotic monsters undoubtedly look like Paullinus' rowboat nappers, but one may well doubt their authentic nature as crabs. In fact, large crabs can only run on the bottom of the sea; they do not swim[22]. How could they rise to the surface?

As we shall see in the next chapter, the Scandinavian story of the *krabben* or *kraken*, identical to the above legends, does not refer to a crab, but to a cephalopod. The "cancres" of Paullinus, the crabs of Arab travellers and Siamese folklore, and the "spiders" of Eskimo fishermen probably all have the same origin.

The octopus and the crab, with their round body surrounded by many

---

[22] *Only very small flat crabs like Neptunus and Callinectes are light enough to be good swimmers. One can catch them far at sea near the surface.*

legs, have enough similarity that unfamiliar witnesses may have confused them. At the time of the Renaissance, cephalopods and crustaceans were often confused, as one may see on the woodcut which illustrates the chapter *De polypis* in the work of Olaüs Magnus. The text clearly refers to the octopus; there is a reference to a "fish with many feet" which lives in caves and changes its colour to that of the substrate where it finds itself. Its body is small compared to its feet. The locomotive tube, although poorly placed, is also mentioned: the author says that the animal has "a flute on his back, which he uses to move in the sea, turning it now to the left, now to the right." However, the illustrator has without hesitation drawn enormous lobsters. As the honorary bishop of Upsala had mentioned that around the Orkneys these "fish" sometimes drown swimmers by keeping them underwater, the figure shows us a poor sailor grabbed by one the crustacean's claws and being pulled off the deck of his boat.

In 1539, Olaüs Magnus had published a map in Venice, on which he had had drawn all the monsters of the nordic regions. It was called *Tabula terrarum septentrionalium et rerum mirabilium in iis ac in oceano vicino* (Map of the northern lands and the marvels which they and the neighbouring ocean contain). It was the prototype of many maps, famous today because of their picturesque drawings: among which, that which appeared in 1556 in the *Cosmographia universalis* of Sebastian Munster, and that published in 1572 by Antoine Lafréri. In all of them the incident ends up as a picture showing a combat between a man and a lobster. Thus are the pitfalls of illustration!

All these stories about pseudo-crustaceans attacking boats or swimmers should have been compared to the old legend of Skylla, which never failed "with each one of its six gaping jaws" to take a man from those vessels careless enough to defy her. The homeric fable was to be brought up to date in 1640 in the monumental zoological encyclopedia of Ulisse Aldrovandi: the kidnapping monster was by now clearly described. It was octopus which the prolific Bolognese naturalist said had been seen "attacking ships at sea and pulling from them men which they had enfolded in their arms to pull them violently into the sea."

The only problem with the probability of this story was that the octopus, although a better swimmer than the crab, is nevertheless also, like it, a bottom dweller: it never comes up to the surface, especially in the "high sea"...

The time has come to remember that the octopus has cousins, expert swimmers, which are not only capable of swimming, but normally do so: these are the squids. They even occasionally leap out of the water and fall onto the bridges of ships!

Since the days of Aristotle, it was known that there do exist extremely large squids. If one could have read between the lines about the story of the famous "octopus" of Carteia, as related by Pliny, one could have even suspected the existence of truly gigantic specimens. A critical study of the sources of the legend of the good-smelling *cetus*, often mistaken for an island, would also have supported these suspicions.

THE MEDIEVAL FABLE OF THE ISLAND BEAST

So then, the *hafguse* of the old *Royal Mirror* of Norway, dug up by Worm and Bartholin, the monstrous crabs of Paullinus and of Arab navigators, the ship-wrecking octopus of Aldrovandi, weren't all these bogeymen giant squids? It would have been logical to suppose so. Especially since during the Renaissance nothing was known about abyssal free-swimming octopus. Alas, since there was little attention paid to the observation of various cephalopods and to the comparison of their specific behaviour, nobody in the XVIIth century asked that capital question. To ask it would have been to answer it.

This was, nevertheless, the time when, in England, Sir Thomas Browne was trying to install the reign of reason by warring against gross errors and was asking himself strange questions about what song the sirens sang. It takes time for a revolution to take place. The giant squid was still for a whole century going to remain in the thick fog of legend because no one had been able to dissipate it. Perhaps also because no one had been ready to face the terrifying reality.

Chapter 6

SCIENCE LOOKS
AT THE KRAKEN

———

"We then saw the most marvelous phenomenon which the secret seas have to this day revealed to man. A vast and doughy cream-coloured mass many hundreds of meters in length and width floated at the sea surface. Innumerable arms radiated from its centre, rising and twisting like a nest of snakes apparently ready to blindly catch anything in reach."

HERMAN MELVILLE, *Moby Dick*

In the XVIIIth century, the Age of Reason, as it was to be called, the mystery of the island-beast became particularly irritating because this accursed monster was truly thought to be the largest animal in all creation. It so troubled the peace of naturalists as to come to deserve in their eyes the qualifiers of diabolical and malevolent, which mariners already used for it. After Anton van Loewenhoek had, with his microscope, open a porthole on the world of invisible organisms, and while Reaumur was introducing his contemporaries to the habits of the smallest invertebrates, then all lumped under the name Insects, wasn't it rather frustrating to know so little about such a monumental creature?

However, while the European great powers fought like hungry dogs over the shreds of a world significantly expanded by explorers, the need to put a little order in the animal kingdom was becoming quite acute. During the Middle Ages, and even during the Renaissance, the animal kingdom had gradually became so chaotic that Mother Nature herself would have lost her way in it. Beasts recently discovered in the New World were put cheek by jowl with the dusty and disparate monsters of the Greek pantheon or with the avatars of Vishnu. Various teratological products, like the two-headed snake, the five-legged sheep and the white negro, were lumped together with the familiar animals of the barn yard. That was one of the reasons why, in the middle of the XVIIIth century, three scientists, and not the least, were simultaneously to focus on the problem of the island-beast. If they were not to agree about its nature, all three were nevertheless to recognize its existence.

The first one was the Swedish botanist and physician Carl Nilsson

THE KRAKEN AND THE COLOSSAL OCTOPUS

Ingemarsson, already famous at the time under his Latin pseudonym Linnaeus; the second was a learned Danish prelate, great amateur of natural sciences, bishop Erik Ludvigsen Pontopiddan; the third and last was a German anatomist and botanist, professor Karl-Augustus von Bergen, of the University of Frankfurt.

The light was soon to come from the North!

It is on the faith of commentaries by, among others, Bertholin on the *hafguse* "more like an island than a beast", and Paullinus on the *see-krabbe*, on the back of which a whole regiment could maneuver, that Linnaeus included the embarrassing creature in his fauna of Sweden (*Fauna suecica*) in 1746. He classified it under the name *Microcosmus*, meaning "a small world", with the testacean worms, a heterogeneous group in which he lumped all invertebrates with a shell or internal test. He only commented briefly on the animal:

"It is said that it lives in the Norwegian Sea; as for myself, I have never seen this animal."

Soon afterwards, Linnaeus included the enigmatic Microcosmus in the sixth edition of his *Systema Naturae*, a fundamental work on the systematics of living beings. It is impossible to understand the success of this book on the basis of Linnaeus' classification of animals, which is clearly retrograde compared to that implied in Aristotle's work. The book also contains a number of coarse and comical errors, which Buffon criticized with great verve. Snakes are classified as amphibians, pigs and shrews are included with beasts of burden, an "anthropomorph" category includes the three-toed sloth and a scaly lizard. Linnaeus was a rather mediocre zoologist, and his prestige rested mainly on the cleverness of his botanical classification, based on the reproductive apparatus of plants. What was most important in the *Systema Naturae* was that Linnaeus introduced in this work a binary system of nomenclature and, especially, rules for its application.

Since that time, it has been agreed that, following Linnaeus, each plant and each animal should bear two latin (or latinized) names: a noun, denoting the genus, and an adjective (or noun used as an adjective), denoting the species. The initiative of the great Swedish botanist led, in addition to a classification scheme, to the adoption of an international terminology which eliminated unfortunate confusion.

SCIENCE LOOKS AT THE KRAKEN

For all naturalists in the world, man would become *Homo sapiens,* the hedgehog *Erinaceus europaeus,* and ivy *Hedera helix[1].*

The island-beast received the official name *Macrocosmus marinus.* But what the hell did Linnaeus think that this animal looked like? That he put it in the class Worms (*Vermes*) does not tell us much since that group also included, besides molluscs, zoophytes, batrachians, reptiles, and even plants, the Lithophytes. By specifying that Microcosmus was to be classified with *Testacea*, the testaceans worms, Linnaeus suggested that it should be related to shelled molluscs, or echinoderms. That was rather vague. Besides the generic name MICROCOSMUS, he wrote down the following definition, or rather lack thereof:

Animal -- --- ( ).

All he added was this laconic diagnosis: *Tegmen ex heterogeneis compilatis*, meaning that the skin of the animal was covered by, or even composed of other organisms.

This does not contribute much to our knowledge, and still leaves us at the mercy of gratuitous conjectures. One may suppose that the proverbial resemblance between the *hafguse* and a vegetation-covered island had led Linnaeus to assume that it had a rocky or rigid envelope, of which the naturalist found the equivalent in other marine animals, such as the urchin or the oyster.

With this in mind, it is rather disconcerting to find next to the scientific name *Microcosmus marinus* in the *Systema Naturae* the vernacular German name *Meertraube*, which means "sea-grapes". It has long been known that clusters of gelatinous capsules called by that name are the eggs of the cuttle-fish.

---

[1] *All creatures which did not appear in the second edition of the Systema naturae (1758) had to be subjected upon discovery, or simply upon precise description, to the rather informal ceremony of scientific baptism. Of course, it often happens that the same plant or animal is described under different names by many naturalists. In that case, priority rules: only the first name is valid; all the others are rejected. It may also happen that an author classifies a creature in a genus where it is later found not to fit. For example, Linnaeus called the black rat Mus Rattus and thus classified it in the same genus as the mouse (Mus musculus). In 1867, the Austrian Fitzinger showed that, because of some characteristic anatomical traits, rats deserve to be placed in a disting genus, Rattus . He renamed the black rat Rattus domesticus. However, the priority rule plays in favour of the specific name defined by Linnaeus: the name Rattus rattus was then adopted.) The laws of taxonomy are rather complex and constitute a real code. Difficult cases are referred to a kind of upper court, the International Commission on Zoological Nomenclature, which rules on them. It has become the custom to add to the name of living creatures that of the naturalist who baptised it first. For instance, the Pyrenean desman of the Pyrenees having been first described by Etienne Geoffroy Saint-Hilaire, one writes Galemys pyrenaicus E.Geoffroy. However if the generic name proposed by the original descriptor has later been found invalid, the latter's name is then put in parenthesis: for example, Rattus rattus (Linnaeus).*

117

THE KRAKEN AND THE COLOSSAL OCTOPUS

Rather than being helpful, this detail leaves us completely confused[2].

To reach some understanding of how Linnaeus imagined the animal, which he admitted never having seen, it is absolutely necessary to go back to his sources of information. Unfortunately, the Father of Systematics had the bad habit of quoting them in rather less than explicit terms.

Here is how he quoted his sources in the case of *Microcosmus*, mentioninng for each work the name given by its author to the animal:

Bart. cent. 4 p. 284: Cete vigesimus secundus

Rhed. vivent. t. 22 f. 1.4.5: Microcosmus marinus

Ephem. Nat.Cur.ann. 8. obs.51: Singulare monstrum

Act. lips. 1686. P.48. t.48: Microcosmus marinus

A real puzzle for the researcher! As we already know about the *Anatomical History* of Bartholin, where there is made mention of the "22nd whale" (the *Hafguse*), and Paullinus's observation, published in the *Ephemerids of those Curious of Nature*, we can readily decipher the first and the third reference. However, it took me months to discover what was hidden behind the elliptic abbreviations of the second reference. After an exhausting and fruitless search, a stroke of luck put me on the right track. My lack of success was not surprising, since there was a spelling mistake in the author's name. By *Rhed.*, Linnaeus was referring to the famous Italien naturalist Francisco Redi! The work denoted by *vivent.* was that in which Redi discredited the theory of spontaneous generation by showing by experiment that intestinal worms are born from eggs and reproduce sexually, with full title *Osservazioni interno agli animali viventi che se trovano negli animali viventi*, published in Florence in 1684[3]. One really had to be already familiar with it to recognize this reference! But what did that work on intestinal worms have to do with our giant marine animal?

As soon as I had Redi's little book in hand, I could tell that Linnaeus had completely gone off track in linking the island-beasts of Bartholin and Paullinus to the marine Microcosm described by Redi. The Italian naturalist effectively speaks of a bizarre beast from the depth of the sea, which

---

[2] *It has often been incorrectly claimed, following Louis Figuier (1860) and later Alfred Moquin-Tandon (1865), that Linnaeus had classified the kraken in his works under the name Sepia microcosmus. This is completely false. In any case, in Linnaeus's mind, the Microcosmus was certainly not a cephalopod, for he would have then placed it, like the genus Sepia, among the zoophyte worms and not among the testaceans. As we shall see later, it is the German Oken who gave the scientific name Sepia microcosmus to the legendary kraken.*

[3] *As the last reference refers to lengthy commentaries of the same work of Redi, published in the Acts of the Learned in Leipzig, we need not be concerned with it.*

118

resembles a rock covered with corals and other concretions, forming miniature hills and valleys; he also describes the latter as being covered by bushes and small trees inhabited by small organisms - millipedes, worms and the like, - he depicts this creature as looking like a whole world in itself, hence its name; but, on the very first line, he points out that it is an *animaletto,* meaning a very small animal, not a large *animalone!!!* Linnaeus had to be completely incompetent in Italien to confuse this small creature with a giant of the sea.

Examining the figure published by Redi of the somewhat Y-shaped creature, one can easily recognize - especially if one is an amateur of Provençal cuisine - a vioulet or bitotche, or sea-fig. This animal, a tunicate, is covered with a hard crust on which is attached every imaginable marine organism: truly the parasites' paradise. The Vioulet has kept to this day in the scientific nomenclature the name *Microcosmus*, but even the best documented indices seem to ignore that it was Redi who so baptized it. Blainville appropriately called it in French "l'Ascidie petit-monde", i.e. the "small-world ascidian".

We don't know whether Linnaeus realized his mistake, whether he grew disgusted of not being able to define the marine Microcosm with any more precision, or whether he began to doubt the existence of such an elusive giant. In any case, he expurgated it without further comment from later editions of his fundamental work. One might speculate that it may have been on the advice of his friend and collaborator Peter Artedi that Linnaeus performed this act of cleansing. Cold and critical, Artedi often advised Carl - Kalle as he called him - to eliminate from his work all that was based on legend or anecdotes. And as Artedi was the real zoologist of the pair, the imaginative enthusiast that was Linnaeus followed without complaint. In this case, he would have been better advised to leave some room for dreams, for at one blow there disappeared, from a catalogue that had the pretension of being exhaustive, not only the largest existing invertebrate but also the innocent vioulet, so common in the Mediterranean.

As for Artedi, he showed that he was in no position to chide Linnaeus for his lack of attention. One night, in Holland, just after he had visited the cabinet of curios of the famous pharmacist Seba, in Amsterdam, he was so lost in his thoughts that he walked straight into a canal and drowned.

While Linnaeus had been very reserved and discrete on the subject of the marine Microcosm, bishop Erik Ludvigsen Pontopiddan turned out to be quite prolific about the *kraken*[4], in which we quickly recognize the island-beast of Paullinus and Bartholin.

The legend of the Kraken is surely very ancient; the fact that it was so widespread among Nordic people, the faith that was put in it, and the degree of detail which accompanied it, all vouch for its antiquity. However it

---

[4] *One should strictly write: the Krake, but this noun usually appears in the original text with its suffix n, which corresponds to the definite article, and usage has consecrated the form: Kraken.*

THE KRAKEN AND THE COLOSSAL OCTOPUS

is in the very thorough *Natural History of Norway (1752-53)* of the Danish bishop that one first finds an extended written version and some attempt at explanation. One should also note, to justly appreciate the status of Pontippidan as a zoologist, that most of the names which he gave to the birds of Scandinavia are still in use today.

The learned prelate, who was bishop of Bergen from 1746 to 1764, devoted a large part of his large, two-volume, work to the study of fabulous beasts of northern seas, whose existence did not seem to him without some ground. After having examined the case of the Mermaid, and that of the Sea-serpent, he arrives at what he called "without contest the greatest marine monster of the world". One suspects immediately that we shall once more hear about the animal which we already know under the names *hafguse* and *sae-krabbe*.

"It is called Krake, Kraxe, or as some say, Krabbe, [crab] which is the preferred name. This latter name seems to fit better with the description of the creature, which is round, flat and full of arms, or branches. Some also call it Horve, or Sae-horve, and others Anker-trold [anchor-troll]. All the foreign authors, both ancient and modern, which I have consulted on this subject do not seem to know very much about this creature, or even have a correct idea about what it is."

That is quite clear. In the middle of the XVIIIth century - only two hundred years ago - entire populations believed in the existence of an animal of prodigious, extraordinary size, but nobody had any idea about its identity...

What was known about it? Only what was told by common northern folk:

"Our fishermen, wrote Pontopiddan, unanimously affirm, without the least variance in their stories, that when they go many miles offshore, particularly during calm and warm summer days, and expect at their location (which they know by sighting many points on shore) to find a depth of 80 to 100 fathoms [145 to 180 m], they often find only 20 to 30 [35-55 m], and sometimes less. In such places, they usually find the greatest abundance of fish, especially cod and ling. As soon as they cast their lines, they say, they bring them back loaded with fish, from which they conclude that the Kraken is at the bottom of the water. They say that it is that creature which causes the shallowness mentioned above and stops their sounding weights. Fishermen are always happy when this happens for they see it as the augur of a good catch. Sometimes more than twenty boats assemble and fish next to each other. They then observe with their lines whether the depth remains constant or decreases. When the latter happens, they know that the Kraken is coming up to the surface and there is not time to waste; they immediately stop fishing, row vigorously and leave as quickly as possible. As soon as they reach a place where the depth returns to normal and where they feel out of danger, they stop rowing and a few minutes later they see the enormous monster appear at the surface. It shows itself clearly, although there never appears the

totality of its body, which no human eye has ever contemplated (except for the young of the species, of which we will speak later). Its back, or upper part, which seems to be about a mile and a half [2 km] in circumference (some say more, but I chose this number for more certainty), at first sight looks like a group of islets surrounded by something which floats and undulates like marine algae. Here and there, one sees near the sea level sand banks, on which various kinds of little fish leap without cease until they all sink back in the water. Finally, many points and shiny horns appear, which grow in thickness as they rise above the water surface; some times they are as wide and tall as the masts of medium size boats. It seems that these would be the arms of the creature and it is said that if they managed to get a hold of even the largest warship they could drag it to the bottom.

"After a bried stay at the surface of the water, the monster begins to sink slowly again, and at that time the danger is as great as before, for in going down the beast produces such an upwelling and a whirlpool that it carries everything with it, as does the Maelstrom [...]

"As this enormous animal must, in all probability, be classified in the genus Polyp or Starfish, as I shall prove it more fully later, it seems that those parts of its body that are seen to be raised at will and which are called horns or arms must actually be its tentacles or feelers. The animals use these to move around and to collect their food.

"In addition, to help foraging, the Creator has endowed these animals with a peculiar and powerful perfume which it can emit at some times and which it uses to lure and attract other fish to itself.

"This animal has another strange characteristic, which numerous seasoned fishermen know by experience. They have noticed that during many months the Kraken or Krabben eats continuously and that during other months it continuously rejects excrements. During this evacuation, the surface waters are discolored and appear thick and turbid. It is said that this murk is so pleasing in smell and taste to other fish that they immediately gather above the Kraken; it then opens wide its arms, or horns, gulps down its welcome guests, and transforms them by digestion, over a sufficient time, in a lure for more fish of the same kind."

Such is the legend of the *kraken*. We have quickly recognized in it the island-beast of northern seas, the ambiguous *cetus of* medieval bestiaries, the Jasconius of St. Brendan, the *hafguse* of the Norwegian *Royal Mirror* and of Bertholin, the monstrous crab of Paullinus and the pseudo-*microcosm* of Linnaeus. Thus, we have seen the various elements of this legend gradually merge over the centuries: the insular appearance that the animal shares with the cetaceans; the rarity of its surface appearances; the delicious perfume which attracts fish from all around; the laborious digestion alternating with feeding frenzies; the presence of multiple tentacles; and finally a strength sufficient to drag to the bottom the largest ships. There is only one truly original variant in Pontopiddan's story of the *kraken:* it no longer attracts its preys through perfumed burps, as Thomas Bartholin related; it is its excrements which exude an exquisite smell. Their use as a lure for preys which appear on the menu of the following meal - proof of the admirable concern for efficiency manifested by the divine Providence - is the final touch of genius which completes the picture. It is not improbable that the discovery of floating mass of ambergris might have something to do with this last embellishment.

"I only relate here what many have already stated, points out bishop Pontopiddan, but I can't vouch for that last trait with as much certainty as I can about the very existence of this surprising creature, although I don't see in it anything contrary to nature."

It wouldn't make sense to imagine that the bishop of Bergen, an erudite and undoubtedly intelligent man, would have taken upon himself to state with certainty the existence of this unlikely creature merely from tales told by a few fishermen. It appears from his text that he had carried out, over many years, a detailed enquiry among numerous people and over a wide area, in part with the assistance of the pastors of the various dioceses falling under his ecclesiastical authority. Thus, through the help of the reverend Friis, assessor of the consistory, pastor of Bodo, in the province of Nordland, and vicar of the College for the Advancement of Christian Knowledge, he had even managed to lay his hands on a report, less than a century old, about the discovery of the corpse of a *kraken*. As there couldn't be many chances of examining such an enormous animal live, that would have been, in Pontopiddan's mind, a unique opportunity to observe its entire body:

"In the year 1680, a Kraken (perhaps a young and careless individual) ventured into the waters which surround the rocks and cliffs of the parish of Alstadhaug [in the Gulf of Ulwangen], although that kind of creature usually stays many leagues from land, which is also why they must usually die offshore. It so happened that the long outstretched arms, or antennas, which this creature seems to use as does the snail, grabbed in their motion some trees which were growing near the water, which they could easily have uprooted. However, in addition, it also got stuck in some rock fissures and crevices and become so firmly wedged that it could not free itself and died and rotted in place. The carcass, which during its slow putrefaction filled a large part of the narrow fjord, made the latter impossible of access because of its unbearable stink."

Another point which argues in favour of the good faith of Pontopiddan and his informers: the *kraken* was not pictured as some fabulous bogeyman, although it was thought to have enough strength to pull down the largest ship:

"The Kraken never had the reputation of being very dangerous, unless some have taken the life of people who could not complain about it. I have heard of only one case of agression, which happened a few years ago near Fridrichstad, in the diocese of Aggerhuus. It is told that two fishermen had quite accidentally, and to their great surprise, fallen upon an area where the

water was full of thick mud, almost similar to a swamp. They immediately tried to get out of that place, but they did not have time to turn around fast enough to avoid one of the horns of the Kraken, which crushed the bow of their boat. Although the sea was as calm at it can be, it is only with great difficulty that they escaped on board their wreck."

Far from finding anything wrong with the mysterious beast being mistaken for an island, bishop Pontopiddan saw in that the explanation for another mystery: ghost islands. He recalls that, in his description of the Faroe Archipelago (*Faeroe and Faeroes reserata*), published in 1673, the Danish pastor and topographer Lucas Jacobsen Debes had spoken of islands which suddenly appeared and no less suddenly disappeared. Two other works of the end of the XVIIth century also alluded to similar phenomena: the *Mundus mirabilis tripartitus* of Eberhard Werner Happel, also called Happelius, and the *History of Norway* by Thormod Torfeson, called Torfoeus. The latter in particular noted the apparition in 1345, in the Breidenfjord, on the coast of Iceland, of an island never seen before.

Nobody to that day had been able to understand such phenomena, Pontopiddan pointed out. Thus, it was not surprising that simple marine folk should consider such islands, whose appearance and disappearance defied calculation and multiplied the pitfalls of navigation, as the abode of evil spirits. It was quite obvious that one could not imagine finding, at the surface of rough seas capable of crushing the most sturdy ships, floating islands such as sometimes form in calm and stagnant waters. That's why sailors saw in those found in northern waters the result of some devilish intervention. However, the learned bishop pointed out with good sense, that "one should not so lightly incriminate the fallen angel":

"I would rather believe, he said, that the devil which so quickly builds and destroys those floating islands, is no other than the Kraken which some call soedraule , or soe-trold, meaning sea oger [in the sense of "marine evil"].

He brought, in support of his statement, the recent adventure of baron Carl Grippenhielm, who had sought in vain, off Stockholm, an island called Gummars-öre, indicated on the chart of geographer Buraeus. One day however, he ended up by finding three points of land rising among the waters and asked the pilot of the boat whether this was Gummars-öre. To which the sailor answered that all he knew was that this apparition heralded either a storm or a large abundance of fish. For Gunnars-öre, he said in the sinister tone which one can imagine, is a pile of reefs at sea level where the *soe-trold* lurks.

"Who will not immediately understand, adds Pontopiddan, that this eclipsing island with its pointy parts, and the promise of a great abundance of fish, is no other than the soe-trold, the Kraken itself!"

The bishop of Bergen nevertheless claims that he is not one who is easily fooled; he treats as a "nototiously fabulous and ridiculous novel" the story of the whale taken for an island which dives into the sea when a fire is lit on its back; he even accuses his Swedish colleague Olaüs Magnus of gullibility for have spread that tale, two centuries earlier. The courageous biographer of the *Kraken* (and of the Sea-serpent) did not suspect that two centuries later he would in his turn also be the object of nearly universal ridicule.

We already known what Pontopiddan thought about the zoological nature of the incredible Kraken. In his description of the legend, he had not been able to keep himself from stating his opinion on that subject. As the animal was surrounded by many arms, it had to be some kind of "polyp" (octopus) or star fish. One might perhaps be surprised to see a mollusc and an echinoderm so closely associated, as if they were of the same genus. However, anyone who has seen a crawling and writhing Ophiurid starfish, with its prehensile arms, will not have failed to be struck by its astounding resemblance to an octopus. Let's not forget that in Pontopiddan's days, zoological classification was still in limbo. Linnaeus had just only published the first editions of his *Systema naturae*. In his grouping of the invertebrates, whose internal anatomy was poorly understood, the great Swedish naturalist had relied on external characteristics, whose similarity could sometimes be attributed to some convergence phenomenon. Thus, all invertebrates not covered by an articulated shell (Insecta) had been limped in the worms (Vermes).

In any case, from the vague description given by the fishermen, it was just as legitimate to see in the *Kraken* a giant ophiurid as a giant cephalopod. It is actually that first identification which Pontopiddan prefers. He uncritically accepts at face value Pliny's text about the *arbor,* so large that it cannot pass through the Pillars of Hercules: he sees in it an obscure allusion to the *kraken* and a confirmation of its gigantic size. In addition, he mistakenly associates this passage with that which follows it, about the *rota,* with its eight spinning arms. He imagines that both comparisons, one with a tree, the other with a wheel, relate to the same animal. The transformation of arms into branches, suggested by the first passage, leads him to link the fabulous beast of Scandinavian waters with the star fish "called *Zeesonne* (sea-sun) in Dutch and *Stella arborescens* by Rondelet and Gesner" and which he himself called "Gorgon's head" (it is the *Asterophyton arborescens* of modern zoologists, which we mentioned earlier). The Scandinavian Bishop quotes, in support of his idea, the fact that the inhabitants of the Norwegian coast believe that the Gorgons' heads are baby *Kraken,* or perhaps even its eggs.

Later, more appropriately, Pontopiddan draws a resemblance between the attractive perfume of the *Kraken* and that of the musky octopus mentioned by Pliny under the name *Ozoena*. This incursion into Pliny's *Natural History* of course brings our episcopal commentator to mention the story of the monstrous "polyp" of Carteia. He then concludes that among the various species of polyps and starfish, which he includes in the same genus *Kors-trold* (cross-shape evil troll, meaning a beast with radial symmetry) "there are some which are of much greater dimensions than others and even, if appearances are to be trusted, larger than the largest inhabitants of the ocean."

Even though bishop Pontopiddan's diagnostic is rather unfortunate, his conclusions are nevertheless quite pertinent. After all, one can only blame him of having believed without restriction the information pertaining to the extraordinary size of the animal. A mile and a half in circumference? That gives a diameter of more than 600 metres! Even with the best of good will, this is a little too much for a single animal...

In the light of what we know today, anatomists, physiologists or engineers could easily demonstrate that such an animal is *mechanically* impossible. It would probably fall apart under the stress of a storm. In any case, not having nervous fibers 300 metres long, it could never manage to coordinate its movements, and it is hard to imagine that its heart could be strong enough to pump blood as far as its body's periphery, so as to nourish and oxygenate all its cells. To survive, it would have to have many brains, many hearts, innumerable gills and kidneys, etc... It would have itself to be a combination of many organisms, which brings us back to the same conclusion: such a size is *too much for a single animal*.

Even if Pontopiddan, in his day, could not possibly carry out such a line of argument, common sense should have warned him. Well versed on the writings of the authors of antiquity, he had often noted that legend always exaggerates the size of marine "monsters". It was already known that whales, far from measuring hundreds, even many thousands of meters, as some had claimed, were no longer than a few tens of meters. That is why he had so ridiculed the island-whale of the gullible Olaus Magnus.

The most rigourous commentator of the *kraken* was undoubtedly professor Karl-August von Bergen (1704-1760) of the University of Frankfurt. He also, after life-long study, had carefully compared all old Scandinavian stories pertaining to the mysterious monster of the northern seas. However, somewhat more critical than his predecessors, he expressed amused astonishment about Linnaeus's unjustifiable identification and the "far fetched" diagnostic of Mgr Pontopiddan.

On the subject of Linnaeus's identification, he indulged himself in some sarcasm; "Everyone knows in what sense man is sometimes called a microcosm, but even naturalists don't agree on the identity of the marine animal to which they give this name." He went on to chuckle over the casualness with which Linnaeus had brought under the same generic name two organisms of such different dimensions than Redi's small animal and the island-beast of Paullinus and Bartholin, merely on the basis of the heterogeneity of their carapace.

To try and identify the Scandinavian giant, there was, according to the German scientist, no better basis than Pontopiddan's in-depth study. He had shown that it was to be classed in the family of "polyps", and in that, he might be right, although one should be careful not to rely "only on the presence of tentacles and antennas". In fact, most of the limbs of this animal, which no man had even seen in its entirety, always remained hidden below the surface.

According to von Bergen, there is nothing in the stories of the fishermen that says that the arms of the unknown creature have as many branches as those of a Gorgon's head, with which Pontoppidan tries to link it. And if the Gorgon's heads, which are found in large numbers on the shores of Norway, were actually the offspring of the *kraken*s, these should be more numerous and would be seen more often.

It was even less likely that the enormous beast should belong to the class of shelled molluscs, as Linnaeus claimed it did. As the German naturalist had verified with a large murex, these animals are rather heavy, and if they were to reach a very great size, they would never manage to float. In this respect, the animal was more likely to belong to the class Crustaceans.

Truly, concluded von Bergen, its exact structure will be known only on that day when fishermen harpoon one and drag it ashore. All that could be said, for the time being, about this beast was what various authors had been unanimous about: that it is the largest of all marine animals, that it has never been seen in its entirety, for it only rarely allows its arms and tentacles to be seen; that the surface of its back, covered with algae and other marine plants, looks from afar like small islands or reefs near the surface; that it exudes, like the musky octopus, a powerful smell which attracts some fish; that, with its multiple feet, it is certainly not a whale; that it fills and empties its belly only once a year; that it is during calm conditions and during the summer months that it shows up at the surface; that it always moves in a vertical direction; and finally that its reproductive cycle must be very long.

Without drawing formal conclusions, professor von Bergen took pleasure in emphasizing everything that linked the Kraken with cephalopods. He even expresses surprise that Pontoppidan did not give more weight to the episode of the "polyp" of Carteia, quoted by Pliny. The identification which he suggests is thus an excellent approximation. Nevertheless, at that time, the elements which would have permitted a more precise identification of the beast were already available.

Pontoppidan, rather than to laugh at the story of the island-whale as related by Olaüs Magnus, would have been well advised to link the legend of the Kraken with another passage of the *Historia gentibus septentrionalibus,* where there was mention of "*the horrible monsters which are found on the coast of Norway*" (book XXI, chapt.V). The latter were described in great detail:

"In the sea off Norway are found a kind of strange and monstrous fish, the name of which is unknown (so large that they are sometimes counted among whales), which are extremely frightening to look upon, and also seem very cruel. They are built in a most frightening shape and manner. Their head is square, full of sharp points in all directions, and of long horns, in such a way that they look like the roots of freshly uprooted trees; it is 10 to 12 cubits long [6 to 7 m]. Their colour is black; their eyes are very large, being 8 to 10 cubits around [5 to 6 m]; the pupil is very red and shining, one cubit wide [60 cm], and it is visible far under the waves at night by fishermen at dark times. Their hair is made like goose quills, very long and thick, and hangs like a beard. The rest of the body is small in comparison to the head, which is

square and large, for it is at most 14 or 15 cubits [8.5 to 9 m]. One of these marine monsters can pull to the bottom a large, fully loaded ship, even though it may be filled by good and strong sailors."

The habitat of this animal, the horns which surround its head, and the strength which allowed it to pull down large vessels, all indicate quite clearly that this text, from 1555, relates to the monster which will later be known as the *Kraken*. The abundance of details helps in recognizing its exact nature.

The crown of sharp horns, arranged as the roots of a tree, betrays without the shadow of a doubt a cephalopod mollusc[5]. The presence under the head of long and thick "hair", similar the the quills of geese, assures us that this cephalopod belongs to the group Decapods (we recall that the squid probably owes its French name, "calmar", in part, to the presence of these two extra tentacles somewhat in the shape of goose feathers). Finally, the enormous, and rather exaggerated, eyes suggests that the animal is a squid rather than a cuttle-fish.

The pertinence of this diagnostic can be verified today from the dimensions quoted, which are in the same proportions as those of most squids. In a giant specimen, the body might reach 9 m in length; the head, arms included, might reach around 7 m; the pupil of an eye measures nearly 60 cm. However the circumference of the whole eye would not exceed 3 m. An error must have slipped in in the transcription of the dimensions attributed to the circumference of the eye: one should probably read "eight to ten feet" (2.40 to 3 m) rather than "eight to ten cubits". This is indeed confirmed by a glance at the original Latin text of 1551.

Thus, the agreement between the various measurements is such that one is led to believe that they were taken from one or more dead individuals; they are then quite likely to be correct.

According to Olaüs Magnus, Erich Falchendorff, "archbishop of Nisdrose" (actually Nidaros, old name of Trondhjem, at that time capital of Norway), was supposed to have sent to Pope Leo X, in 1520, a detailed epistle pertaining to "this strange and novel fish monster". The letter was even accompanied by the head of "another monster, also quite fearsome, pickled". It was that of a walrus[6].

Squids 14 to 16 metres in length, without including the tentacles? This is a long way from the *kraken*, 600 m in diameter, but equally so from the small squids of 50 cm and even of the so-called giants of 2.50 m mentioned by Aristotle in the Mediterranean! All of a sudden, the story of the

---

[5] *That this monstrous "fish" had been thought to be a "whale" is no real objection. In those days, most marine creatures were called "fish", and we know that the name cetus, used in the original Latin text, was applied to all large marine animals.*

[6] *It may be less surprising to find a prince of the Church keeping the Pope informed of the marvels of nature when we recall that Leo X was Giovanni de' Medici, great protector of letters, science and the arts. Raphael has left of him a famous portrait, to be found in Florence, in the palazzo Pitti.*

pseudo-octopus of Carteia becomes more probable. One can understand how the sight of such giants, probably smelly, as are many cephalopods, could have spawned or maintained the medieval fable of the *cetus*, partly confused with a whale. The possibility that such monsters might be capable to sink at least small ships now becomes less of a laughing matter. And the incredible legend of the *Kraken* suddenly becomes clearer...

Let us recall, on the one hand, that squids are often gregarious. That fact is well known today for some species, but naturalists, even in antiquity, perhaps already suspected it, since Pliny mentioned that *lolligines* sometimes leaped out of the sea in such numbers that they could sink a boat.

On the other hand, Pontopiddan insisted upon the fact that no one had ever been able to look at the entire body of a *kraken*, except for a "young one" stranded in the gulf of Ulwangen in 1680. What fishermen always saw, over a wide area of the sea, was "a number of small islands surrounded by something which floats and undulates like marine algae", and "here and there sand banks at the water surface", among which finally appeared "many points or shiny horns" sometimes as large and tall as the masts of ships.

When we bring these two facts together, the gregarious nature of squids with the discontinuous appearance of the "ogre of the seas", the solution of the mystery strikes us. The *Kraken* is not a̲ sea monster, but a whole group of sea monsters, a *school of giant squids*. One should have thought of it long ago, since 600 metres in diameter is much too much for a single animal[7].

The most astounding part of this whole story is surely the fact that some modern scientists revived, only a few years ago, one of the most fantastic aspects of the old legend of the Kraken: that about the way in which the monster causes the apparent depth of the sea to change by placing itself between the bottom and the fishermen's sounding log.

Here are the facts. In 1946, the American navy revealed that four years earlier, during the war, three of its specialists in acoustic sounding, Eyring, Christensen and Raitt, had detected in the ocean, between depths of 300 to 450 metres, a mysterious "layer" which reflected sound waves. The original discovery, off the coast of California, referred to a 500 km wide strip, but in later years it was discovered to this strange phenomenon occurred nearly everywhere in the deeper part of the oceans. (In some areas, two, three, and up to six reflective layers were discovered, at different levels.)

It was first assumed, of course, that the echo had to be the result of some discontinuity between waters masses of different temperature, density or chemical composition. However, already in 1945, the biologist Martin W.Johnson, of the Scripps Oceanographic Institution, had detected the first indication of the true nature of this "false bottom". He had observed that the

---

[7] *If one may thus conclude that the Kraken of legend is no other than a gigantic squid, it certainly has nothing to do with the octopus. It must be for Germans that the problem of the giant Cephalopods must seem messiest: they have had the unfortunate idea to use the word kraken for the octopus!*

deep scattering layer, or E.C.R. layer, so called after its three discoverers, shifted vertically according to a daily rhythm; at night, it approached the surface, to fall during the day to greater depths. It deliberately avoided high light intensities[8]. Impossible not to think of it as an accumulation of living creatures.

Three hypotheses have been put forward about the identity of the sound scattering creatures. According to the first, they are small planktonic shrimps, similar to those eaten by baleen whales. It is a fact that many planktonic organisms perform vertical migrations, regulated by the light of day. However many wondered whether even a dense throng of these minuscule creatures could significantly affect sound propagation.

This is why some prefer the second hypothesis, according to which the E.C.R. layer might be formed by schools of fish, whose swimming bladders are known to be excellent sound reflectors. Their daily ascent and descent would follow those of the plankton on which they feed. However, this explanation is contrary to observations which show that, in the ocean, fish populations concentrate in well defined zones where they find abundant food. Nothing suggests a uniform horizontal distribution of a fish-rich layer similar to the deep scattering layer.

The third hypothesis is bolder, and at first sight, seems rather improbable. Its rare supporters believe that the "false bottom" is made of an enormous concentration of squids. One is first tempted to protest against such an opinion. If it is difficult to accept that there exists at a certain depth a continuous layer of fish, a group as immense as it is diversified, how could one suppose the existence of a similarly broad distribution of squids alone?

Actually, these pelagic and nocturnal cephalopods are found in all of the world's oceans, from the glacial waters of the polar seas to the tepid waters of the Equator. Rachel Carson reminds us that:

"Squids are known to be the sole food of the sperm whale, found in the open oceans in all temperate and tropical waters. They also form the exclusive diet of the bottlenosed whale and are eaten extensively by most other toothed whales, by seals, and by many sea birds. All these facts argue that they must be prodigiously abundant. It is true that men who have worked close to the sea surface at night have received vivid impressions of the abundance and activity of squids in the surface waters in darkness."

The author of *The Sea Around Us* quotes on this matter personalities such as John Hjort, Thor Hejerdahl and Richard Fleming, who all had the opportunity of seeing immense schools of squids gather at night at the

---

[8] *In 1954, Dr. Brian P.Boden and Dr. Elizabeth K.Boden of the Scripps Institution showed that, during the day, the light intensity was only 3/10,000th of a footcandle inside the reflecting layer. In other words, the organisms of which it is formed seek a light intensity corresponding to that of a 100 watt light source at a distance of 800 metres.*

surface. She concludes however that "most people find it difficult to believe in the ocean-wide abundance of squid."

As a zoologist, I cannot but indulge in a few comments on this topic.

Of all marine invertebrates, cephalopods are by far those with the highest level of organization; research has revealed that even at a the mental level they occupy a rather high rank in the animal kingdom. Professor John Zachary Young, of the University of London, says of the octopus that: "It is the most intelligent of the lower animals, and, of all invertebrates, it is the one that has the largest and most developed brain." As in vertebrates, the latter is even enclosed in a cartilaginous box. Numerous experiments have shown that cephalopods are capable of rather complex reasonings and learn easily. Professor Henri Pieron has shown among other things that an octopus, faced with a bottle containing crabs was ingenious enough to remove the cork and extract its preys.

In by-gone days cephalopods dominated the world during tens and perhaps hundreds of millions of years, longer than any other group. While we know today about 400 living species, paleontologists have succeeded in identifying more than 8,000 fossil species, which of course represent only a puny fraction of their actual abundance.

One need only reflect on the past splendor of the nautilus group, of which five or six living representatives remain, on the extraordinary diversity of the shapes of ammonite and belemnite shells, two completely extinct groups. Among them were the first giants of the sea; in Ordovician beds were found calcareous sheaths of Nautiloids, 3 metres long. The coiled shell of the Ammonite *Pachydiscus seppenradensis*, with a diameter reaching 2 metres, is to our knowledge the largest shell that ever existed. Even in the period which followed the Ordovician, the Silurian, the largest known Arthropods, the eurypterids or giant sea scorpions, apparently did not exceed 3 metres in length. Never did the crustaceans, nor their extinct cousins the trilobites, ever posess a combination of defence and attack skills as highly perfected as those of the octopus, or could move at the speed of squids. In the world of invertebrates, cephalopods remain to this day the undisputed champions, as much by their strength and their speed as by their size.

Even the emergence of fish has not completely taken away the prestige of the multi-tentacled molluscs. They can hold their own against the most dangerous and voracious of them all, the sharks. It also seems that some species manage to surpass in mass even the largest fish. It is then not surprising that they can put up a good fight against the most powerful and best armed of all marine mammals, the sperm whale.

Besides, squids, with their great speed, have such a varied ration - molluscs, crustaceans and fish - that they must be able to find enough food almost anywhere. On the other hand, their group is adapted to the widest range of temperatures and to considerable changes in pressure. There are species of squids in all of the world's oceans, near the surface as well as in abyssal depths: *Chiroteuthis lacertosa* was fished out of a depth of about 5,400 m.

Although we may hesitate, given the enormity and diversity of the fish group, and of the gigantic size of whales, to call the squids the Kings of the Sea, we must recognize that they occupy in the ocean a place which is at least the equal of that of rats, sparrows, flies, and...people on land. Capable of adapting to all conditions, they are cosmopolitan creatures par excellence.

Moreover, in some places, the extraordinary density of squid populations has been observed *in situ* thanks to exceptional circumstances. It so happens that because of some mysterious poisoning, to which we shall return, the sea is sometimes strewn with an uninterrupted blanket of their corpses. On Jan 10th, 1858, at 43° 12'S and 37° 15'E, the crew of the Dutch merchant vessel *Vriendentrouw* (Captain de Greevelink) found, during two hours of navigation, as far as the watch could see, the sea covered with dead squids. Further, my friend André Capart, director of the Institute of Natural Sciences of Belgium, who has participated in numerous oceanographic expeditions, in particular to the South Atlantic in 1948-49, told me that no matter where he found himself at sea, it always sufficed, at night, to stop the engines and to turn the projectors onto the sea to immediately see hordes of squids.

The excellent naturalist Ivan Sanderson provided an excellent summary of the situation:

"Most people don't know what a squid is; nevertheless, the collection of all these animals is undoubtedly a greater mass of pure animal matter than that of the sum of any other pair of creatures. Squids exist in innumerable numbers of seemingly infinite schools in all the sea and oceans of the world, and nearly three-quarters of this planet are covered with water, of depths on the average of 4,000 m. In the midst of this vast volume of liquid, there are probably more squids than anything else."

Thus, if there exist in the oceans creatures which might have some chance of being distributed nearly uniformly and continuously, it is indeed the squids. Anyway, if it were confirmed that the deep scattering layer was composed mostly of squids, and thus that it was true that the *Kraken* interferes between our soundings and the bottom of the ocean, would that not be a wonderful victory for legend?

Part Four

THE COMPLETE STORY
OF THE SUPREME SQUID

"The vast ocean makes itself admirable to all nations by the strange and diverse kinds of fish which it produces, which are also most admirable not so much for the infinite quantity with which they are found in the sea, so numerous that one would think that there are less stars in the sky than
there are fish in the sea, than for their excessive and terrible size, their monstrous form and shape..."

OLAÜS MAGNUS, *History of the Northern People*

Chapter Seven

THE FIRST PIECES
OF EVIDENCE

As early as the XVIIth century, a Sherlock Holmes of Zoology could have deduced, on the basis of legends and anecdotes alone, the existence in the North Atlantic of squids of a tremendous size, comparable to that of whales. To establish what Science would still not recognize for another two hundred years, all he would have had to do would have been to use that same method which I use in my research on cryptic animals: to scrutinize the texts of ancient authors, perhaps prone to exaggerate, but also full of naive honesty; to track the sources of various fables; to sieve through reports and laymen's testimonies, and thus, altogether, to exploit the least item of available information, without ever rejecting *a priori* even the most implausible statements. What is true is not always plausible!

These giant squids, were gradually surrounded by an aura of terror and metamorphosed into myths, and it is of some interest to ask ourselves what may have been at the origin of those old folk stories, which sometimes turn out to be surprisingly accurate. It may, of course, have been the fleeting and defective observation, by navigators, of live individuals appearing at the sea surface. But it must have been mainly the leisurely examination and perhaps measurement of dead specimens. As there is little chance that such oceanic titans would have been harpooned, such an examination could have occurred only occur thanks to accidental strandings.

If one digs into old chronicles, one finds some of their traces. When, in the second half of the XIXth century, zoologists could at last hold in their hands the material proof of the existence of these monsters, namely fragments of their bodies, they held up these ancient texts in triumph. But it should have been *before* putting the remains into jars that these stories should have been consulted, so as to guide the investigations. *Afterwards*, they are only of historical value.

Thus is revealed the exact nature as well as the utility of the cryptozoologist's task.

Let us cross through the enchanting, but distorting fog of legend to try and arrive at the truth beyond: for that purpose, we will, for the time being, stick to concrete and well documented facts.

As I mentioned it earlier, the report of Trebius Niger on the "polyp" of Carteia is by all appearances the first known document about the stranding of a giant squid, namely, on the Atlantic coast of Spain. It is probable that a number of specimens, stranded on the Norwegian coast, contributed to Olaus Magnus' extremely significative description of the "monstrous fish", accompanied by rather precise measurements. The first text is however ambiguous, and the second does not refer to any single specific incident.

One finds, in Conrad Gesner's *Historia animalium* (1555-1560), a passage which, although better dated than the previous one, remains vague and seems at first glance unrelated to our topic:

"In the year 1530 of our era, in the month of January, there was brought to Venice a monstrous serpent, of which this is a faithful reproduction [our Fig. 43]. It came from Turkey. The king, to whom it was presented as a gift, exclaimed: `This monster, there is no doubt about it, is the omen of some horrible cataclysm.'

No one would think of seeing in this incident the discovery of a giant squid if it weren't that the illustration is in some ways quite revealing. It is clearly an artistic interpretation, still quite medieval in style, of a description, probably coarse and redundant, of the more or less well preserved remains of an animal. No creature has that many heads, so that they must stand for something else. However, nothing is most likely to be compared to the jaws of snakes than the arms of a squid, studded with toothed suckers similar to a series of mouths. As in the case of Homer's Skylla, the necks of the hydra would represent the arms, reduced to seven because of the magic of that number. The elongated body could well be that of a squid. All this, however, remains conjectural.

It is in the pages of an Icelandic chronicle, the Annals of Björn of Skardsa (*Annalar Björns á Skardsa*), that one finds, for the year 1639, the first detailed mention of the stranding, on the north coast of the island, of what is indisputably a giant squid.

"(1639). In the fall, a curious creature, or marine monster, was cast onto the sands of Thingöre, in the land of Hunevand[1]. Its body, as long and wide as that of a man, had seven tails, each seven ells in length [1.20 m]. These tails were covered with buds similar to eye balls, with golden eye lids. In addition to these seven tails, there also came out above them a much longer one, 4 to 5 toises [4.95-5.50 m] in length. There was no bone, nor cartilage in the body, which to the eye and to the touch looked like the belly of the female of the lumpfish[2]. There was no head to be distinguished except perhaps through the presence of a couple of openings located near the seven tails.

---

[1] *The icelandic text speaks of a* sio-skrimsel, *meaning a "sea-ghost". Skrimsel is the catch-all word given to all unknown and fearsome creatures, and corresponds to our "monster" in its most popular sense.*

[2] *The text says:* grasleppu-hveliu-kvidur, *meaning approximately: "The soft-one, female of the Shield". This is the fish with a lumpy and naked body comonly called in France mollet or gros seigneur and in England lumpfish (*Cyclopterus lumpus*).*

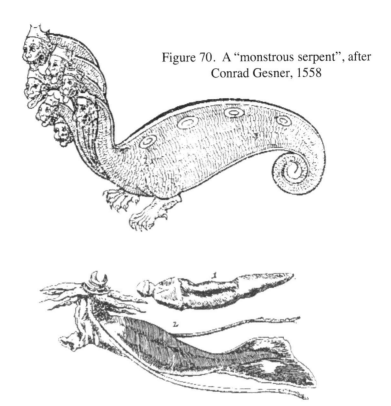

Figure 70. A "monstrous serpent", after
Conrad Gesner, 1558

Figure 71.
1. The marine monk of Enckhuizen (Netherlands)
2. Squid from the Gottorf Cabinet of Curios.

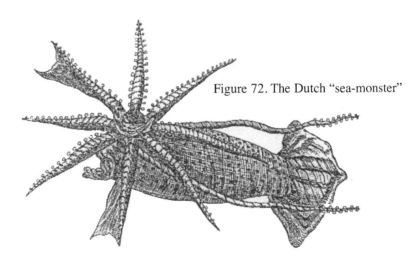

Figure 72. The Dutch "sea-monster"

Figure 73. Dom Pernetty's squid, 1763-64

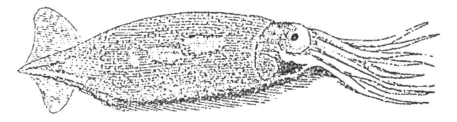

Figure 74. Section of a
human eye (left) and that
of an octopus (right)

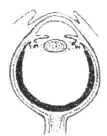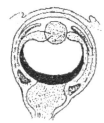

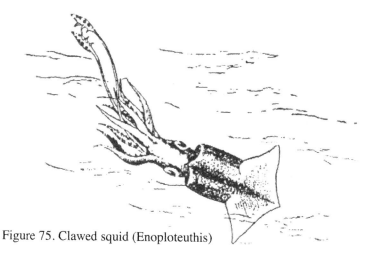

Figure 75. Clawed squid (Enoploteuthis)

Figure 76. The double-headed squid of Dingle-I-cosh,
1673.(*Leabharlann Náisúnta
na h'Eireann*)

Figure 77. Struggle between a
giant squid and a sperm whale,
after Frank T. Bullen.
(Drawing by Arthur Twindle,
Monsters of Land and Sea,
1913, Cassell & Co.

Figure 78. A more recent depiction of a fight between a gigantic squid and a sperm whale

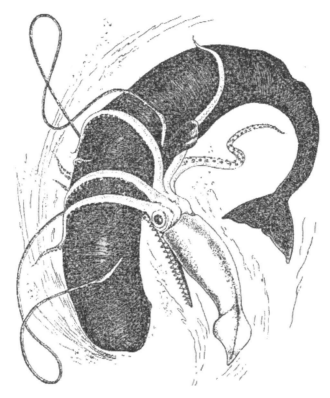

Figure 79. An 18 metre-long sperm whale just about to be cut-up
(Photo Saunders, F.Lane-Rapho)

Figure 80. A reconstitution of the original ex-voto of Saint-Thomas' chapel in Saint-Malo

Figure 81. The colossal ocotpus of Pierre Denys de Montfort (1764-1821), as shown in the *Suites á Buffon*, 1801-1802

Figure 82. The incident of the attack of a ship by a colossal octopus off Angola, as it appears on teh *Naturalist's Library*, 1860

Figure 83. The same incident, freely dramatized in *Monsters of the Sea*, by John Gibson, 1897

Fig (84) A modern version of the same incident by the illustrator Gino d'Achilles (Aldus Books, 1975)

Figure 85. A detail from the attack off Angola incident,, in one of Philippe Hettingen's works of the 1920's

Una barca di pescatori è stata assalita al largo di Jersey (Isola Normanna) da un polipo gigantesco i cui tentacoli misuravano sette m di lunghezza. Durante la spaventosa lotta sostenuta col mostro rimasero uccisi due pescatori (Disegno di Beltrame)

Figure 86. In September 1922, the sensational Italian weekly *Illustrazione del Popolo* reported that a fishing boat had been attacked off Jersey by an octopus with 7m long arms. Although the incident had claimed the lives of two fishermen, it had never left a trace in the police records.

THE FIRST PIECES OF EVIDENCE

" A number of reliable men observed this prodigy, and one of the tails of the monster was carried to the old abbey of Thingore to be examined."

The first well-informed people who consulted this manuscript had no difficulty in recognizing the monster described in it. When, in the middle of the XVIIIth century, Eggert Olafsen and Dr. Biarne Povelsen explored the interior of Iceland (at that time a Danish possession), under the auspices of the *Videnskaberne Selskab* (Learned Society) of Copenhagen, they had the good luck of finding this priceless text. In the account of the expedition, published in 1772, Olafsen commented, quite sensibly[3]:

"It seemed to us immediately that the chronicler had made a mistake, confusing the forward and hind parts of the animal: in that case, the bud-studded appendices are not tails, but arms or tentacles; and since only seven are mentioned, it would seem that the eighth had been torn off.

"Who cannot tell now that this animal was just a very large squid. (Sepia) But of which species? That, what we cannot determine, since the structure of the stomach is not described, nor is that of the mouth, although it would appear that it had been damaged and distorted. The description of the buds and the cups is rather curious, but nevertheless carries an air of authenticity because of the precision with which some details, such as the colours, are described."

In squids, the stemmed suckers, with their edges finely lined with small teeth, are irresistibly reminiscent of clusters of eye balls, fringed by large eye lashes: a choice motif for surrealistic painters! The old chronicler's comparison was quite opportune. So was Olafsen's conclusion when he used the above text to link the stranded beast with a fabled Icelandic bogey-man. According to popular belief, this horrible monster lived in Grimsey Fjord; it chopped off the heads of seals and sank ships:

"The kind of fish, wrote Olafsen, which is seen here as a monster [soe-spoegelse, literally sea-ghost, this time in Danish], perhaps because of its size, will undoubtedly be recognized by learned naturalists as belonging to the marine worms."

At that time, all molluscs were placed within the catch-all group of Worms. The presence in the mutilated monster of Thingöre of one very long tentacle, in addition to a series of smaller ones, shows that it was a cephalopod of the group Decapoda, but there is no detail in the text which would allow one to decide whether it was a squid or a cuttle-fish. The diagnostic expressed by the Icelandic explorer was thus as accurate as it could be at the time.

In any case, we now have an unambiguous document establishing the existence in the icy Arctic Ocean of ten- armed cephalopods reaching up to

---

[3] *From the French translation of 1802, by Gauthier de Lapeyronie.*

4 m in length, not including their long tentacles, and with a total length of at least 7 m. We are still a long way from the dimensions of the "polyp" of Carteia or of the "monstrous fish" of Olaüs Magnus. However, the seven-tailed phantom of Thingöre is nevertheless much greater in size than the "great squids" of Aristotle, which at 2.50 m only already passed for the giants of their time.

Iceland was a long way away; it must have seemed further in those times than Mars does to us today. What happened there usually remained unknown, even among the most learned of the continent. One had to wait for the report of Olafsen and Povelsen expedition, a century and a half later, for the Thingöre monster to receive some publicity.

In 1661, it was on the coast of the Netherlands, then one of the intellectual centres of the western world, that a similar incident occurred. The event, which coincided with the grievous visit of Charles II of England, was thought worthy of recording for posterity. The animal was sketched live, or at least soon after its death, and its description, as well as the circumstances of its capture were reported in a tract, published in German, which seemed to have enjoyed quite a broad distribution. I have translated it as follows:

Representation of a horrible marine monster
as it was drawn from the sea at the end of 1661
in Holland, between Scheveningen and Katwyk.

"The above marine monster was captured in Holland, between Scheveningen and Katwyk, at the place where were moored the English ships which had come to fetch His British Majesty. After having been caught, it lived for another three hours, but it behaved, when captured, in such a manner that the fishermen thought that the devil itself was caught in their nets; it surrendered only after a spike had been driven through its body and it had been thus immobilized. It is approximately three and a half feet [1.05 m] long, and has a marvelous head, on which lies a star with eight branches about 1 foot long [30 cm], two of which have skin flaps, like the wings of a bat. On the whole area of the star, there are numerous small buds covered with a small crown, and when the animal was still alive, these small buds sparkled like tiny mirrors. From the star emerges an eagle's beak which can open and close; below this there is a mouth like that of a pig, with a tongue inside it. Between the mouth and the star are the eyes which, in the live fish, were of such a terrifying appearance that they were scary to look at. When the eyes had been taken out for embalming, it was noticed that the internal core of the eye ball had the size and the shape of a pearl. The eyes themselves were of the size of those of a large calf; for a single one of them, the fisherman was offered the sum of two hundred Dutch florins. Under the star, or crown, which the monster has on its head, there arise two long arms about two fingers thick. There are two orifices through which it takes food, but there is no outlet; there were no intestines found in its body, but only a liver and some grease. Consequently, physicians and learned people who have travelled widely cannot identify this monster with any species found either in Italy, or

THE FIRST PIECES OF EVIDENCE

*Turkey, or India, and consider it as a marvelous and unique creature, whose
significance is known only to the Almighty."*

This rather curious description calls for some comments. The
precision of the details which accompany the figure, in itself quite excellent,
is such that two centuries later the great Danish zoologist Steenstrup will not
hesitate to create, on the basis of this document, a new species of slender
squids of the northern seas, *Ommastrephes pteropus*. However some of the
interpretations of the original commentator are completely erroneous. What
he takes for a mouth, with a tongue in it, is evidently the opening of the
mantle cavity, from which projects the siphon. We have known since the
days of Aristotle that the latter is the real outlet of all cephalopods. There was
then no justification to attribute to the monster two mouths and to deprive it
of its anal orifice. All this under the pretext of making it an ambassador from
the Almighty, charged, without doubt, with the mission of demonstrating the
extraordinary versatility of a Creator capable of inventing a beast exempted
from the most prosaic servitudes of existence: this devil was indeed nearly an
angel! All the same, the Dutch commentator vies for poetic license with
Victor Hugo, who disgustingly combined the two natural orifices of the
octopus.

The comment on the appearance of the internal nucleus of the eye
balls is most interesting. The eye of cephalopods is structured, except for
some details, very much like that of the higher vertebrates: the principal
distinction being in the spherical shape of the crystalline lens. The latter's
resemblance to a pearl in shape and appearance has long been known;
archaeological digs have shown that at the time of the Incas, Peruvians used
the lenses of large cephalopods as ornaments and that the ancient Egyptians
used them as eyes for their mummies. More recently, natives of the Sandwich
Islands sold them as real pearls to naive Russian travellers[4].

In this case, the figure leaves no doubt as to the identity of the
animal captured; it is undoubtedly a squid and not a cuttle-fish. The cylindrical
form of the body and the bilobate nature of the fins are significant in this
respect.

The size of the Dutch squid - about 1.35 m from the end of the tail
to the tip of the arms - is much less impressive than that of the Icelandic
specimen. Nevertheless, at the time, the incident was sufficiently sensational
that the carcass was made into a royal gift. As one learns in 1710 from
Johann Lauerentzen, assessor of the Copenhagen consistory, the remains of
the fateful "fish" were brought back from the Netherlands by the illustrious
Hannibal Sehested, who had the honour and the grace of offering them to
Frederik III, king of Denmark and Norway. Thus, they came to contribute to
the collections of the Royal Museum, of which Holger Jacobaeus, president
of the University of Copenhagen, was to compile the catalogue in 1696.

In the month of May of 1662, eight days after the feast of the
Ascension, another cephalopod, of an even more modest size, was captured in

---

[4] *Cf. J.E.Gray (1830) and Boulenger(1935*

the Elbe, at Holstein, near Hamburg. This time, according to the description published in the local chronicles, it was only a 90 cm long cuttle-fish, tentacles included, i.e. a specimen hardly larger than normal. Nevertheless, the incident attracted much attention, and the animal was described as "a particularly extraordinary fish". Perhaps this was because there had been some much publicity about the monster found on the Dutch coast: multi-tentacled marine demons were becoming popular.

These creatures attracted the attention of Adam Oelschlaeger, also known as Olearius, a learned German which Frederik, duke of Holstein, had sent from 1633 to 1639 as his ambassador to the great-duke of Muscovy and the King of Persia, and which he established as his private librarian and curator of the Cabinet of Curios of his residence in Gottorf. Among the marvels of that collection, where Egyptian mummies were piled up next to bezoars and stuffed crocodiles, Adam Olearius kept the carcass of a 30 cm long squid, which he thought was still young and could have grown to the size of a man and even longer. What led our curator to this opinion was, one might have guessed, the recent capture on the Dutch coast of the "horrible sea monster", 1.35 m long, as well as the catch of a 90 cm "particularly extraordinary fish" near Hamburg. On the basis of their descriptions, Olearius had immediately linked these two creatures and his own small squid:

"Who cannot see, following the descriptions and the published illustrations, that all three fish belong to the same genus? Although the Dutch specimen and ours are longer and somewhat rounder than that found near Hamburg, both kinds are known to scientists, and may be classed among Mollia (meaning Molluscs)."

He then proposed to classify the latter among *Sepiae* (cuttle-fish), and the former among *Loligines* (squids).

Before this particularly accurate diagnosis, Adam Olearius had had no qualms about laughing at so called "physicians and scientists" who, unable to recognize the animal captured in the Netherlands, had seen in it a "marvelous creature", an ill omen for the landing of the English troupes. They should have read Aldrovandi and Jonston, he said.

Let us not leave the *Gottorfische Kunstkammer* and its astute curator without having a look at one of the most curious pieces of its collection; it will play a decisive role in this story. In the catalogue of strange treasures of the duke of Holstein, its description immediately precedes that of various monsters known to be cephalopods. The text describes it as a "fish called *Monachus marinus*, because its upper part looks a lot like a monk." Adam Olearius adds that this bizarre creature had been "captured alive at Enckhuizen", a small town on the shores of the Zuidersee, on the eastern shores of East Frisia.

The curator of the Cabinet of Curios did not of course draw any parallels between this grotesque creature and the cephalopods which he then described. However, it would seem that he might have had ulterior motives when he grouped all these illustrations on the same plate. Because, for anyone who cares to notice, it looks like the famous "sea monk" might be no other than the mutilated and shrunken body of a squid, closely resembling that of the Dutch coast, which is drawn beneath it. Undoubtedly, the crown of "hair" which surrounds the head of the monk fish indicates the attachment point of the arms which were torn off it.

Olearius adds that Aldrovandi also described a *Monachus*, but that his specimen had scales and did not resemble that of Enckhuizen. Nevertheless, in spite of some clear differences between the two monks, professor Steenstrup would later not hesitate to see in Aldrovandi's monster a giant squid, also earlier described by Guillaume Rondelet. We shall return to this at the appropriate point.

The next episode in the story of the giant squids is much more sensational than its predecessors; we are immediately immersed in the strange atmosphere of Bertold Brecht's *Threepenny Opera*.

Towards the end of 1673, a character about whom nothing is known except that he was called James Steward and that he was a merchant, began to wander through the towns and villages of Ireland, raising in his footsteps a reaction of curiosity or horror, amazement or incredulity, but never indifference. Everywhere he went, he carried with him a mysterious oblong box, taller than himself, and an equally bulky roll of canvas. After having set up his tent on the fair ground or at the market place, he invited the passers-by to listen to the true story of the double-headed monster which he had caught with his own hands at Dingle-I-cosh, in County Kerry. To shore up his words, he displayed numerous letters of trustworthy witnesses who had seen the beast's carcass and were aware of the circumstances of its capture. Then, in exchange for a "modest" contribution, the good people were invited to come inside and contemplate the "image of the fish, drawn in its exact proportions on canvas".

On the canvas, finally unrolled, the curious, seduced by the hawker's harangue, could gasp at the naive representation of a squid, nearly six meters in length, whose oval body seemed wrapped in the loose folds of a scarlet mantle. From the latter stuck out the pointed end of the body, graced with what looked like the two lobes of a caudal fin.

The head had two enormous eyes and eight short hands came out of it, studded along their whole length with a double row of suckers. At the centre of this Gorgonian hair-do there were two longer arms, as large as the others, but smooth and naked, and ending in a point[5].

The most astounding part of this portrait was that the beast was fitted with a kind of large spherical outgrowth, bearing mandibles and adorned with a second pair of roughly drawn eyes: a second head, no less, of a smaller size.

Besides the beast's image, an explanatory text provided further information. The hawker read it in a pompous tone, rolling his r's for emphasis. With the end of his long wand, he underlined the most striking

---

[5] *Those are of course the two tentacles. They were perhaps contracted, or reduced to a mere pedoncule following some mutilation.*

THE KRAKEN AND THE COLOSSAL OCTOPUS

words, simultaneously pointing out the corresponding details in the picture:

"This monster was taken at Dingle-I-cosh, in the County of Kerry, beeing driven up by a great storm in the Month of October last, 1673; having two head, one great head, (out of which sprung a little head two foot, or a yard from the great head) with two great eyes, each as big as a pewter dish, the length of it being about nineteen foot [5.80 m], bigger in the body than any horse, of the shape represented by this figure having upon the great head ten horns, some of six [1.30 m], some of eight or ten [2.40 to 3 m], one of eleven foot long [3.35 m], the biggest horns as big as a man's Leg, the least as his wrist, which horns it threw from it on both sides; And to it again to defend it self having two of the ten horns plain, and smooth, that were the biggest and middle horns, the other eight had one hundred Crowns a peece, placed by two and two on each of them, in all 800 crowns, each Crown having teeth, that tore any thing that touched them, by shutting together the sharp teeth, being like the wheels of a watch, The Crowns were as big as a man's thumb or something bigger, that a man might put his finger in the hollow part of them, and had in them something like a pearl or eye in the middle; over the Monster's back was a mantle of a bright Red Colour, with a fringe around it, it hung down on both sides like a Carpet on a table, falling back on each side and faced with white; the crowns and mantle were glorious to behold: This monster had not one bone about him, nor fins nor scales, or feet, but had a smooth skin like a man's belly. It swoom by the lappits of the mantle; The little head it could dart forth a yard from the great one, and draw it in again at pleasure, being like a hawk's beak, and having in the little head two tongues, by which is thought it received all its nourishment; when it was dead and opened the liver weighed 30 pound.[13 kg]"

That this extraordinary creature had actually existed, the brave Irish couldn't deny: in a long box, two of its arms, 2.50 m long, as well as the small extra "head" could be seen marinating under their very eyes. However a careful examination of that "head" would have convinced them that the eyes, which the showman had represented on the painting, were pure embellishment. In a letter from "a very Sober person in Dublin" which describes the animal in detail, there is no mention of such eyes:

"The head was not soe bigg as my fist, the mouth and two hard shells upon it very black and shap'd somewhat like to an Eagles Bill, but broader; In the mouth there was two tongues, and (as the Man declared that tooke this monster) the Beast had naturall power to draw this head in or putt it out of the Body as necessity required."

In this squid, the beak was in some way hoisted at the end of some retractable horn: what was described as a second head was no other than the pharynx!

The double appearance of the tongue should not surprise us. In all cephalopods - in all molluscs, in fact - the tongue is a large fleshy and bilobate bud, where the larger posterior lobe is covered by a kind of continuously growing rasp, the *radula*. The latter is used to break down into small pieces the pieces of flesh torn up by the beak.

Another striking detail in the beast's description is the mention of teeth which covered the "crowns" of the arms, in other words, the suckers. It is a fact that among squids, the suckers are studded along their entire periphery with small horny teeth; this is especially noticeable among the largest species, where they look like trepanation tools or hole-piercing bits. Their sucking power is of course enhanced by the scarification of the flesh into which they encrust themselves.

Thus, with this kind of large squid, one draws much closer to Victor Hugo's unlikely octopus, which blows out the skin and make the blood spurt out "under its disgusting pressure".

We know of the circumstances of the capture of the animal by the hawker, thanks to a letter which Mr Thomas Hooke, of Dublin, sent on 23 December 1673 to his friend Mr John Wickins, of London:

"That in the month of October last, I think about the 15th day he was alone riding by the sea-side, at Dingle-I-cosh and saw a great thing in the Sea, which drew his eye towards it, and it came just to him; when he discerned the horns it began to look frightfully, he said he was sometimes afraid to look on it, and when he durst look on it, it was the most splendid sight that ever he saw; The Horns were so bespangled with these Crowns, as he calls them; they shewed he saith like Pearls or precious Stones; the Horns it could move and weild about the Head as a Snail doth, all the ten; the two long ones it mostly bore forwards, the other eight mov'd too and fro every way; When it came to shore its fore parts rested on the shore, and there lay; He got help after awhile, and when he saw it stirred not to fright them, he got ropes and put them about the hinder parts, and began to draw it on shore, and as it stir'd not to hurt them, they grew bold, and went to pull with their hands on the Horns, but these Crowns so bit them, that they were forced to quit their hold: the crowns had teeth under every one of them, and had a power to fasten on anything that touched them; they moved the Horns with handspikes, and so being evening they left it on the shore, and came in the morning and found it dead."

The enterprising showman with his "two-headed beast" finally reached Dublin, with the intention of presenting his trophies to the Lord Lieutenant, the chief of the local magistrature. What the latter thought of this, the story does not tell. The whole incident created quite a sensation and has left various traces: a manuscript letter and two broadsheets, one printed in Dublin, the other in London. The first is probably a specimen of the prospectus distributed to the visitors of the fair; the second is a kind of tract which includes, besides the animal's portrait and description, a reproduction of three letters vouching for its authenticity.

These precious and picturesque documents were gathered by a London book-seller, Thomas Thorpe, a collector of documents about the history of

Ireland. They finally ended up among the shelves of the library of the Royal Society of Dublin, from which they were dug up only in 1875 by the naturalist A.G.More, on the occasion of the reprinting of a description of the stranding of a giant squid on the coast of Ireland.

We learned already through Erik Pontopiddan, bishop of Bergen, that in 1680, a "young" *kraken* was so embarrassingly wedged in the reefs of the gulf of Ulwangen, in Norway, that it perished and that for some time the stink of its corpse made all traffic impossible in that fjord. No detail was provided on this undoubtedly enormous animal except that it possessed a multitude of arms. But we can be sure that it was a squid, not an octopus. Good as it is at crawling, an octopus would have had no difficulty in extricating itself from this situation, while the squid on the other hand is lost as soon as it ventures in shallow waters.

So far, we thus have proof that during the XVIIth century alone, five strandings of squids of more or less gigantic size took place on various European coasts: in Iceland, in the Netherlands, in Germany, in Ireland and in Norway. It is to be expected that these few cases which were documented and transmitted to us may well not have been the only instances of strandings. We can thus wonder how Science should have so long remained silent or circumspect about the existence of these giants of the sea. News did not travel fast, nor very far in those past centuries. These isolated incidents, whose details became distorted through successive transmission, did not contribute to the development of an accurate body knowledge about titanic cephalopods. On the contrary, a legend gradually wove itself around them, for romantic imagination always fills in the gaps of knowledge. The exaggerations of the myth, drawing as it does from supernatural sources of wonder, finally generate an incredulous and opposite reaction.

Thus, when Dom Antoine-Joseph Pernetty, abbott of the abbey of Burgel, member of the Royal Academy of Sciences and Letters of Prussia, corresponding member of the Academy of Florence, and librarian to his Majesty the King of Prussia, accompanied Bougainville on the frigate *l'Aigle* during his trip to the Falkland Islands[6] in 1763-64, the prestige of his titles and of his functions, as well as his erudition were not sufficient to convince his readers of the truth of the rumors which he brought back from far away seas. Men of letters were willing to accept without argument his outrageous statements on the stature of Patagonians, but they merely smiled at his stories of giant squids:

---

[6] *These islands, located in the South Atlantic, not far from Tierra del Fuego, were called at that time Isles Malouines by the fishermen from Saint-Malo whose ventures carried them far from home. Having passed under British ownership, they were baptized Falkland Islands, which is the name which they have kept to this day (although still called Malvinas in Spanish). In all fairness, they were discovered between 1591 and 1593 by the famous English navigator John Davis, in honor of whom they were first called the Davis Southern Islands. Sir Richard Hawkins, another English explorer, later also landed on them and called them Maiden Islands.*

THE FIRST PIECES OF EVIDENCE

"According to the feeling of the mariners of the South Sea [related the abbott], the squid is the largest fish in the ocean. It seizes its prey using the mobile hooks which he has at the end of its muzzle. These sailors also say that it attaches itself to boats, using those same hooks, and climbs up the rudder; if it does this at night without being noticed, it can tip the ship on its side by its enormous weight, until it causes it to tip. Thus, a vigilant guard is kept, with axes and other sharp instruments, to cut off the hooks of this fish as soon as he lays them on a ship. Our captain [the sieur Duclos-Guyot, from Saint-Malo] and his brother, who have travelled many times into the South Seas vouched for the veracity of this fact; however they also added that they had eaten one weighing only 150 pounds [66 kilos] or thereabouts and that it was an excellent fish. To judge by the small one which I have illustrated, it must be very delicate to the taste. The cone which holds its body and the fish itself were nearly diaphanous."

The illustration which accompanies this text confirms the identity of the animal: it is an excellent drawing of a squid.

The brothers Duclos-Guyot could only guarantee the existence of specimens weighing around 75 kilos, but this was already a lot: the total length of an individual of such a mass must be around 3 metres. When one considers that the squids described in the learned treatises of the day did not exceed 10 kilos, one must recognize that the benedictine father's report did contribute something new.

At the very same time that Dom Pernetty was finishing writing his *Histoire d'un voyage aux îles Malouines,* an indisputable fact occurred which was to confirm, although indirectly, the pertinence of his claims.

In 1768, Captain James Cook had undertaken his first trip around the world in the east-west direction. Two naturalists travelled with him: one was the British maecenas Sir Joseph Banks, the other Daniel Solander, one of Linnaeus' favorite disciples, which he had sent, like many others, to explore the world in search of yet undiscovered animal and plant species. The wish of the Father of Systematics was soon to be granted in a most sensational fashion.

After rounding Cape Horn, the *Endeavour* was sailing towards the new islands of the South Pacific when there occurred, by 30° 44' South latitude and 110° 33' West longitude, a significant event which the captain was to enter into his log:

"Mr Banks found a great Sepia which appeared to be slain by the birds; its mutilated body floated on the water; it was very different from the cuttle-fish which are found in the European seas, for its arms, instead of suckers, were armed with a double row of claws, very sharp, resembling those of a cat, and which it could, like that animal, withdraw at will."

Are we to be spared no horror about those viscous scarecrows! Fragments of this mollusc were brought back to England in 1781 and entrusted to the Museum of the College of Surgeons, in London. When the famous anatomist

THE KRAKEN AND THE COLOSSAL OCTOPUS

Richard Owen examined them later (in 1841), he reported the following:

"The fins have a rhomboidal form, which allowed the animal to swim both forwards and backwards. Comparing the dimensions of this Cephalopod, from the available arms, with complete adult animals of the same species, but of a smaller size, one finds that its body must have been at least 4 feet [1.20 m] long and the whole animal, including tentacles, must have been over 7 feet [2.10 m]."

The description of the animal's fins, which did not form a continuous and uniform hem around the body, convinces us that the animal was not a cuttle-fish, but a squid. Even though Aristotle had already clearly defined the differences between these two types, a regrettable confusion again brought them together at that time.

The strange clawed cephalopod encountered by Cook's companions was soon to be in the news again and once more to be incorrectly identified in the same manner. In his *Natural History of Chile*, published in 1782, the great Chilean writer, father Juan Ignacio Molina, mentions a *Sepia unguiculata*, a clawed cuttlefish, similar to Sir Joseph's. At the beginning of the next century, Pierre Denys de Montfort will bring the level of confusion to a peak by calling this same mollusc an "unguiculated octopus".

It will be only with the publication of the important monography on cephalopods by Ferussac and d'Orbigny (1835-48) that the truth will be clearly established. In fact, Alcide d'Orbigny had recognized, on a drawing showing one the arms of the "great cuttle-fish" of Sir Joseph Banks, the characteristics of a genus of Mediterranean squids, which he had described under the name *Enoploteuthis*, meaning "armed squid". In honour of father Molina who, according to him, had been the first to name the larger Pacific species, although with an inappropriate generic name, he named the animal *Enoploteuthis molinae*[7].

The size of the clawed squid caught in the Pacific by Mr.Banks was far from being as impressive as that of some squids stranded here and there on European shores. Two metres long, including tentacles, was not much compared to the dimensions of the "polyp" of Carteia, to the 14 to 16 metres of the "monstrous fish" of Olaüs Magnus, to the 7 metres of the squid from the old Icelandic chronicle, and to the 6 metres of the "two-headed monster" of Dingle-I-cosh. Nevertheless, in his comments on that mollusc, professor Owen emphasized its frightening character:

---

[7] *According to the rule of priority, d'Orbigny should, in all fairness, have called this species* Enoploteuthis unguiculata, *since the specific name bestowed by Molina remained valid. In fact the generic name given by d'Orbigny also did not stand: in 1882, professor Steenstrup recognized that the species described by Molina was so different from other armed squids that a new genus had to be created for it. Thus the clawed squid of Chile is now called* Cucioteuthis unguiculata.

THE FIRST PIECES OF EVIDENCE

"The natives of the Polynesian Islands, who dive for shellfish, have a well-founded dread of these formidable animals."

The claws, which in these animals replace the suckers, must make for particularly atrocious encounters even with specimens of moderate size. The situation must be similar to an assault by a herd of enraged leopards.

One fact is certain however, after having studied the fauna of Chilean waters, father Molina did not believe in the presence in that area of the Pacific of really very large squids, clawed or not. Regarding an animal which he calls *Sepia tunicata* (actually a species of *Ommastrephes*, and thus a slender squid), he notes that:

"Sailors exaggerate the size and the strength of this animal, but it is certain that those caught in Chilean waters weigh no less than 150 pounds [66 kilos]"

That is the very same weight quoted by Dom Pernetty's sources. The whole sentence actually sounds like a response to the latter's allegations. Squids weighing 150 pounds are fine, says Father Molina, but to turn them into the largest creatures of the sea is a sailor's tall tale[8].

Bragging or otherwise, sailors were not the only ones to tell such tales. In his *British Zoology* (1777), the English naturalist Thomas Pennant, also author of an interesting *History of Quadrupeds*, provided some astonishing details on a cephalopod which he calls, following Linnaeus, *Sepia octopodia*, and which is no other than the octopus:

"In hot climates, some are found of an enormous size. A friend of mine, long resident of the Indian isles, and a diligent observer of nature, informed me that the natives affirm that some have been seen two fathoms [3.65 m] broad over their centre, and each arm nine fathoms [16.50 m] long. When the Indians navigate their little boats, they go in dread of them; and least these animals should fling their arms over and sink them, they never sail without an axe to cut their arms off."

That was rather vague, but the information could pertain as well, if not better, to squids than to octopus.

In 1784, more accurate facts were brought to public attention which strongly argued for the good faith of sailors and people of India. In his study on the origin of ambergris (1785), Dr. Schwediawer reported the surprising revelations made by a whaling captain from New England whose honesty was above all suspicion. This upright and sensible man, who had been introduced to him personally by Sir Joseph Banks, had been occupied for many years with the "south fishery", i.e. whaling in the south seas. About ten years earlier,

---

[8] *Poetic justice, Molina in turn was to be considered unreliable, first by the German Schneider and then by the French Ferussac.*

THE KRAKEN AND THE COLOSSAL OCTOPUS

"he had hooked a spermaceti-whale that had in its mouth a large substance with which he was unacquainted, but which proved to be a tentacle of the Sepia octopodia, nearly 27 feet in length [8.25 m]; this tentacle however did not seem to be entire, one end of it appearing in some measure corroded by digestion, so that in its natural state it may have been a great deal longer."

"When we consider, continued Dr.Schwediawer, the enormous length of the tentacle of the Sepia here spoken of, we shall cease to wonder at the common saying of the fishermen, that the cuttle-fish is the largest fish of the ocean."

The name *Sepia octopodia* (eight-legged cuttlefish) proposed by Linnaeus to describe the octopus seems really non-sensical, since the cuttle-fish is a decapod, a ten-footed cephalopod. However, Linnaeus classified nearly all cephalopods in a single genus. In this case, the term *Sepia* was not such a bad choice. There was a good chance that the said arm could be that of a decapod, squid or cuttle-fish, rather than that of an octopus. The sperm whale and the squid are hereditary enemies. And why should this mutual hostility between two carnivores come as any surprise? They compete in the same environment; they are both pelagic, which is not the case for the octopus, which huddles at the bottom.

Squids and sperm whales engage, in the depths of the ocean, in homeric battles. The numerous remains of squids found in the digestive tubes of harpooned sperm whales and the uncountable rows of round scars left on the skin of these cetaceans by the serrated suckers of their opponents are sufficient evidence of these struggles. A few people even have had the rare privilege to witness certain phases of these murderous duels. Among them, one should note the British writer Frank Bullen, who in 1875, at the age of eighteen found himself on board a whaling ship. In his famous *Cruise of the Cachalot*, he left a striking description of what he says was "the strangest sight I ever saw". It happened at eleven at night, at the mouth of the Strait of Malacca:

"There was a violent commotion in the sea right where the moon's rays were concentrated, so great that, remembering our position, I was at first inclined to alarm all hands; for I had often heard of volcanic islands suddenly lifting their heads from the depths below, or disappearing in a moment, and, with Sumatra's chain of active volcanoes so near, I felt doubtful indeed of what was now happening. Getting the night glasses out of the cabin scuttle, where they were always hung in readiness, I focussed them on the troubled spot, perfectly satisfied by a short examination that neither volcano nor earthquake had anything to do with what was going on; yet so vast were the forces engaged that I might well have been excused for my first supposition. A very large sperm whale was locked in deadly conflict with a cuttle-fish, or squid, almost as large as himself, whose interminable tentacles seemed to enlace the whole of his great body. The head of the whale especially seemed a perfect net-work of writhing arms - naturally, I suppose, for it appeared as if the whale had the tail part of the mollusc in his jaws, and, in a business-like, methodical way,

148

was sawing through it. By the side of the black columnar head of the whale appeared the head of the great squid, as awful an object as one could well imagine even in a fevered dream. Judging as carefully as possible, I estimated it to be at least as large as one of our pipes, which contained three hundred and fifty gallons [1590 litres]; but it may have been, and probably was, a good deal larger. The eyes were very remarkable from their size and blackness, which, contrasted with the livid whiteness of the head, made their appearance all the more striking. They were, at least, a foot in diameter, and, seen under such conditions, looked decidedly eerie and hobgoblin-like. All around the combatants were numerous sharks, like jackals round a lion, ready to share the feast, and apparently assisting in the destruction of the huge cephalopod."

In the November 1952 issue of the magazine *Le Chasseur Français*, the high seas captain Max. P. Robin relates how he witnessed, not very long ago, a similar fight between a sperm whale and "an enormous octopus, with a span of 8 to 10 metres".

An octopus? One first suspects a mistake. King of the free open spaces of the ocean, the sperm whale hardly seems to have the opportunity to engage in a combat against octopus, which hide in the cracks of coastal reefs or in the depths of the ocean. But, isn't the great toothed whale in the habit of diving, or sounding as whalers would say?

It has recently become known that sperm whales indeed dive to incredible depths. Hardly a quarter of a century ago, all naturalists claimed that it would be impossible for cetaceans to dive beyond about a hundred meters. Some did accept, with some reluctance, that they might be able to stand the enormous pressures of the abyss. But all agreed that if they descended into it, they would die of the bends during their ascent, of the same caisson disease which strikes divers which return too quickly to the surface[9].

Cetaceans breathe air at atmospheric pressure, and survive under water entirely on the oxygen supply carried in their lungs. It thus seemed impossible that they should be able to dive to great depths. There were two possibilities: either they could return slowly to the surface and die of asphyxiation before reaching the surface, or they returned quickly and died of an embolism on the way up.

A single zoologist did not share this opinion. This was the German explorer-zoologist Willy Kükenthal, then a professor in Breslau, who had bluntly stated in 1900 that: "The sperm whale can dive down to 1,000 metres." If sperm whales were not so keen to eat squids that they sometimes mistake submarine cables for their tentacles, we might still be calling Kükenthal a dreamer, if not a nut.

---

[9] *Under high pressure, more nitrogen dissolves into the blood. This wouldn't matter if it weren't that under rapid decompression, associated with a sudden ascent, this gas is released in the form of small bubbles, as does carbon dioxide in a suddenly opened pop bottle. These nitrogen bubbles are responsible for the damage caused to capillary vessels.*

THE KRAKEN AND THE COLOSSAL OCTOPUS

A few years ago, the skipper of the cable ship *Guardian*, captain Harne, had a damaged cable pulled up, 500 metres from the coast of Ecuador, in a depth of 512 m. Imagine his surprise when he brought up, at the end of his make-shift jigging line, a 10 m sperm-whale: the cable was wrapped around its lower jaw!

Rather than simply admit that the large cetacean had actually dived down to a depth of more than 500 metres, a number of far-fetched hypotheses were put forward. According to the least extravagant, it would have been entirely by chance that the corpse of a sperm whale had become tangled in the cable, while sinking to the bottom.

In April 1932, a similar incident occurred, this time in even more extraordinary circumstances. A break in the cable which links Balboa, in the Panama Canal Zone, to Esmeralda, in Ecuador, had been detected. When the cable ship *All America* pulled up the cable, off Columbia, from a depth of 987 m, it was found that a dead sperm whale, 15 metres long, was entangled in it, like a kitten in a ball of wool. The cable was wrapped not only around its lower jaw, but also around its whole body and tail fin. To be so tangled up, the hungry cetacean had to have struggled mightily...

By 1970, we already knew of about fifteen incidents of this type; one of them occurred in a depth of 1133 metres!

Once it was established that sperm whales could actually dive to depths in excess of 1,000 metres, scientists began to wonder about the mechanisms which make such a feat possible.

Thanks mostly to the research of the American physiologists Irving and Scholander, it has been shown that the sperm whale can dive for more than 75 minutes, i.e. an hour and a quarter! That it manages to do so is not because its relative pulmonary volume is greater than that of a man -it is actually smaller by a half - but because it renews more completely the air of its lungs when it comes to breathe at the surface (up to 90%, compared to only 10 - 15% in man). Moreover, its respiratory centre is less sensitive to the presence of carbon dioxide, which gradually accumulates in blood which is not re-oxygenated; its breathing does not accelerate, the way ours does, and it doesn't consume more oxygen when the latter's concentration decreases.

Furthermore, besides storing oxygen in its blood, it also stocks it in its muscles, wherein there is found a respiratory pigment, myoglobin, similar to the hemoglobin of red blood cells. In this way, it can come back to the surface slowly without fear of suffocation. On the other hand, it is less threatened than a hard-hat diver by gaseous embolism because its air is not renewed, and it thus carries with it only the limited quantity of nitrogen present in its original breadth. In addition, during its dive, the pressure of the water on its flanks pushes into its bronchial tubes, its trachea and its nasal cavity most of the gases not absorbed by the blood.

Finally, the acceleration of its heartbeat during ascent favours the rapid elimination of dissolved nitrogen[10].

In spite of all this, it has been shown that marine mammals may not be entirely immune to the bends. It thus makes sense to ask what might drive the sperm whale to sound so dangerously. It is certainly not seeking security, which could be found at 100 metres, or at most at 500 meters, in the eternal night of the abyss. It can only be hunger. The two cases of damaged cables prove it. When we recall that sperm whales feed almost exclusively on cephalopods, we can't but draw a comparison between the cables and the tentacles of their prey, and to imagine a fatal mistake, easily excused by the darkness.

Only the lower jaw of the sperm whale is studded with teeth - terrifying ivory daggers, 10 cm long - which fit into corresponding grooves of the upper jaw. One supposes that, when hunting, the cetacean hangs its toothed jaw down, like a lure. Perhaps, in the shadows of the depths, occasionally illuminated by phosphorescent organisms, the white teeth are the only visible part of the big black cetacean and voracious squids lunge upon it as Mediterranean octopus onto olive boughs dangled by fishermen...

The incident of the telephone cables leaves one pondering. It would seem to suggest that the sperm whale drags its jaw along the bottom, like a dredge, in the hope of catching other preys. It is possible that it might in this manner be seeking squids, which perhaps sometimes rest on the bottom, but it must also happen that it catches in such conditions benthic cephalopods, i.e. octopus. That has indeed been shown by the analysis of the contents of its stomach. This should make us prudent in our conclusions. Our knowledge of the most gigantic cephalopods is based to a large extent, as we shall see, on the examination of the debris vomited by harpooned sperm whales in their agony or subsequently found by whalers in their stomachs: large quarters of

---

[10] *These various safety mechanisms are common to all cetaceans. Except for the sperm whale and thebottle-nose whale, most of them do not seem capable of diving beyond 300 to 350 metres. The exceptional performance of the sperm whale might thus not be completely explainable through these mechanisms. Some have suggested that the most distinctive feature of the large toothed cetacean, its enormous cylindrical head, containing a tonne or two of* spermaceti, *might be the clue to the mystery of its performance. It is thought that this oily liquid allows it to close its blow-hole in a completely water-tight fashion, even under the greatest pressure. It is also possible that because its density increases in the cold deep waters or through voluntary contractions by the cetacean, spermaceti may assist in accelerating its descent. The sperm-whale's head would then behave as a hydrostatic control organ. The importance of the role played by spermaceti in deep diving is confirmed by the fact that the other cetacean which has a reservoir of it, the bottle-nose whale, is reputed to dive down to 1,300 m. It is interesting to note that the latter also feeds on cephalopods.*

gelatinous flesh, enormous tentacles, monstrous beaks. Knowing the similarity in habitat of the squid and the sperm whale, one might be tempted to identify all these remains as squid. In fact, they could occasionally be those of an octopus, with important consequences on the total dimensions of its live owner. To be specific: for a common octopus, an arm is about 5/6 of the total length, while for an ordinary squid, it is only about 1/3 (excluding tentacles). Thus, a 1 metre long arm, prior to identification by a competent person, might be supposed to come from a 3.50 m long squid or a 1.2 m long octopus. Quite an appreciable difference!

Chapter Eight

PIERRE DENYS de MONTFORT, OUTLAWED MALACOLOGIST

People soon grow weary of a process which they find confusing. They tire of following us. They then invent their own, and if we don't follow it, they resent us [...] It is dangerous not to conform to the idea that people have of us, for they don't readily change their views.

JEAN COCTEAU, *La difficulté d'être*

"Some circumstantial evidence carries a lot of weight, as when you find a trout in the milk", wrote in his journal the great American writer Henry David Thoreau.

The discovery, in the mouth of a sperm whale, of the eight meter long arm of a cephalopod, is an event of the same order. It forces one to reflect on the size that tentacled molluscs can attain. One would think however that most people deny any value even to the most strongly based inference, for someone was to lose his reputation, and perhaps also his life, for drawing logical conclusions from this incident. The report from Dr. Schwediawer, which mentioned it, in 1783, struck the imagination of a young French naturalist enthralled with the wonders of nature. His name was Pierre Denys de Montfort, though he was to become citizen Denys-Montfort during the Revolution.

Born in 1764, this dreamy and enthusiastic young man had, from the most tender age, felt a strong attraction for natural sciences.

"I travelled from my early childhood on, he once told the professors of the Paris Museum; ripened early by the observation of nature, all my faculties naturally turned to the study of natural history; to this day, it has been my bliss, and to it I owe my most joyful moments."

Alas! Young Pierre belonged to a generation sacrificed by history.

THE KRAKEN AND THE COLOSSAL OCTOPUS

He grew up in a crumbling world, from which another was to be born after a stormy gestation; as the monarchy gradually collapsed in France, revolt grumbled among the masses. Pierre was twenty-five years old when the Bastille was taken: the age at which one looks for a job. The Republic was proclaimed; in the sadly famous words of Fouquier-Tinville, who sent Lavoisier to the scaffold, "it did not need scientists". In fact, what it needed at that time was soldiers, policemen, judges and executioners. While the Terror reigned on French soil, the revolutionary armies, eager to share with the rest of the world the benefits of the New Regime, sacked and bloodied a good part of Europe. Of an adventurous bent, either because of his passion for the new ideas, or merely wishing to draw a veil on his insufficiently modest origins, citizen Denys-Montfort participated with enthusiasm in what he called the "seven campaigns for freedom". These certainly gave him the opportunity to devote himself to his country. Of course, they did not leave him much time for study, but after leaving the army, he remained, "too happy to have retained, in the bustle of military encampments, his innate taste for natural history."

At that point, the geologist Barthélemy Faujas de Saint-Fond took an interest in him and hired him to be his assistant for a while, before he was appointed to a similar position at the Jardin des Plantes, in Paris. The young man's exceptional qualities were soon noticed at the Museum of Natural History. During General Bonaparte's expedition to Egypt, the geologist Sylvain Gratet de Dolomieu, trying to ensure the support of a competent assistant, asked Denys-Montfort to accompany him. At the same time, the famous Daubenton also requested his services and asked him, among other things, to accompany him to Germany. Our hero gave up the mirages of Egypt: he preferred to work with the great anatomist, who had been Buffon's principal collaborator and who, he said, personally honoured him with a "fatherly friendship."

His knowledge of a number of foreign languages caused Denys-Montfort to be particularly appreciated at the Museum, where he rendered great services as a translator. The professors of the eminent institution were even going to make him the official mineralogist of the round-the-world expedition proposed by captain Baudin. Alas!, circumstances did not permit this voyage, which would certainly have fulfilled the wishes of one who, in his youth, had already seen three continents.

When Daubenton died in 1799, and thus caused a vacancy in the chair in mineralogy, Denys-Montfort immediately presented to the professors of the Museum a request that they should call to the chair the unfortunate Dolomieu, for whom the Egyptian expedition had turned into a disaster: by some incredible misfortune, he had fallen into enemy hands... This generous initiative was, to speak truly, not entirely disinterested, for our assistant-naturalist asked at the same time that he should be made interim holder of the chair, until Dolomieu could occupy it himself.

In any case, the attempt failed. Probably because, having been too busy fighting the enemies of the Republic, Denys-Montfort, at the age of thirty-six, had, by his own admission, "published very few things". He was about to fill this serious gap, for Sonnini de Menoncourt had not hesitated to charge him with writing that part of the *Suites à Buffon* (follow up on Buffon's work) devoted to molluscs.

Denys-Montfort had launched himself into this work with a kind of passionate fury. Enamoured of broad syntheses, while concerned at the same time not to forget essential details, he had begun to dig with the patience of a benedictine monk through all the scientific literature of his time. That is how he came upon Dr. Schwediawer's note on the origin of ambergris. He immediately perceived the opportunity of sensational revelations on a subject which was rather discredited at the time: the existence of giant cephalopods.

The era of the great compilers - Pliny, Aelianus, Vincent de Bauvais, Gesner, Aldrovandi, Jonston, Wotton - was over. Under Buffon's prestigious thrust, zoology had matured, and zoologists were henceforth determined to credit only precise observations and concrete contributions. Needless to say, most of them put the Scandinavian *Kraken* and all its imitators in the category of marine superstitions .

Might not the monstrous arm waived by Dr. Schwediawer - Dr. Swediaur, as he was called in France - reverse this opinion? That is certainly what Denys-Montfort, well versed in all the ancient literature, thought desirable. However, wishing to gather a body of convincing testimony before jumping to conclusions, he decided to conduct a systematic enquiry among the American whalers settled in Dunkirk.

Some years before, Charles Alexandre de Calonne, at that time General Controller of Finances, had worried about the collapse of the French whaling industry. The Basque pionneers had, over the centuries, gradually been displaced by the Dutch, who had learned from them all the tricks of the trade; they in turn had been robbed of their monopoly by the English, which the Americans, after declaring their independence, were soon to expel from the rich waters of the New World. In the hope of reviving a large scale whale fishery in France, Calonne had recruited shipowners among the whalers of Nantucket and had made them interesting offers, inviting them to settle in his country with their families.

There, thought Denys-Montfort, are the people who should be consulted to find out more about the enormous cephalopod, whose tentacle was found in the mouth of a sperm whale.

Our impetuous malacologist's enquiry was crowned with success. He could then report in great detail what two captains of the Dunkirk whaling fleet had told him:

"One of these captains, by the name of Benjohnson, told me that he had harpooned a male whale, which in addition to its extremely prominent penis placed on its belly, appeared to have another one coming out of its mouth; being, with the rest of his crew, quite astonished at this sight, he had them, as soon as the whale was tied up along his ship, stick a hook through this elongated and roundish mass of flesh, which they then solidly secured with a number of slip knots [...] They could hardly believe their eyes, when they saw that this fleshy mass, truncated at both ends, the thickest of which was as

big as a mast, was nothing but the arm of an enormous octopus [sic], whose sunken suckers were broader than a hat; the lower end appeared freshly severed; the upper end must have been cut in some struggle which had preceded its capture by some time, for it was scarred and extended by a kind of extension of the size and length of a man's arm. This limb of an enormous (quid) octopus, accurately measured with a fishing line, turned out to be seven fathoms, or thirty-five feet long [10.65 m], and the suckers were arranged in two ranks, as in the common octopus. What must have been the total length of the arm which, at its already severed upper end still had a diameter of more than six inches [15 cm]? "

According to Denys-Montfort, there would have been about ten feet missing from the arm at its upper end "to go from a thickness of six inches to that of the tip of a needle", and from ten to twenty feet at its lower end, clearly cut off below the umbrella membrane[1]. This gives about 18 metres for the total length of the arm, which is probably an exaggeration, since nothing proves that this particular cephalopod had a well developed umbrella, as is indeed the case for most squids. If one pays attention to the word *quid*, which Denys-Montfort reports rather unfaithfully (it was obviously a *squid*), the American whaler thought that the arm belonged to a squid, and not to an octopus!

The other testimony quoted by our bold malacologist is even more astonishing:

"Another one of these American captains, called Reynolds, told me, in the presence of Benjohnson and of a third mariner employed in the same fishery, that, one day, after having harpooned a whale which was simultaneously hit by two harpoons, one of which penetrating behind its ear did not allow it to engage in a long struggle, so that it was almost instantaneously at bay, there was seen floating at the water surface a long fleshy body, red and slate in colour, which they took for a sea-serpent, and the sight of which even frightened the sailors which rowed the whaleboats."

One man, bolder than the others, noticed that the presumed reptile did not have a head and seemed quite inert; the captain had the long and fleshy body hoisted onto deck with a capstan. They then noticed that "what he and his crew had taken for a sea serpent was nothing else than the truncated arm of an enormous octopus which they recognized through the large suckers which covered it on one of its sides, and which were the size of a plate at their thicker end, the other end ending in a sharp point."

With the great flensing knifes used to carve the whale's blubber, the

---

[1] *The umbrella is the more or less prominent membrane which connects, usually at their root, the arms of cephalopods. When the animal opens its arms to grasp its prey, one has the impression of seeing an open umbrella. The resemblance is almost perfect in some species, where the umbrella stretches to the end of the arms. That is the case, among others, in* Cirroteuthis, *which, lacking a radula, probably feeds on plankton and falling organic debris, which it must gather by spreading its umbrella.*

captain had a few slices of that arm cut off, but as he found the flesh quite tough, he ordered most of it thrown back to sea, not without having had it carefully measured.

"The base was two and a half feet in diameter [75 cm], and thus seven and a half feet in circumference [2.25 m]; its total length was nine fathoms, or fourty five feet [13.7 m]. "

The incident of the red and slate pseudo-serpent ends on a rather prosaic note which adds a final convincing touch of authenticity:

"Some sailors carved up a few roundels from a piece which had been kept; by dint of beating them up and rincing them in sea water, they managed to make a good meal out of it, which after a few days of hanging became so much better than they regretted what they had thrown overboard, thinking that the entire arm must have weighed over a thousand pounds."

Here again, it is almost certain that this reddish arm of a frightening length and size - perhaps the greatest ever reported - belonged to a squid. Already, we have seen that the "two-headed monster" of Dingle-I-cosh was covered with a scarlet coat. That was not just an embellishment on the part of the hawker. We will gradually notice that this colour is common to most exceptionally large squids[2].

That is of course a fact that Denys-Montfort could not have been aware of. Are we then to excuse him on that account for having featured exclusively octopus in this business of gigantic cephalopods? I don't think so.

Emboldened by the revelations of the American whalers of Dunkirk, he had not hesitated, in his *Natural History of Molluscs*, published in 1802, to speak of cephalopods as the "most enormous animals existing in the world". However, he recognized among them only two species of phenomenal dimensions: the *colossal octopus* and the *kraken octopus*.

Nevertheless, he had taken the trouble of consulting with care and attention to detail the ancient literature devoted to this question and thus had in his hands all the pieces of the puzzle which could have allowed him to ascertain the existence of giant squids. Even when he had heard that for the sailors of Bayonne "the squid is the smallest as well as the largest animal in

---

[2] *In fact, it is impossible to define the colour of cephalopods, since due to effect of the chromatophores which cover their skin, they can change it with surprising ease. Thus, the octopus, which at rest are of a grayish-green colour, with reddish spots here and there, become coloured with many shades of red when they are excited: all hues of red, purple, mauve and blue wash over them in a flash and they sometimes take on highly contrasting marbled patterns. Speaking of a Mediterranean slender squid, Jean-Baptiste Verany wrote: "When alive, this cephalopod is of a slightly transparent livid white, tinted with blue, green and silvery iridiscent rose... When it has lost all vitality and the interplay of chromatophores has ceased, its colour is a uniform brick red." It is thus not surprising that giant squids found dying on a beach or at the sea surface - or even their mutilated remains - should have been described as being of a more or less bright red.*

the sea", he took it for a mistake, and blinded by his obsession, imagined that they were really talking about the octopus.

However, it is not for that particular bias that the French naturalist was blamed; it was for having invoked in the defense of his thesis the most fantastic stories. This practice branded him for ever in the scientific and even in the popular literature with the reputation of being a bragger and a dupe.. Is it really deserved? Today, one can well doubt it.

On what evidence did he then hope to base the existence of his *colossal octopus*? First of all of course, on the old reports in Pliny and Aelianus, about the salt fish thieves of Carteia and Pozzuoli, then on the vague rumours retold by Gesner, Aldrovandi and Jonston, about sailor-snatching octopus. He gave especially great weight to the discovery of enormous tentacles in the mouth of sperm whales and to a number of first-hand reports.

The first of these reports is anonymous, and perhaps exaggerated, but as it is most probably based on a real incident, which captured the imagination of a whole town, it is worth looking into:

"One may see, in Saint-Malo, in the chapel of Saint-Thomas, to whom sailors of this country turn to in their most extreme dangers, an ex-voto painting, which represented the great danger which had nearly caused the loss of a ship of this harbour, on the coast of Angola, where it was trading in slaves, ivory and gold. The ship having completed its commerce, and the crew all having returned on board, most fortunately, as we shall soon see, the captain was thinking of raising the anchor and setting sail to leave that coast for the American islands, when all of a sudden, by calm weather and in full daylight, there rose from the depth of the sea, which it caused to boil over a large area, a marine monster of frightening dimensions which climbed over the ship's deck, hung on to the vessel, wrapped itself into the sheets and the masts to their very top, with arms as long as they were flexible and frightening; hanging all its weight in the rigging this monster caused the vessel to list until it lay on its side, threatening to sink into the abyss. In this extreme peril, each one listening only to his own counsel, all crew members ran for weapons; scared to death by this sudden and strange invasion, each sailor, each man on board, spontaneously grabbed any means of defence at hand, and attacked together this terrible ennemy, each for himself, with axes and cutlasses.

The magnitude of the danger endowed the most cowardly with courage; no one sought refuge at the bottom of the hold, but all fought vigorously for the common good. However, almost despairing in their efforts, the ship listing ever more and lying on its side, no longer hoping for salvation, these brave sailors from Saint-Malo put themselves in the hands of the patron saint of their harbour, and they made a vow to Saint Thomas, promising a pilgrimage if they survived this struggle through his help, and returned on the sea to find once more their homes and their families. The confidence which they put in celestial assistance may have, by enhancing their forces, been of some help and cooperated in their victory; but one might also believe that their sharp knifes were even more useful, since without their assistance this miracle would not have taken place. With great blows of their axes and with the edges of their sabres, these sailors finally managed to cut the arms of this horrible animal, which was no other than an enormous octopus; when they succeeded in separating the arms from its body, the latter fell to the bottom; the ship, no longer being pulled to the side nor in danger of being submerged, straightened up on an even keel, with the tips of its masts pointing to the sky, at the desperate moment when, by all appearances it was to be pulled down by this mollusc, whose weight was sufficient to sink the ship.

"Religiously faithful to a vow expressed in a moment of extreme danger, deeply struck by the great peril which they had faced in that occasion, the entire crew, upon its return to Saint-Malo, hurried to fulfill their vow in its entirety, as soon as they arrived in port; brushing aside their wives, refusing to embrace their children and to indulge in the sweet emotions of love, friendship and fatherhood, the sailors tied up their ship, came ashore, nearly naked and bare foot, in a solemn procession to the chapel of the saint, and offered in his sanctuary the expressions of the gratitude which they owed him for the special and significant protection which he had so generously bestowed upon them. Not satisfied with this first solemn ceremony, these sailors also wished to carry to posterity the proof of their gratitude towards saint Thomas by commissioning a painter to represent on canvas, as well as he could, their terrible struggle and the pressing danger which they had faced at that disastrous moment when they thought they had reached the end of their existence. It is thanks to their pious fervour and religious fidelity that this tradition has come to us, with its representation, which we now seize as evidence for natural history, which takes advantage of all materials whose authenticity and evidence cannot be doubted."

Denys-Montfort's gravest error was undoubtedly to publish in his work a copy of the famous ex-voto: the painter had naturally produced a very personal and most emphatic representation of the incident[3]. This naive image is in fact the only part of the story whose "authenticity and evidence" is clearly questionable. Had the naturalist reproduced it only as a matter of curiosity, it might have been found enjoyable, but he seemed to attribute to it a definite documentary value, since he specified in regard to the copy which he had made:

---

[3] *I have been told that the original ex-voto burned during the war of 1940 following one of the numerous bombings of Saint-Malo. However, Frank Buckland already reported in 1891 that one of his correspondents, from Dinard, had looked for it in vain, and had not even been able to find a church or a chapel dedicated to Saint Thomas. He had heard however that such a picture was to be found in Marseille, where it commemorated a combat between a monstrous octopus and a crew from Saint-Malo. Actually, Armand Landrin speaks of such an ex-voto which, according to him, was still in 1867 within the church of Notre Dame de la Garde, in Marseille: however, it commemorated an incident which had taken place off the coast of South Carolina. In spite of the assistance of my friend Jean Delaborde, who was then chief quarter-master of the French navy, I could not get a hold of this precious relic.*

THE KRAKEN AND THE COLOSSAL OCTOPUS

"We did not draw the struggles of the crew because, in so little space, it would have been difficult to depict this combat in a way which preserved real proportions; a man alone being to that octopus what a nutshell was to a large ship."

In fact, one finds in the text only a single indication of the size of the monster: its long arms wrap around the rigging and reach to the top of the masts. In this kind of ship, the masts may reach 30 metres in height. That's a lot, but that would not be *very* exaggerated for tentacles of giant squid. In contrast, if it had been a colossal "octopus" which had arms as large as those shown on the ex-voto - nearly 3 metres in diameter - the sailors would never had been able to hack them off, even with boarding sabers!

The attack of the octopus having taken place off Angola, Denys-Montfort took as confirmation the advice of de Grandpre, author of a *Voyage in Africa*[4]. The latter, he said, had "assured him of the existence of these monstrous octopus on the coast of that country, some distance from the shore" and had repeated, regarding the Saint-Malo ex-voto that "the incident was undeniable, fully recognized, and had never been doubted."

"The negroes of the coast of Africa, his informant had told him, are extremely afraid of this octopus, for he fairly often throwns its arms on their boats and canoes, dragging them to the bottom; on the coast of Guinea, this mollusc is called ambazombi: an evil or sorcerous fish; they regard it as an evil spirit."

If these "monstrous octopus" are found "at some distance from the shore", their zoological identification is likely to be incorrect. An octopus proper never wanders far from the bottom; in any case, it never rises to the surface, unless it be at the shore itself. A cephalopod encountered at sea can only be of the swimming type; in all likelihood, a kind of squid. That is supported by the most detailed testimony reported by Denys-Montfort, who then seems to ignore the pertinent details revealed by his source:

"Captain Jean-Magnus Dens, a respectable and reliable man, after having made a few trips to China for the Gothenburg company, had finally come to rest from its maritime expeditions in Dunkirk, where he lived, and where he died a few years ago, at an advanced age. He told me that during one of his voyages, lying at about 15°S, some distance from the coast of Africa, off St.Helens island and cape Nigra, he was becalmed for a few days, and he decided to seize the opportunity to clean up his boat and to have its hull scraped. Accordingly, he lowered a few boards alongside, on which the sailors stood to scratch and clean the hull. The sailors were carrying out their tasks when one of those squids, in Danish ankertroll[5] , rose from the bottom of the sea and threw its arms around the body of two sailors, which he pulled off, with their scaffolding, and threw in the sea; he then threw a second arm onto

---

[4] *That author's full name is Louis Marie Joseph count O'Hier de Grandpré and his book* Voyage à la côte occidentale d'Afrique, *was written in the years 1786 and 1787.*
[5] *Meaning of course* ankertrold, *the anchor-troll.*

another crew member, who was trying to climb up the mast and was already at the first rungs of the shrouds. However, as the octopus had also seized on the large sheets of the shrouds and wrapped itself up in them, he could not pull off his third victim, who started screaming for help. The whole crew went to his assistance; some, using harpoons and spikes threw them into the body of the animal, which they pierced deeply, while others, with knifes and hatchets cut off the arm which held the unfortunate sailor, which had to be prevented from falling in the water, for he had completely lost consciousness.

"Thus mutilated and pierced in its body by five harpoons (some of which built so as to slide into a transverse position and hang in the body of the animal by a broadening and with two sharp hooks) this terrible octopus, with two crew members, tried to return to the bottom of the sea by its enormous weight alone. Captain Dens, not having given up on saving his men, let the lines attached to the harpoons uncoil; he held one himself, which he allowed to run when he felt some resistance; when they had almost reached the ends of the lines, he ordered them pulled in, which worked for a while, as the octopus let itself drift; they had already pulled in about fifty fathoms of line when the animal quashed all his hopes by pulling again on the lines, which he once more let uncoil. They took the precaution of tying them solidly to the ship at their end."

"At that point, four of the lines broke and the harpoon of the fifth lost its hold and came out of the body of the animal with a jolt which shook the ship. Thus this brave and honest captain had to regret the loss of two men, which became the prey of a mollusc about which he had often heard in the north, which to this time he had considered as completely fabulous, but in whose existence he had to believe after this sad adventure. As for the man who had been squeezed in the folds of one of the arms of the monster, and to whom the ship's surgeon had ministered all available succor from the very beginning, he opened his eyes and recovered his speech; however, having been nearly choked and crushed, he was in horrible pain, fright had affected his senses, and he died the following night.

"That part of the arm which had been cut off from the body of the octopus, and which had remained tangled in the shrouds, was as broad at its base as the foreyard, ended in a very sharp end and was studded with buds, or suckers as large as a ladle; it was 5 fathoms, or 25 ft long [7.60 m], and as the arm had not been cut at its root, since the monster had not even shown its head out of the water, the captain figured that the whole arm could have been 30 to 40 feet long [10 to 15 m].

"He held this adventure as one of the greatest dangers which he had faced at sea, and did not doubt for a moment that, should that terrible and monstrous animal have thrown all its arms on deck, it might even have sought to drag the ship under, which could well have happened, for they would have been in a greater danger from the fact that they were navigating in full peace and did not have their weapons at hand, feeling that they had no need for them and going about their work in the greatest feeling of security. The tragic event spread terror among crew members, and for the five days that they stayed becalmed, they did not dare to sleep, the watch people being

THE KRAKEN AND THE COLOSSAL OCTOPUS

continuously alert, in the fear of a return appearance of their dreadful enemy."

This kind of incident had been reported from other areas than the coast of Africa. The naturalist George Mercer Dawson related in the English magazine *Nature* that the Chimsgau Indians, in British Columbia (Canada), told that at the beginning of the XIXth century a two-master with oriental crew had been seized by a giant squid in Milbank Sound (at a latitude of about 52°N). The Indians added that the evil influence of the beast had caused the ship to be wrecked on the coast at some point further south.

One should not be surprised that Denys-Montfort, who insisted in calling an octopus what captain Dens still called a squid, also made the giant cephalopod of the north, the mysterious *Kraken*, into an octopus. Nothing in the descriptions that he quoted justified such an identification:

"Among the great number of Norwegian captains which I have seen arrive from Christiania and Bergen, loaded with stockfish (dried fish), as well as fir planks and beams, many told me that they had an intimate knowledge of all the facts that one could draw on to prove the existence of these great octopus in the north, and that in addition to traditional knowledge, they had some of their very own. One of them, Andersen by name, told me that he had seen on the rocks near Bergen, two arms of a very large octopus, which he took to be the Kraken, still tied to each other by a part of the mantle, or membrane which links them near their base; that these arms were so large that he could hardly put his arms around them, and that they were each ten steps long, which, counting two-and-a-half feet per step would give twenty five feet [7.6 m]. Andersen added that during a few days, these octopus arms were adrift and were finally carried elsewhere by the waves, and that it was not very rare to find such remnants and debris of great octopus."

I have shown in the previous chapters that, already at the end of the XVIIth century, one could have established the likely identity of the *Kraken*: it could logically only be that species of super-giant squid which was seen now and then in the NOrth Atlantic. Thus, Louis-Augustin Bosc, son of a physician of Louis XV, who was writing up the *Suites à Buffon* devoted to Worms, was much closer to the truth than his colleague Denys-Montfort when he wrote:

"The famous Kraken, which northern sailors claim can tip a ship, seems to be no other than a cuttle-fish. Their reports, stripped from all exaggeration, show that, at the very least, there exist specimens large enough to be able to snatch men from their small boats with their arms and carry them off to sea."

Denys-Montfort indeed knew about his colleague's opinion, but quoted him merely to show that he was not the only zoologist of his time to believe in the existence of gigantic and aggressive cephalopods...

If our bold malacologist took the kraken for an octopus of gigantic size, how did he distinguish it from the *Colossal Octopus*? Through the difference in geographical distribution of the two monsters? Not at all. Rather through their different temperament. Thus, he considered the *Colossal Octopus* as an evil creature, "whom Nature had endowed with a propensity for destruction and slaughter". He blamed it for the disappearance of ships which "sink all of sudden and in a manner which seems like a miracle". He even thought that it could stop a ship underway, and held it responsible for those dramatic incidents which ancient writers blamed on a much smaller fish, the remora. "It would seem, on the contrary, that the *Kraken* has much more peaceful and quiet habits."

Did this opinion, clearly inspired by a highly qualified comment of bishop Pontopiddan, justify the separation of the *Colossal Octopus* and the *Kraken Octopus*, whatever their true nature might be? One may well doubt it. Shouldn't Denys-Montfort perhaps have attributed the apparently greater agressivity of the southern giant cephalopod to the more powerful imagination of southerners?

The declarations of Pierre Denys de Montfort received in the scientific world a rather icy reception. In fact, the rashness of his words was to discredit him forever. Open any encyclopedia today, and you will not find the least biographical mention of the French malacologist.

Already in 1830, Cuvier, in the bibliographic index of his *Animal Kingdom* (2nd edition) labeled him briefly as a "strange man, who claimed to be the former naturalist of the king of Holland". Mistaking the first part of his patronymic as a given name, he distorted it and called him Montfort (Denis). Usage was to perpetuate this error. Who cares about the name of a "strange man " anyway?

Whenever one speaks today of Denys de Montfort, it is always to insult or ridicule him. "He had to be really sick in his mind", spoke of him at the end of the last century professor Félix-Archimède Pouchet, of the Natural History Museum of Rouen. Only a few years ago, an eminent malacologist of the British Museum, Dr. W.J.Rees, called Denys de Montfort "an unscrupulous scoundrel at one time in the employ of the Paris Museum". He accused him of having "created" a gigantic octopus by which he managed to sink several English men-of-war, England being then at war with France." This unwarranted calumny, which Dr.Rees was merely repeating, shows that no dishonest maneuver was spared to discredit the poor naturalist.

Here is how this astounding story was presented in some popular science articles. In 1782, six French ships had been captured in the West Indies by admiral George Rodney, and taken to the nearest port under an escort of four British ships. Afterwards, all ten ships diasppeared in circumstances so mysterious that Denys de Montfort hypothesized that only an attack by a colossal octopus could explain the disaster!

It is true that our rash malacologist presented this particular explanation for the tragedy, of which he had only heard a distorted report. But that he should have seriously thought of having his favorite cephalopod avenge the French defeat does not make any sense. It would have been indeed a rather strange manner of serving French interests to have half a dozen of its own ships sunk by a *deus ex machina* capable of more discernment.

THE KRAKEN AND THE COLOSSAL OCTOPUS

It was claimed in England that when the British admiralty had given a more prosaic interpretation of the event, Denys de Montfort, finding himself discredited, desperately tried to salvage his scientific reputation. Incapable of doing so, he had abandoned the natural sciences and started on a criminal career which took him to the galleys, where he died. The truth is quite different.

Pierre Denys de Montfort deserves better than insults and oblivion. In addition to the first four volumes of *The Natural History of Molluscs*, he contributed to zoology a *Conchyliologie Systématique* in two volumes which, according to professor Léon Vaillant (one of the rare scientists, perhaps the only one, who treated him with fairness) "is still authoritative, for a number of genera which he [Denys-Montfort] created are still part of the nomenclature; one finds no less than twenty-five in Paul Fischer's *Manuel de Conchyliologie* (Paris, 1887), one of the most complete treatises on the matter."

The work of Denys-Montfort, continues professor Vaillant, is prefaced by a Preliminary Discourse where one sees clearly how deeply the author had tried to delve in his subject and the care which he had brought to the presentation of the work, carrying his zeal to the point of drawing and engraving himself all the figures, thinking that others could have done it more artistically, perhaps, but none more truthfully."

All this has been scornfully rejected, condemned to permanent oblivion.

At the time, a single naturalist accepted without argument the conclusions that Denys de Montfort drew regarding giant cephalopods: the clever German zoologist and philosopher Lorenz Oken. In his famous *Lehrbuch der Naturgeschichte* (Manual of Natural History) (1815), he did not hesitate to attribute to the two oversized octopus described by his French colleague scientific names conforming to Linnean directives: the colossal octopus thus became *Sepia gigas* and the kraken *Sepia microcosmus*. However one had to admit that the impetuous Oken, by his real name Ockenfuss, self-taught and vehemently verbose, was also considered by the scientists of his time as somewhat "exalted", sometimes as a "scientific fraud" and at least an "excentric".

In his own country, the audacious Denys de Montfort met mostly with sarcasm and insults. It is significant that his two main works remained incomplete. Of the *Natural History of Molluscs*, only the first four volumes, which include generalities and Cephalopods are of his hand; the last two were entrusted to Félix de Roissy. Of the *Conchyliologie Systematique*, two volumes only were published, in 1808 and 1810, specifically those dealing with univalve; the third never appeared. It is more probable that the mistrust and the growing hostility of the scientific world prevented the completion of these nevertheless important works by Denys de Montfort.

All this had the effect of hardening the resolve of this enthusiastic and obstinate spirit. One day, not long after the publication of his work on Molluscs, which some were already ridiculing, Denys de Montfort met his colleague Defrance, and told him quietly, perhaps alluding to Pliny's *arbor*:

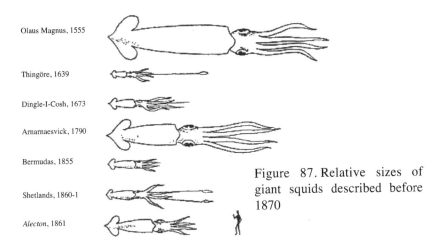

Olaus Magnus, 1555

Thingöre, 1639

Dingle-I-Cosh, 1673

Arnarnaesvick, 1790

Bermudas, 1855

Shetlands, 1860-1

Alecton, 1861

Figure 87. Relative sizes of giant squids described before 1870

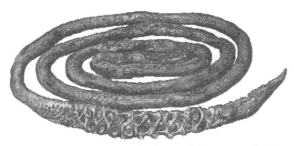

Figure 88. Tom Piccot's trophy: the end of one of the super-giant squid's tentacles

Figure 89. the various pieces of the beak of *Architeuthis*

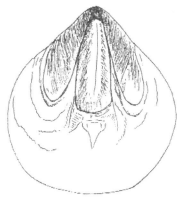

Figure 90. Mouth piece of a super-giant squid, seen from above (after Harting, 1859)

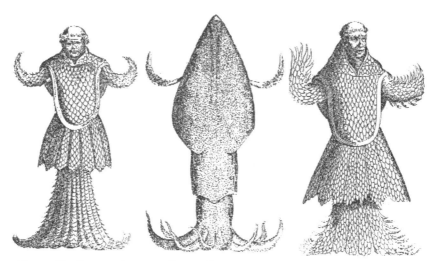

Figure 91. Sea monk, according to Rondelet (1554), on the left; the same, after Belon (1551), on the right, compared by Streenstrup in 1855, to a giant squid, in the middle.

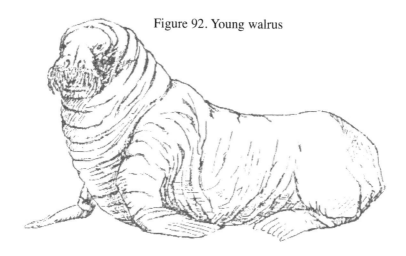

Figure 92. Young walrus

Figure 93. Isn't the walrus perhaps the real sea-monk?
(Photo Björn-Atlas)

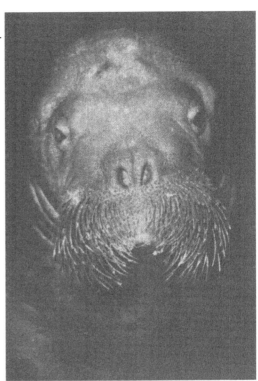

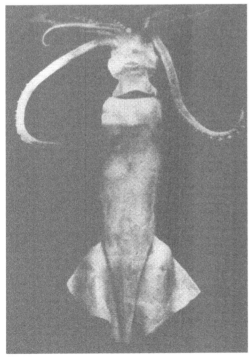

Figure 94 Slender squid
(*Ommastrephes sloani*),
with a characteristic diamond-shaped caudal fin.
(Photo Popper-Atlas)

Figure 95. The *Alecton* squid (1861), after the original drawing of Ensign E. Rodolphe

Figure 96. The *Alecton* squid after Louis Figuier, 1866

Above: Figure 97. Daniel Squires and Théophile Piccot, attacked by a super-giant squid in their boat in Conception Bay, Newfoundland, 1873

Below: Figure 98. The legend soon spread that enormous octopus can attack ships and pluck crew members from their decks

Figure 99. Johan Jepetus Steenstrup (1813-1897), the scientific god-father of the super-giant squid *Architeuthis* (Painting by August Jerndorff, 1885)

Figure 100. The *Alecton* squid after Arthur Mangin, 1864

"-- If my Kraken is accepted, I will have him spread its arms across the Strait of Gibraltar."

Another time, in front of the philosopher Jacques-Joseph Champollion-Figeac, the older brother of the famous orientalist, our jokester told his old teacher Faujas de Saint-Fond:

"If my colossal octopus is accepted, in the second edition I shall have it sink a squadron!"

Having been told of these jokes a few decades later, Alcide d'Orbigny was shocked and saw in them evidence of an extravagant and perverse mind. The least we can say today is that the perpetual secretary of the Académie des Sciences (d'Orbigny's post at the time) did not have much of a sense of humour.

Denys de Montfort's *Kraken* did not have the opportunity of stretching its arms across the Pillars of Hercules, for he was not really accepted. There was no second edition of the *Natural History of the Molluscs* edited by Sonnini, because, precisely, the *Colossal Octopus* was not accepted; it was considered more prudent to hand over the preparation of this opus, from the fifth volume on, to a more stable zoologist.

Pierre Denys de Montfort was not beyond all blame. Like many passionate scientists, his critical sense often failed him, and he had a tendency to blindly endorse all those facts which served his purpose. To compound the situation, a rather unfortunate incident had at the time endowed him in the broad public with a solid reputation of gullability.

A young foreigner, who called himself Nadir-Mirza Shah, son of Persian potentate Charok Shah, had come to settle in Paris, after having had to leave his country because of civil strife. Not knowing French, the exiled had placed in the *Moniteur* an advertisement inviting people who spoke Persian, Turkish, Russian, German or Chinese to come and visit him. Denys de Montfort, distinguished polyglot, had answered his invitation and received from his mouth the story of romantic adventures which had not failed to seduce him.

The news of the presence of the oriental prince had spread quite rapidly following an article published in the *Gazette de France*. However, a citizen by the name of Olivier, who pretended to be well informed of political events in the Middle-East, had immediately replied that the so-called Nadir-Mirza Shah could only be an impostor. Part of the press, always on the look-out for a scandal, supported his arguments, and there was soon nothing but a concert of vicious barking. It was said, among other things, that the king, of whom Nadir-Mirza Shah pretended being the son, was a eunuch, which made his fatherhood rather improbable. Others accused - supreme insult- the young man to be a Jew. Polish, said some; Dutch, said others.

During that troubled period of French history, to be considered with suspicion was dangerous, and the poor immigrant ran the risk of being thrown in jail. So, Denys de Montfort, indignant, decided, for the sake of justice, to defend the young man, whose misfortunes had touched him. He was certainly not the only one who believed in the authenticity of the Persian prince. Mr. de Varenne, who had lived for twenty-two years in Russia and knew the Middle-East well, has also vouched for his good faith. The botanist and traveller Michaux, who had journeyed through much of Persia, was convinced, after a long conversation with the foreigner, that he was "the son of a general or of the governor of some Persian province."

Armed with such guaranties, Denys de Montfort had published in 1801 a kind of defence of the noble refugee: *The Life and Political Adventures of Nadir-Mirza Shah, Prince of Persia, now in Paris*. If one is to believe the bibliographical dictionary of Quérard, this turned out to have been a most unfortunate initiative:

" *This so-called Persian prince was nothing but an impostor, who knew neither the history, nor the geography, nor the language of Persia. His adventures were a web of lies, just as absurd as they were unlikely. This charlatan managed to fool a number of people in Paris, among whom Mr. de Varenne and Mr. Denys de Montfort.*"

This business being somewhat remote from the subject of this book, I have not taken the trouble to search through all the chronicles of the time to find out the truth. However, Quérard's judgment seems rather improbable. Perhaps Nadir-Mirza Shah was not the son of the Shah of Persia and was indeed of a more modest origin: I would not be able to judge.

However, I would be surprised that a person who "knew neither the history, nor the geography, nor the language of Persia" could have fooled such erudite travellers as Michaux and Varenne. A man who could speak five languages, among which Russian, Turkish and Chinese - if he hadn't he wouldn't have dared advertise that he could - was not just anyone. If he was truly an impostor, he was truly a master, able to fool even well-travelled, intelligent and sophisticated people. It is not without a certain ill will that one can use this incident as a basis for disparaging the "gullible character, or the romanesque imagination" of the unfortunate Denys de Montfort.

Of course, this episode was not likely to enhance the prestige of the defender of the colossal octopus. It was only natural to wonder if the latter might not come from the same barrel as the controversial Persian prince. It was no less natural, however, for a mind fascinated by the marvelous to have been just as seduced by the lively adventures of an oriental potentate as by the possibility of monstrous cephalopods.

After that time, Denys de Montfort no longer held any official position in scientific agencies or in teaching. In spite of the sarcasm, he tackled with courage and doggedness his methodical classification of shells, but it would seem that the material conditions faced by a scientist condemned to work independently did not allow him to complete his task. After the publication, in 1810, of the second volume of the *Conchyliologie Systématique*, he continued until 1816 to write articles for the *Bibliothèque physico-économique, instructive et amusante*, but his malacological work was to remain unfinished. He first withdrew into the country, where life was easier than in Paris, and devoted his time to more prosaic and more profitable efforts than conchyliology, notably agriculture. In 1813, he even published a booklet entitled *Ruche à trois récoltes annuelles, fortifiée, économique, et son gouvernement, ou Moyen de mettre des abeilles à couvert contre les attaques de leurs ennemis. (A Hive with three annual harvests, fortified and economical, and its management, or On the means of putting bees under cover against the attacks of their enemies)*.

Neither the sale of that booklet, nor bee farming itself seemed to have improved his situation, for, a year later, he published an even more modest brochure for the only purpose of making a little money. This was *A Small vocabulary for the use of the French and the Allies, containing the names of some of those things most essential to life in many languages: French, Latin, Hebrew, Belgian, Dutch, German, Prussian, Hanoverian, Badish, Tyrolese, Swiss, English, Scottish, Irish, Spanish, Italian, Venitian, Roman, etc*

There were nineteen languages listed on the cover, but one could find inside the 16-page booklet only 7 vocabularies, many of the languages listed being only dialects. In reducing the study of foreign languages to 100 or 150 commonly used words (numbers, every-day objects, food, etc) Denys de Montfort was a precursor of our modern methods of practical instruction.

After 1816, Denys-Montfort shrank into complete obscurity, and we would probably not know what happened to him, were it not for an anecdote related by a learned malacologist of the last century, professor G.-Paul Deshayes. We have it from professor Vaillant, who heard it from his mouth.

Having come at an early age from Nancy to Paris, in the first years of the XIXth century, Deshayes started early in the science in which he was to become a master: conchyliology. Like many people simply taken with their beauty and strangeness, he had a passion for shells. But he soon became more than a dilettante, and he amassed a collection which was to become world famous. In Paris, he gathered fossil shells in the city's surroundings, and he faithfully frequented the stalls of commercial naturalists, seeking specimens missing in his collections.

One day, in one of these stores, where boxes of butterflies were stacked next to amethyst crystals, stuffed birds, and frogs in bowls, Deshayes saw the door open and a person enter "of miserable appearance, with rough beard, torn clothes, wearing frayed pants, and a patched jacket but who, in spite of his ragged appearance stepped in with confidence."

This strange kind of hobo pulled from under his clothes a canvas bag, the contents of which he spread on the counter under the merchant's eyes:

- Monsieur, I return your shells, he rasped, in a surprisingly distinguished tone."

On the counter lay about twenty shells, including some very rare ones.

THE KRAKEN AND THE COLOSSAL OCTOPUS

- Very good, answered the naturalist, who carefully counted them, and who then, after rummaging in the cash drawer, gave a twenty sous piece to the strange individual.

- Thank you. You...have no others to give me? the latter asked in his scratchy voice, where there now pointed some hint of some anxiety.

- No, but come back in a few days; I might have some by then.

The stranger folded his canvas bag with a sigh, stuffed it under his jacket, and quickly took his leave. Deshayes, torn between pity and curiosity could not help ask the merchant about the identity of the strange character.

-What? Don't you know him? he answered. Yet he has a name famous in conchyliology: it is Denys de Montfort.

Denys de Montfort! The young man couldn't believe his ears.

Yes, it was Pierre Denys de Montfort, fallen into the blackest misery "one could even say abject", added the merchant. There was no longer for him any hope of ever escaping it. His only livelihood was to make the rounds of commercial naturalists, who occasionally gave him some shells to identify. Even though this occupation required a rare and exhaustive knowledge, it hardly sufficed to earn a man's daily bread.

Of the coin he had just earned, the miserable would devote a few *sous* to buy bread and some sausage. The rest would buy *eau-de-vie*.

Holding his meager supplies tightly under his arm, as if carrying some fragile treasure, or the source of all life - wasn't this after all what these supplies represented in his mind? - he returned home, in a hurry, to one of these Parisian hovels the like of which there are still too many remaining in the old city: a room only a few square meters, under the roof, whose only piece of furniture was a large piece of paving stone, taken from the street. This stone had multiple uses: during the day, it served the poor man as a seat; at night, it helped to tilt his bed. His bed? It was the door to his cubby-hole, which Denys de Montfort took off its hinges every night and on which he lay without any bedding.

Once at home in his hovel, one can imagine him, squatting on his paving stone, eating with tears in his eyes, the bread and sausage awaited for many days, wetting his meal with a few swallows of alcohol. As he has nothing left but his despair, and that he is perhaps shivering with cold, he will continue to drink until hell burns his guts and he finally sinks into the marvelous ocean of his dreams, of those dreams which undoubtedly precipitated his fate.

On his rude and disappointing awakening, made more doleful by the greyness and dampness of this sordid garret, what would you want him to do but to continue, like Poe, to kill "a worm which does not want to die"? He will only stop drinking when there is no a single drop of alcohol left in the flasks which litter the floor. Then his steps hesitant, hie eye feverish, his breadth foetid, his throat horribly dry, he will go and beg the shell salesmen for more work.

The end of this sorry tragedy is found in the biographical index of Cuvier's *Animal Kingdom*:

> *...died of starvation in a Paris street in 1820 or 1821.*

In the street, just like Poe. Even more horribly than Poe, who still had time to be carried to the hospital before breathing his last. On the paving stones. After all, he must have felt at home...

Thus died, alone and unhappy, misunderstood and ridiculed, even scandalous, a great French zoologist, gifted writer, guilty only of having boldly ventured in the wake of the sea monsters.

As if this miserable end hadn't sufficed in making him atone for his indubitable failings - his overly confident attitude, his stubborness, his fatal inclination for fantasy - Denys de Montfort has not even known, if one might say, the posthumous satisfaction of a rehabilitation. All he got for his funeral eulogy were the snickers and squeaky homilies offered in charity to the poor and weak-minded.

Yet, our damned malacologist was not without some merits. A sharp and particularly penetrating intelligence burns in all his writings like a consuming fire. After having perused all the literature devoted to giant cephalopods, I can vouch that, at a time when literature surveys were much more difficult than today, nobody managed to gather all the existing data on this thorny problem better than he: his culture must have been immense. His investigations were to be a useful guide to researchers for the whole period preceding the XIXth century, and for that alone, he deserves much gratitude. I could quote many who, having impudently pillaged his works, have expressed their debt of gratitude only through insults.

If one may smile at the gullability of Denys de Montfort, the incredulity of his opponents and detractors was even more outrageous than his own naivety, and would deserve even more sarcasm. Even in its most fantastic form, his colossal octopus is closer to the truth than the two-meter span octopus which many have until recently seen as a giant among cephalopods.

Above all, one should acknowledge in this French naturalist that so many have enjoyed ridiculing, a courage the like of which has rarely been seen in his field. For myself, I will proudly associate myself, for their meaning as well as for their style, with those lines with which he prefaced his study of gigantic cephalopods:

> *"Placed between two reefs which both threaten us with shipwreck, it is with the candor of courage that we will proceed: on the one hand, the facts that we will present will seem so strange and foreign to what is known to us that, in spite of their authenticity, it will take some time for us to become used to them, refusing perhaps at first to believe them; it is thus more than*

probable that this first impression will lead us to reject them as fabulous and metaphorical stories typical of the Orient. Expecting the worst, the author hesitates and find himself in a difficult situation: for if, on the other hand, he leaves undescribed and buried in obscurity, through his timid and voluntary silence, facts which have already been quoted and acknowledged by naturalists, his work will remain incomplete, his goal unfulfilled, for he will not keep up with the frontiers of science: avoiding the difficulties which should stimulate and enhance his determination, he leaves to others the opportunity of completing his work. In this peculiar situation, there is but one choice possible: if one is not to be blamed, one must have the courage to seek the truth, face with boldness the legitimate incredulity with which it will be received, put forward and document the origin of the best arguments to prove one's allegations, and finally, and by degrees, transform doubt into the solid conviction which invariably follows the exposition of the truth."

Chapter Nine

THE MONSTER IS IDENTIFIED

While the intrepid Denys de Montfort was pursuing his enquiry among the American whalers of Dunkirk, insistent on attributing to octopus deeds which were more probably those of enormous squids, some of the latter continued to attract attention throughout the world, in the north as well as in the south, the east as well as the west.

Thanks to the investigations of Olafsen and Povelsen, we already know that in 1639 an enormous squid, of the size of a man, had been cast ashore on the northern coast of Iceland. The work of another Icelandic explorer, the naturalist Sveinn Paulsson (or in Danish style, Svend Paulsen), has brought us a testimony concerning a similar incident which occured one hundred and fifty years later in the same region. This time, we are dealing with a specimen of truly formidable proportions.

From 1791 to 1793, Paulsson had explored some of the more remote corners of his native island, still at that time under Danish control. Of these travels, there has remained only the manuscript of the traveller's *Dagböger* (*Journal*), preserved since that time at the *Islandske litteraire Selskab Bibliothek* in Copenhagen. That is where the famous Danish zoologist Steenstrup found it in the middle of the XIXth century; it contains an extremely interesting section regarding very large cephalopods.

In Paulsson's *Dagböger*, under the entry for February 1792, one finds a letter which the prefect of the northern district, C.Thorarensen, had sent him on the preceding 29th of December from the cloister of Mödrevalle. Through this letter, the honorable civil servant wished to inform Paulsson of some remarkable events about natural history, in particular the following:

"In November or December of last year (thus in 1790) in Arnarnaesvick, in this very community, there emerged a creature which one could call a Kolkrabbe, for it was apparently completely similar to the animal of this name, except for its size. Its longest tentacles were more than 3 and a half fathoms in length [5.50 m], while its body, starting from its head, was three-and-a-half fathoms long [6.4 m] and was so big that a man could just span it with his outstretched arms.

"The beast was discovered intact, but I heard about it only after it had

completely spoiled and decomposed and had, as usual, been cut up as bait for cod, for fishermen particularly appreciate it for that purpose. Men, who for the most part had taken pieces from that creature, reported, without complete agreement, that there were four long tentacles, and ten altogether, as one finds in the genus Sepia. But it seemed to me, following Olafsen and Povelsen, that there were probably only two long tentacles. This cuttle-fish was probably quite similar to the strange animal stranded in 1639 on the sands of Thingöre, about which the travellers speak in their Voyage, page 416; its presence proves that although some of these creatures are found of a small size, mostly in the Mediterranean, there also exists another, much larger species."

Professor Steenstrup asked himself whether, instead of 3.5 *fafnen* (fathoms), one should not read 3.5 *alen* (cubits) for the length of the body, which would bring it down to about 2 metres. He suggested this correction, arguing that the measurements given made the body much too long with respect to the tentacles. However, it would seem to me that it is difficult to defend such a shrinking of the body. First of all, had it been only two metres long, it would not have been so large and an adult man would have had little difficulty embracing it (*omfavne*); but the text refers to it being difficult to span it (*overspaende*). It would seem to me that the choice of a verb with the prefix *over* rather than *om* (around) accurately reflects the thinking of the author in this regard.

It was more likely the length of the tentacles which was too short with respect to that of the body! This would be more readily explained by the fact that the latter are often mutilated in fights with sperm whales and other toothed whales. If that were so, as I suggest, this would also account for the strange report by the fishermen to the effect that there were *four* tentacles which were longer than the others. These probably were what was left of the two long tentacles proper, together with the two longest arms.

Thus, if the dimensions soberly reported by prefect Thorarensen are accurate, this squid was at least a dozen metres long, without its tentacles. It was not far from the dimensions of the "monstrous fish" of Olaüs Magnus, whose length ranged from 14 to 16 metres.

As exceptional as it may seem, such a length is not impossible. One-and-a-half century later, a giant squid will be measured to be almost as large as that of Mödrevalle, although it shall then be seen as a champion.

In 1802, the scene shifts to the antipodes of Iceland, in the South Pacific, where we now hear more about giant squids. Off the coast of Tasmania, one of them was encountered by the French expedition of Nicholas Baudin, send to the austral regions by First Consul Bonaparte and greeted back by Emperor Napoleon. Having boarded the corvette *Le Géographe* on October 19, 1800, the young and brilliant naturalist Francois Péron, at twenty-five already a correspondent of l'Institut, witnessed the incident and left a detailed description:

"This same day [January 9, 1802], we saw in the water, not far from

the ship, an enormous kind of cuttle-fish, probably of the genus Calmar (Loligo, Lamarck), as large as a barrel; it rolled noisily in the waves and its long arms stretched at the surface waved like as many reptiles. Each one of these arms was no less than 19 to 22 decimetres long, with a diameter of 18 to 21 centimeters. It is probably to a specimen of this species that Dom Pernetty does not hesitate to attribute dimensions so enormous, and a weight so great that when it manages to hang onto the yards of a ship by climbing along the rigging, it can cause its loss and make it sink. A childish tale perhaps, a revolting exaggeration, which probably finds it source in the observation of some monstrous animals of this kind."

There is no doubt that, after returning to France, Péron consulted the *Histoire naturelle des mollusques* of his colleague Denys de Montfort, published in 1801 and 1802. However he did not intervene to set the record straight and render onto the squid what belonged to the squid. Perhaps he no longer had the strength. Exhausted by a fever contracted during the *Géographe* expedition, which lasted from 1800 to 1804, he died on Dec 14, 1810 in his native town of Cérilly. He was only thirty-five.

A few years later, another French expedition, that of *l'Uranie* and *la Physicienne*, under the direction of Louis Claude de Saulces de Freycinet led to an observation very similar to that of the unfortunate Péron. In the account of their voyage around the world, Jean Quoy and Joseph Gaimard, the two zoologists on board *l'Uranie*, described the circumstances:

"In the Atlantic Ocean, near the equator, by calm weather, we picked up the remains of an enormous squid; what the birds and the sharks had left behind still weighed over 100 pounds [44 kilos], although it was only one half of the animal, without any tentacles, so that one may surmise, without exaggeration, that the weight of the whole animal was over 400 pounds [167 kilos]"

In his *Manuel de l'histoire naturelle des mollusques et de leur coquilles* (Manual of the natural history of molluscs and their shells), published in 1829, Paul Karel Sander Rang, who was both an eminent malacologist and a senior officer of the Corps Royal de la Marine, relates a third such encounter, also in the Atlantic:

"We met, in the middle of the ocean, a species quite distinct from the others, of a dark red colour, with short arms and the size of a barrel."

Somewhat less explicit, but certainly in the same vein, is the report of admiral Cécile, whom Ferussac and d'Orbigny quote in their *Monographie des Céphalopodes*. During his voyage on the *Héroïne*, he had seen, according to his commentators, "an enormous cephalopod passing near his ship."

In those troubled times, when political regimes followed each other without resembling each other much, the French seemed to travel a lot. Did the company of Pacific cannibals seem more attractive, to the most enlightened, to that of their compatriots, enamoured with revolutions and cleansings? In any case, they did meet a large number of giant squids, which are not likely to be encountered in the woods of bucolic France.

If, in the middle of the XVIIIth century, the existence of very large squids was known only to a few lucky eye- witnesses and could be suspected on the basis of old writings only by a few enquiring and ingenious types, by the middle of the following century, it should have no longer been in doubt for anyone. As we shall see, that was still far from being the case.

At that time, actual material remains were no longer extremely rare. Many museums, in the whole world, had remains of more or less enormous cephalopods.

In France, besides the remains of the large squid found by Quoy and Gaimard, preserved in the Museum of Natural History in Paris[1], there was also the material proof of the existence of a rather large squid in the Mediterranean. Among the collections of the Faculty of Sciences of the University of Montpellier, there was a specimen of slender squid (*Ommastrephes*) 1.82 m long, caught by fishermen near Cette around 1845. That was the creature which Paul Gervais, then professor in that university, identified with Aristotle's *teuthos*.

From 1851, on the eve of the Second Empire, there was also available in France the reliable testimony of Jean- Baptiste Vérany, director of the Cabinet of Natural History of Nice, although that city and the worthy curator where still at that time under the rule of a foreign sovereign, the king of Sardinia.

In his monograph on *Mediterranean Molluscs*, Vérany asserted that the squid "attains a weight of 10 kilos."

"I examined one, he said, which was more than a metre long, without including its tentacular arms, and I have had a dorsal pen 65 cm in length. Usually, these large specimens are stranded on the beach, where they are found dead."

Elsewhere, the Nice malacologist added that it was the todare squid[2] which reached such a dimension (at least in the Mediterranean):

"I have never met one less than two decimeters in length, and I have measured one at 16.55 decimeters, including its tentacular arms; it weighed 12

---

[1] *These remains were still at the Jardin des Plantes in 1824; that year, the Dutch naturalist J. van der Hoeven saw them in a jar bearing a label as coarse as it was inaccurate: "Various parts of the intestines of an enormous squid found dead at sea near the equator, weighing nearly 400 pounds, by Mr. de Freycinet."* (Cf. Algemeene Kunsten Letterbode, *1856, p.246.) In 1862, Crosse and Fischer searched in vain for these precious relics at the Museum*

[2] Loligo todarus *Delle Chiaje. D'Orbigny had placed this species among the* Ommastrephes, *the slender squids, under the name O. todarus. However, finding this species too far from the type species O.bartrami, Steenstrup created for it the new genus* Todarodes.

kilograms. One was caught in Nice weighing 15 kilograms, and fishermen assure me that they have seen some stranded of a greater weight."

Anyway, that species seemed to be found everywhere in the Mediterranean, since there was an Adriatic specimen preserved in the Trieste Museum, also weighing about fifteen kilos, found stranded on the Dalmatian coast. As the prefect of Mödrevalle had rightly remarked to Sveinn Paulsson, the giant squids of the Mediterranean were but small fry besides their oceanic and northern cousins."

A number of Dutch museums could also be proud of their impressive trophies. At the second meeting of the British Association for the Advancement of Science, held in Plymouth in July 1841, colonel A.Smith had revealed the existence, in the Harlem museum, of "various parts of a gigantic cuttlefish". To bolster his claims, he had shown a drawing of the beak and of some other organs of the monster.

On the other hand, when Pieter Harting was named, in 1859, director of the Museum of Natural History of the University of Utrecht, he discovered a jar which contained the mouth parts of a cephalopod, 12 centimetres at its thickest, as well as a number of suckers in the shape of serrated cups. The latter's dimensions - from 13 to 25 millimetres in diameter - suggested that they had belonged to a specimen of phenomenal size.

Highly excited by this discovery, the new director attempted to ascertain the origin of these surprising remains, but he could find no information except that they came from a collection of natural history objects gathered at the end of the previous century by a local pharmacist, Mr. Juliaans.

Harting presented a summary description of these pieces to the Amsterdam Academy of Sciences. On the occasion of this presentation, his colleague Willem Vrolik revealed the existence of fragments of another gigantic cephalopod in the collections of the Zoological Garden of Amsterdam. There was there a complete pharynx, 13 centimetres high, with most of the oesophagus attached; the terminal piece of an arm, 6 centimetres thick at its base; an enormous eye, 7.5 cm high and over 8.5 cm wide, and thus larger than that of a right whale; and finally a piece of tentacle still studded with hooks, which betrayed the identity of its owner. These various debris had been found in the stomach of a shark caught in the Indian Ocean by the crew of a merchant ship returning from the East Indies to Amsterdam.

In 1860, Harting published a detailed description of the remains of these two cephalopods. He attributed the first set to a squid of the genus *Ommastrephes*, obviously much larger than its Mediterranean cousins, and the second, in all likelihood to a clawed squid, *Enoploteuthis*. The Dutch naturalist went as far as stating that, all things considered, the stories of the Kraken and of the Colossal Octopus of Denys de Montfort were but "exaggerations of an irrefutable truth."

The British museums were almost as rich in this respect as their Dutch counterparts. At the Royal College of Surgeons, in London, were respectfully preserved the remains of the large clawed squid brought back by

THE KRAKEN AND THE COLOSSAL OCTOPUS

Sir Joseph Banks from his voyage around the world with Captain Cook. In the cellars of the British Museum, there marinated, in an enormous jar, an arm at least 2.75 m long, 28 cm in circumference at its base, which narrowed gradually to a point at its extremity. It was studded with two alternating rows of 145 to 150 suckers. The largest of these were 1.25 cm in diameter; their stemmed-glass shape and their serrated edge showed that they belonged to a squid.

The exact origin of that rather long piece of arm was unknown. At best, it was thought to have come from the coast of South America. When the famous English zoologist William Saville Kent pointed out in 1874 the existence of this trophy, he classified it as an *Ommastrephes*, probably of the species *todarus*.

All these squids of a size much greater than the giant squids of the Mediterranean, - I shall then call "super-giants" - justified the creation of a new genus. This initiative was taken by the Danish naturalist Johan Japetus Steenstrup (1813-1897). If the Kraken, as we by now recognize it, could receive a scientific baptism according to the strict rules of nomenclature, it was thanks to some new acquisitions made by the Copenhagen Museum.

The Copenhagen Museum was no richer than the others; rather the contrary. However, an active and enterprising young man worked there, one who had his eyes and ears open wide.

Of course, Dr. Steenstrup hadn't always been entirely fortunate in his conclusions. The discovery of pictures of mammoths, as well as that of some of its bones, split with stone tools and then burnt, had completely discredited Cuvier's dogma that men and extinct animals had ever co-existed. Well informed about what went on in the world, Steenstrup pointed out that remains of mammoths had been found in Siberian glaciers in such a state of preservation that dogs had eaten them: cave men could well have been inspired by such carcasses to sketch their cave drawings and could even have roasted frozen meat and gigantic marrow bones. In order to disprove this rather clever explanation, it was necessary to demonstrate that glaciers had never existed in some of the regions were mammoth bones had been discovered among kitchen relics. Steenstrup was wrong, but one couldn't blame him for what was a truly legitimate objection.

In the matter of gigantic cephalopods, he showed that he also meant, in the footsteps of Denys de Montfort, to ensure that natural history profited from "all materials whose evidence and authenticity cannot be doubted." The rumors which ran all over Scandinavia about the existence of the formidable Kraken had made an impression upon him. He had begun to sift carefully through all the old chronicles, seeking traces of what might have given rise to the legend, or at least kept it going, for it seemed as old as the world. He thus discovered, on the one hand, the *Annals* of Björn of Skardsa, thanks to the travel relation of Olafsen and Povelsen, and on the other, Sveinn Paulsson's *Journal*. He thus learned that super-giant squids had been stranded on the north coast of Iceland in 1639 and 1790.

That was the occasion for the young researcher of a first communication, in 1847, to the Scandinavian Society of naturalists.

THE MONSTER IS IDENTIFIED

Steenstrup's revelations would have probably fallen into oblivion, even into ridicule, just as those of Denys de Montfort had, if a few years later, luck had not come to his rescue of our Danish zoologist, as if to help him bring concrete proof to his allegations.

In December 1853, the sea cast a gigantic squid onto the beach at Aalbaek, in Jutland, right in Danish territory. As usual, the Kattegat fishermen immediately cut-up the beast for bait; the chopped up remains filled up a number of dog carts. Luckily, the marine commissioner Kjelder, of Skagen, picked up the pharynx, as large as a child's head, and ensured that it reached the vigilant hands of Steenstrup.

This indisputable piece, with a beak 11.5 cm long and 8.3 cm wide, was to be the basis of the scientific description of the incredible colossus under the name *Architeuthis monachus*.

Architeuthis, chief-squid, was a most justifiable name. But why *monachus*, monk? Because Steenstrup, ever conscientious and obedient of the rules, thaught that the giant squid had *already* received a scientific name, and that it had to be taken into account in its new baptism.

The deep erudition of the Danish naturalist went beyond old Scandinavian texts. He knew l'*Histoire entière des poissons*, of Guillaume Rondelet from Montpellier, and had been quite struck by the description found therein of a "*Monster in a monk's costume*"[3]:

"In our days, in Norway, there was caught after a great storm a sea monster, which was immediately called a monk by all those who saw it. For it had the face of a man, but rough and coarse, its head shaved and smooth. On its shoulder there was a cap, like that of a monk, with two long spurs in lieu of arms, the end of the bottom ending in a long tail. The middle part was much broader and had the shape of a military cassock.

"The portrait on which I based this sketch was given to me by the very illustrious lady Marguerite de Valois, queen of Navarre, who had obtained it from a gentleman who was bearing a similar one to the emperor Charles the Fifth, who was then in Spain. The gentleman said that he had seen this monster, as described in its portrait, in Norway, cast by the stormy seas on the beach at a place called Dieze, near a town by the name of Denelopoch."

Steenstrup had no difficulty recognizing the town called "Denelopoch" by Rondelet. One should read *den Elepoch* (Ellebogen), ancient name of the city of Malmö, located in front of Copenhagen, across the water[4]. At the same time, the name *Dieze* (in Latin *Diezum*) also became clear. It was obviously *Die Sund*, meaning The Sund, the strait which separates Sweden from the Danish island of Sjaelland.

---

[3] De Pisce Monachi habitu *(p.492 of the Latin edition of 1554). The French text, somewhat terser, is on page 361 of the "vernacular edition" of 1558.*
[4] *On this matter, Paul Gervais exclaimed: "Rondelet should have said Denmark, not Norway!" But this modern author confirms the reputation which the French have for poor geography, for Malmö is in Sweden.*

THE KRAKEN AND THE COLOSSAL OCTOPUS

There remained to find traces of this incident in local chronicles. Steenstrup first managed to trace it in the work of the historian Sörensen Bedel, who had documented the most salient events which occured during the life of Frederik II, king of Denmark and Norway. Under the year 1545, one could read, among others:

"A strange fish looking like a monk was caught in the Sund; it was 4 ells long [2.4 m]"

That confirmed the soundness of Steenstrup's deduction and justified correcting the spelling mistakes made by Rondelet or his publisher. But the date indicated could not be right, for two other more detailed chronicles placed the incident in 1550. These were the *Danmarks Riges Krönike* of Avild Hvitfelds (1595), and especially the Annals of the reign of Christian III, Frederik II's father, written by Nicolas Krag and Stephan Stephanius.

In most of these ancient documents, it appeared that a "Monstrous and remarkable fish in the shape of a monk" had been caught in the Sund in the year 1550. Brought ashore in a herring net, the animal, 4 ells long, had uttered sharp cries when it had been pulled out of the water. Two days after its capture, it was still alive, for the net had been kept in the water.

By the shape of its head and its features, this marvelous animal was reminiscent of a man, specifically a monk, of which it had the bald head. But along with its human head it had a body whose limbs seemed to have been torn and mutilated.

The body of the monkish monster was brought to king Christian III in Copenhagen, who ordered that it be immediately buried "so as not to cause some scandal among the gossipy people, according to his historian".

After having consulted these ancient documents, Steenstrup recalled having heard of another *Monachus marinus*, that mentioned by Adam Olearius in his *Gottorfische Kunstkammer* (Gottorf Cabinet of Curiosities). Comparing its representation, drawn in parallel with the "frightful marine monster" caught between Katwyk and Scheveningen, he understood that it was the mutilated body of an enormous squid. But then, might this not also be the case for the other "sea monk" caught in the Sund?

Having put side-by-side a picture of a squid and the naive portrait of the "monstrous and remarkable fish" reproduced by Rondelet, the Danish naturalist thought he could detect a similarity in their silhouettes. In the folds of the "monastic robe" of the monster, he saw the eight arms of the squid; in the stumps of its arms, two long tentacles, cleverly re-arranged for the purposes of the argument. In his view, the bald and smooth head of the monk was nothing else than the rear end of the squid's body. As to the cries which the beast had uttered, Steenstrup interpreted them as the noise of the locomotory siphon of cephalopods which, in some seashore areas, had been compared to the cries of babies.

As far as I'm concerned, it is difficult, even with the best will in the world, to believe in Japetus Steenstrup's identification. His fundamental mistake was to take, as the basis of his comparison, the portrait of the beast rather than the text of the old chronicles. As was generally the case, the sketch had most likely been drawn from the text, or from oral traditions, rather than from nature; otherwise, it would not have had this extravagant appearance. There is nothing in common between the picture of Rondelet's monk and that of Olearius's *Monachus*.

It is on the text itself that one must rely to identify the monster. In that case, there is no problem recognizing some kind of seal.

The smooth and shaven head, even the coarse and rustic human face, the spur-shaped arms, meaning in the shape of swimming fins, the body ending in a broad tail, the heard-rending cries uttered upon capture, all these traits are reminiscent of a pinnniped. But which? Certainly not the ordinary seal (*Phoca vitulina*), nor the harbour seal (*P.hispida*), which are common in the Baltic Sea and the Kattegat, through which they perform their seasonal migration. The Scandinavians would never have taken one of them for an extraordinary monster. One is irresistibly reminded of a Greenland pinniped, the hooded seal (*Cystophora cristata*) sometimes also called bladdernose seal[5]. This animal, which migrates yearly northwards to the coast of Norway, where the female gives birth in the spring, could possibly venture as far as the Sund. It would rightly be regarded as a rare beast, but it does not strikingly look like a capucine monk. Its various common names come from the shape of its nose, which the male can blow up like a bladder, to the point of giving it the appearance of a hood, which can extend backwards beyond its eyes.

It would seem to me much more likely that the "sea monk" of the Sund should have been a walrus (*Odobenus rosmarus*). The particularly human-like appearance of its face, the folded nature of its skin, and the shape of its forward legs, strongly distinguish it from seals, and it had, in the Middle Ages, a much better chance than them of being taken as a prodigy by Danes and Swedes, already familiar with a variety of more common seals. Let's recall that in 1520, Eric Falchendorff, bishop of Trondheim, had taken the trouble of sending to pope Leo III the pickled head of one of these pinnipeds, which he considered as some kind of "monster".

Normally, walruses do not venture far from the icy waters of the Arctic Ocean, but, rarely, lost individuals have been found in the winter in Great Britain: on the coasts of Scotland and the Orkneys, in 1902, and even further south, in Ireland, in 1897. In 1926-27, a magnificent male was seen successively in Norway, in the Frisian Islands, in Denmark and in Sweden. In 1939, an even bolder, or perhaps even more lost, animal came through the Sund and ended its voyage on the German coast of the Baltic. Perhaps it was only repeating, in performing this memorable feat, the escapade of its remote ancestor, four centuries earlier. This walrus was however not mistaken for an aquatic monk. Probably because we are well beyond the days of the Reformation... The walrus does look like an old hermit, bald and poorly shaven. The numerous folds of its skin might simulate the cowl of a monk on its shoulders. For the Dane, recently converted to Luther's theories, there

---

[5] *There also exists in the Mediterranean and in the Atlantic, around the Canary Islands, a seal called the Monk (*Monachus albiventer*), but it is to the colour of its pelt that it owes its name: when dry, it looks like a burlap cassock.*

must have been quite a temptation to joke at the expense of the minions of the Papacy...

One might object that the walrus's powerful tusks are too characteristic not to have been noticed. But one should remember that they develop only in the adults, and that those of the female are much smaller than the male's. If the monk of the Sund was really a walrus, as I believe, it must have been a young one, because it was only 2.40 m long. The adult male is always more than 3.0 m long and can even reach 4.50 m.

Steenstrup really had to be obsessed with the idea of gigantic squids to see one of them in the imaginative description of the "monster in a monk's habit" and its grotesque picture. Well, that's how he saw it. As Gesner, friend and colleague of Rondelet in Montpellier, had in 1556 translated into Latin as *Monachus marinus* the term *sömonk* used in the Danish chronicle, Steenstrup felt obliged, for the scientific baptism of the squid stranded on the Jutland coast, to propose at least for a specific name the generic name used by the Helvetic Pliny.

The deed was done, and the priority rule established. The first super-giant squid was, and will always be, named *Architeuthis monachus*.

A new incident was soon to give it a somewhat less monkish brother.

In the fall of 1855, while he was passing between Carolina and Bermuda (by 51°N and 76°W), a Scandinavian navigator, Captain Vilhjalmur Hygom, saw, floating at the surface, the inert carcass of a very large squid: it was more than 3.6 m long.

The captain had the fortunate idea of having the monster's corpse hoisted on board, nearly intact, and to extract various organs, which he brought back to Copenhagen and offered to Steenstrup: the horned beak; the teeth (meaning the radula, the grating tongue); a number of arms, as large as those of a man; suckers, 3 cm across, and a number of interesting parts[6]. The dorsal pen - the squid's internal sword - was no less than two meters long and 17 cm thick in its middle.

---

[6] *The Danish text reads:* "Hornkjaeberne, Tandvaebningen, Armene, Sugeskaalene og flere vigtige Dele af en over 6 Alen lang Bloeksprutte. *The word* Tandvaebningen, *composed of* Tand *(tooth) and of the gerund of the verb* vaebnen *(to arm), literally means "dental armature", i.e. "dental system". In 1873, the American A.S. Packard Jr. translated it by hooks, and claime that this creature was a clawed cephalopod. On this point, he imagined that the hooks, which replace the suckers in armed squids, arose from the middle of the latter(!). Another error: he took the length of 6 Alen to correspond to 18 ft, although a Scandinavian ell is only two English feet. This is not merely a wish to exaggerate. Elsewhere in the same article, where he traces the history of the subject, he translates 15 kilos by 15 pounds and thus diminishes by half the weight of the large squid caught near Nice and mentioned by Verany. And when he quotes Dr. Swediaur, he says that it was a 25 foot long squid which his informer had found in the jaw of a sperm whale: that was the length of an arm alone!*

One has to be extremely wary of all the data of that American author: his work is replete with errors of all kinds. To quote another one: he places on the coast of Denmark the strandings of 1639 and 1790 (he writes 1798) which occured in Iceland. All this has caused considerable confusion. Since A.S.Packard, Jr. was the first to mention, in that same work, the important strandings of super-giant squids in Newfoundland in 1871, he has been widely quoted and of course also copied.

THE MONSTER IS IDENTIFIED

As the beak differed slightly from that of the specimen stranded in Jutland in 1853, the Danish zoologist gave that new giant squid the name *Architeuthis dux* (after having had the unfortunate intention, in a preliminary note, to call it *A. titan*).

These unequivocal facts, described in Danish in a scientific magazine published in 1857 in Christiana, now Oslo, were known at the time only to a handful of specialists, among whom Dr. Harting in The Netherlands, professor Paul Gervais in France, professor Addison Verrill and Dr. Packard Jr. in America. The latter had corresponded with Steenstrup and the two scientists had even met.

However, we have seen that trophies similar to those of Dr. Steenstrup were preserved in many of the world's museums. The public was even invited to admire one of them, more impressive than all the others, in the curio collection of the famous Barnum. At least, that is what the French journalist Bénédict-Henry Révoil tells us in his book *Shooting and Fishing in the Rivers, Prairies and Backwoods of North America*:

"An American captain told me that in 1856, while off the shoals of the Bahamas, his ship had been attacked by an octopus which, stretching its gigantic arms, had reached and pulled off into the sea two of his crew members. With a blow of his axe, the chief helmsman cut his arm. This monstrous appendix was 3.5 metres long and as thick as a man's arm. I have seen that curious natural history specimen in Mr. Barnum's Museum in New York, where it is held, folded up and all wrinkled, in an enormous alcohol-filled jar."

There was no absolute certainty that this piece of arm of an extraordinary size came from a squid, but there were good chances that it should. Anyway, it must have belonged to somebody, and that should have been enough to interest zoologists.

Given the rather advanced state of knowledge about giant cephalopods in the middle of the XIXth century, one remains somewhat astounded at the coy silence about them in zoological texts of the time. Let us consult for example the monumental *Encyclopédie d'Histoire Naturelle* of Dr.Jean-Charles Chenu, a work in which this surgeon at the military hospital of Val-de-Grace, also professor of natural history, meant to present a synthesis of the research of all his predecessors. In the volume devoted to *Crustaceans, Mollusques et Zoophytes*, published in 1858 with the collaboration of Eugene Desmarest, we find no shadow of a trace of giant squids. The slender squid (*Ommastrephes*) is mentioned, but there is no reference to its size, and even less to that of really gigantic squids. Yet, one gets the impression, reading the travel accounts of the beginning of the century, that French travellers could hardly venture to sea without meeting one!

Did Dr. Chenu and Mr Desmarest believe, like Denys de Montfort, that giant cephalopods were octopus? Certainly not. Their opinion on the matter of colossal octopus is very reserved:

THE KRAKEN AND THE COLOSSAL OCTOPUS

"One should place in the ranks of fabulous tales what has been said by Aristotle, Pliny, Aelianus, and Aldrovandi, and repeated even recently by serious travellers and by naturalists, such as Denys de Montfort, for example, regarding gigantic octopus, capable of seizing ships, and to grab with their arms, not only men, but also large cetaceans..."

"What seems true, they add with exquisite prudence, is that there exists in the Pacific Ocean a species which is nearly two meters long."

An octopus with a two-meter span! That was the champion of all cephalopods, according to Dr Chenu and his colleague, at a time when there had already been measurements of at least five stranded squids from 3 to 12 metres in length, without including tentacles, and of half-a-dozen arm fragments of lengths between 7 and 14 metres and with a thickness reaching in one case 75 centimetres!

In other countries, the situation was no better than in France, and this state of ignorance was to be perpetuated in an unusual fashion. The excellent German-American naturalist Willy Ley worried about this some years ago, in 1948:

"The so called "manuals" usually fail to list gigantic varieties after listing well known types. Lydekker's enormous Library of Natural History (I have the edition of 1904) does not devote a single paragraph on any of its 3556 large-size pages to kraken. The even more enormous German Tierleben, originally written by one Dr Alfred Brehm, and now a compendium of fourteen volumes, encyclopedia size, has just one page on the topic."

Willy Ley explains this silence - or this excessive prudence- by the paucity of zoological information available even today on giant cephalopods: according to him, they essentially consist only in the knowledge that these animals do exist.

"Body weights of a ton or more, tentacles of a length greatly surpassing 6 metres, staring eyes 25 cm in dimeter, coloration ranging from dark green to bright brick... these are the facts on record."

Even if our knowledge of giant squids consisted in those facts alone, would that not be enough to justify astonished comments? It is not for the lack of subject matter: the length of the chapters devoted in this book to these molluscs demonstrates this clearly. We know a lot less about other marine, and even terrestrial animals, about which textbooks are more wordy.

In my view, the obstinate silence of all general zoology texts of the middle of the XIXth century is mainly explained by the atmosphere of legend which surrounded this whole story. There is a natural tendency to confuse what is legendary and what is purely imaginary. Except for a few rare exceptions, the zoologists on the staff of those institutions where the remains of the prodigious monsters were preserved kept quiet about them, as if held back by some strange decorum. This unconscious conspiracy of silence was responsible for the ignorance of others. Each one kept his cards hidden, so that the various pieces of the puzzle remained scattered, undecipherable...

The ongoing reticence of some modern texts, written after the most striking pieces of evidence had dissipated the fog of mystery, can find no other justification that the distaste felt by many scientists for anything connected with legend.

Let's not forget it: legend is science where fantasy has filled the gaps. Some times, it is no more than some unjustified generalisation of an isolated fact, rare and exceptional perhaps, but fact nevertheless.

Islands vanish following some mistake in cartography or an earthquake. Schools of enormous animals appear before the eyes of navigators like miniature archipelagos. Fantasy associates two phenomena and a legend is born.

A giant squid, voracious like all its congeners, catches a sailor cleaning the hull of a becalmed vessel. Generations of sailors repeat this striking and frightening story, and the squid becomes an object of universal terror, more feared than the shark, which has nevertheless claimed thousands of victims. (But the shark is only one of the *usual* dangers of the sea.) Another legend is born.

Does the fantastic side of these stories take anything away from reality, from the enormous size and fundamental aggressivity of *Architeuthis*?

Chapter Ten

THE INADMISSIBLE SQUID OF
THE ALECTON

"Leave them be: they are blind leading the blind. And if a
blind man leads another, they shall both fall into the ditch."

MATTHEW, XV, 14

On November 17, 1861, the steam frigate *Alecton* had left Cadix, bound for
Cayenne. On the 30th of that month, at the beginning of the afternoon, about
200 km from Teneriffe when a sailor interrupted the daydreams of *lieutenant
de vaisseau* Frederic Marie Bouyer, leaning casually against the rail after a
hearty lunch.

- *"Commander, the watch points to floating debris on the port side."*

The sailors were already speculating on the nature of the object.

- *"It looks like an overturned boat..."*
- *"No... it is a bunch of algae..."*
- *"Come on! It's red; it looks like a dead horse!"*
- *"A horse in the open ocean? There's a good one! It's just a barrel."*
- *"Not at all, he's right, it's an animal. There, we can see its legs."*

The commander immediately had the ship brought closer to the
puzzling object. Here's what he saw, as he was to describe it in a rather
contorted style when he reached the Canary Islands:

"I recognized the giant octopus[1], whose disputed existence seems relegated to

---

[1] *This unfortunate expression "giant octopus", inspired by the legends which circulated among
mariners, was to be the origin of the confusion thrown in everyone's mind by Jules Verne. He
imagined that the words "squid" and "octopus" were interchangeable. We are still paying for this
error, unforgivable on the part of a writer who claimed some familiarity with science.*

the realm of legend. I was thus in the presence of one of those bizarre creatures which the sea sometimes brings from its depths as if to challenge science. It was too much of a piece of good luck not to seize it immediately; I resolved to examine the monster more closely and to try to get a hold of it."

Immediate preparations for boarding! All run about the deck and get ready to fight this new kind of enemy. Harpoons are prepared, guns made ready, slip knots are set to try to lassoo the enormous mollusc.

Alas! There is a strong swell running; when it catches it abeam, the frigate rolls wildly, which is of little help to the hunters aiming at the beast. On the other hand, the animal, while continuing to swim at the surface, tries to avoid the vessel. Not surprisingly, for at each encounter it is showered with a hail of bullets.

This did not come to bother it overmuch though. After each salvo it would disappear under the surface for a moment, only to surface again soon, agitating and twisting its long flexible arms. But each time that it seemed to take some distance, the ship changed its course and followed its movements.

Finally, after hours of this unceasing and merciless pursuit, it was finally brought alongside. Immediately the harpoons flew and one of them stuck deeply in the gelatinous body.

The animal immediately began to vomit a flood of ink, mixed with sticky substances which exuded a violent odour of musk. The sailors took advantage of the animal's disarray to pass a slip knot around its body. They got ready to slip more ropes around it to hoist it on board when a violent spasm of the animal or of the wave-buffeted ship disengaged the harpoon, only lightly held by the soft flesh. At once, the lassoo slipped treacherously on the squid's viscous skin and stopped only at the end of the body, at the widening of the tail fins.

The whole crew tried to lift the enormous animal onto the deck. Already a large part of its body had emerged above the waves when, under the fearful weight of the gelatinous mass, estimated at two or three tonnes, the rope began to cut into the flesh. All of a sudden, it cut through the tail with as much ease as a wire cuts trough butter.

Freed, the monster loudly fell back into the water. But all had seen it closely enough and long enough to be able to draw its exact portrait.

It was a giant squid. It appeared to measure between 4.60 and 5.10 m from the tip of its tail to its terrible parrot's beak, that is without including its eight arms, 1.50 to 1.80 m long. Its total length was thus from 6 to 7.5 m; it no longer had its two long tentacles, probably torn off during a recent fight with some gluttonous cetacean. The relative weakness of the monster's reactions seemed to indicate that it was in a moribund state.

It colour was of that same brick-red that many travellers and whalers had assigned to giant squids and to their arms, vomited by sperm whales. Its eyes were gigantic, of a frightening fixity, and as large as dinner plates. Its mouth, armed with a fearsome beak, was nearly half-a-metre across. Its cigar-shaped body, strongly thickened in the middle, was of an impressive mass. Finally, its fins, located at its posterior end, were in the form of two rounded lobes.

THE INADMISSABLE SQUID OF THE *ALECTON*

Captain Bouyer would have liked to complete the job with the monster, but to tie it up and secure it, it would have been necessary to launch a boat. He did not particularly care to endanger his men's life just "to satisfy a feeling of curiosity, even if this curiosity was based on science". Although pressed by his officers and men, he refused to launch a boat, fearing "that in such a close encounter the monster should throw its long sucker-studded arms on the sides of the boat, tipping it and perhaps choking some sailors in its whips, loaded with electrical fluids."

After a few final efforts at boarding it, the monster, mutilated at both ends, continued to avoid the ship and was abandoned to its fate.

The struggle had lasted more than three hours.

As soon as he arrived in Teneriffe's harbour, at 8:00 a.m. on December 1st, Captain Bouyer, in the company of some of his men, rushed to Mr Sabin Berthelot, French consul in the Canaries, to give him a detailed account of the adventure and to have it authenticated by his companions present. He invited the diplomat on board so that he could examine the tail fragment taken from the monster, which alone weighed 14 kilos. The consul could verify that it was a soft substance impregnated with a smell of musk.

The following day, the commander of the frigate sent an official report of the incident to Field-Marshal Philibert Vaillant, at that time minister of the Navy.

That is why, at the first sitting of the Academie des Sciences of Paris which followed these events, on Monday, December 30, 1861, two communications on the same subject were on the agenda.

The first was by the great physiologist Pierre Flourens, who read the letter addressed by commandant Bouyer to the Minister of the Navy, who had the good grace of sending it to the Academy. The second was by Alfred Moquin-Tandon[2], who read the most pertinent passages of a letter sent to him by consul Sabin Berthelot.

Besides the weight of his report as an official representative of the state, who had talked to many of the witnesses of the incident and examined the meager remains of the monster, Mr Berthelot brought forth another item of corroborating evidence:

"I must add that I also questioned a number of old Canary Island fishermen, who assured me having often seen, far off shore, large reddish squids, 2 metres long and more, which they did not dare to catch..."

---

[2] *This eminent zoologist was to publish in 1865, under the pseudonym A.Frédol, a work of popular science entitled* Le Monde de la Mer, *in which among other subjects he traced the history of giant cephalopods. That very book, excellent in most respects, provided Jules Verne with the essential of the knowledge which he displays, rather clumsily, in* Twenty Thousand Leagues under the Sea. *In its English translation, the work of Horace Bénédict Alfred Moquin-Tandon, alias Frédol, was used a reference for English speaking as well as French authors. It does contain, however, quite a few errors and inaccuracies, which have been reproduced here and there. To keep to the topic at hand, he blames Sonnini, the author of the* Suites a Buffon, *for the exaggerations of Denys de Montfort; he gives Aristotle's cubits a value of 62 cm, which they never had; he attributes to the squid of Quoy and Gaimard a total weight of 50 kilos, which was only that of its meager remains; on the other hand, he gives to the cephalopod mentioned by Dr Swediaur a length of 25 feet, which was that of its tentacle, etc...Therein lies the source of many of the errors of Dr Packard Jr.*

187

THE KRAKEN AND THE COLOSSAL OCTOPUS

Following reading of the two letters, whose details complemented each other, a coloured sketch of the terrible struggle, the work of ensign E.Rodolphe, one of the *Alecton*'s officers, was circulated among the astounded academicians.

To conclude the session, the president, Henri Milne-Edwards, pointed out that the marine animal described seemed to belong to one of the species of gigantic cephalopods whose existence had been mentioned by various authors and whose debris were preserved in a number of museums. He quoted the text from Aristotle on the great *teuthos* of the Mediterranean, and without "tarrying on Pliny's stories or the obvious exaggerations of Olaüs Magnus and Denys de Montfort", recalled the reports by Péron, Quoy and Gaimard, Rang, and the important communications of his Danish colleague Steenstrup (1853) and of his Dutch colleague Harting (1860),

This time, the skepticism of the most reserved scientists was put to a severe test! All reports that could be found on gigantic cephalopods were feverishly assembled. While Bénédict-Henry Révoil received confirmation from Canadian fishermen of the existence of gigantic squids in the high seas, professor Henri de Lacaze-Duthiers collected similar statements from the shores of the English Channel. It was suddenly clear that, after all, there were in hand more than mere presumptions of their existence. It was not only at Barnum's or at other freak shows, suspected of the most outrageous fabrications, that one could see parts of monstrous tentacles and enormous beaks: there were also to be found in the collections of professor Steenstrup in Copenhagen, in the museums of Utrecht and Harlem in The Netherlands, in the Faculty of Science of Montpellier, at the British Museum and at the Royal College of Surgeons in London.

And quite recently - was it in 1860 or 1861? - the mutilated carcass of an enormous squid had stranded on the west coast of the Shetland Islands, between Hillswick and Scalloway. If one was to believe professor Allman, the cephalopod's arms measured 2.43 m in length, its tentacles were twice as long, and its mantle, ending in fins, was 2.13 m long. Altogether, that added up to 4.57 m, without the long tentacles. One of the suckers examined by professor Allman was over 2 cm in diameter[3].

Once verified the authenticity of all these anatomical specimens, it was normal to ask which kinds of squids they could have belonged to. For some of them, of moderate size, their claws made their identification easy: they belonged to the clawed squids (*Enoploteuthis*). As for those squids with diamond-shaped fins, Paul Gervais ranked them with the slender squids (*Ommastrephes*). The stranding of a very similar "marine monster" on the Dutch coast in 1661 had indeed shown that this kind of squid also inhabited the North Sea.

---

[3] *This stranding was reported in the fifth volume of the important* British Conchology *by Gwyn Jeffrey, published from 1862 to 1869. It was not the first time that the super giant squid had been mentioned in that area. Walter Scott, in the documentary notes relative to his novel* The Pirate *(1821), mentioned that the Kraken still had the reputation, at the time of his novel, of haunting the waters of the Shetlands and Orkneys. He added that: "Some years ago, a large object was seen in the beautiful bay of Scalloway, in Zetland, so much, in vulgar opinion, like the Kraken, that, though it might be distinguished for several days [...] yet the hardy boatsmen shuddered to approach it, for fear of being drawn down by the suction supposed to attend its sinking."*

THE INADMISSABLE SQUID OF THE *ALECTON*

In all these species of giant squids, the mantle, meaning the body itself, was never longer than a metre and usually much shorter. What then of the larger ones, those whose mantle measured several meters, and which thus deserved the name of "super-giant"?

Following Dr. Steenstrup, we know that a new genus, *Architeuthis*, had to be created for the super-giant squids. But according to Dr. Harting, there was not sufficient evidence to justify this innovation. In his view, the various fragments collected did not differ significantly from the corresponding parts of *Ommastrephes todarus*, described by d'Orbigny, and until proof of the contrary, should be assigned to that species.

As for the squid hunted by the *Alecton,* two French malacologists, Crosse and Fischer, undertook in 1862, to determine its type by elimination. That is was a Decapod, no one could doubt, although it had lost its two long tentacles. And among Decapods, it did not fit in the *Sepia, Sepiola, Sepioloidea, Sepioteuthis* and *Cranchia*, which were quite dissimilar in shape. Crosse and Fischer soon reached the conclusion that:

"The Ommastrephes, which include some species of very large cephalopods, differ from the Canary Islands squid in the teeth of the upper edge of the suckers of the sessile arms et by the mobility of their eyes. The Cephalopod seen by Mr. Bouyer had, in contrast, eyes of fearsome fixity.

"Everything leads us to believe that our species belongs to the family Loligidae of d'Orbigny and to the genus Loligo of Lamarck; we propose for it the name Loligo bouyeri, a name which will remind naturalists of the name of that officer who gave the most details on the gigantic cephalopod of the Canary Islands."

The elimination of the genus *Ommastrephes* under the pretext that the *Alecton's* squid had eyes of 'fearsome fixity" may appear debatable. It is quite possible to have a frozen glance in a literary sense without having the eye immobilized in its orbit. Besides, it is quite possible that the eyes of a moribund beast might have lost some of their mobility...

One will be no less surprised to see our two malacologists place an animal more than 6 metres long in the same genus as the ordinary squid (*Loligo vulgaris*), twenty times smaller. However, on this point, they justify their decision as follows:

"The examination of very large cephalopods, belonging to well known species of well determined normal dimensions, seems to show that their growth is not limited, as in higher vertebrates (mammals, birds), and that it continues throughout their life[4]."

---

[4] *This statement was actually based on a more humble opinion of Pieter Harting. Speaking about those squids whose gigantic remains were preserved in Dutch museums, he had written: "Are we sure that these large individuals are of different species than other, much smaller, which are well known to science? The size of the body cannot be used as a criterion to distinguish species, especially when dealing with animals which probably grow as long as they live."*

THE KRAKEN AND THE COLOSSAL OCTOPUS

So, when faced with a very large squid, two hypotheses come to mind. Either it is a gigantic specimen of a known species, or it belongs to a different, very large, species. In the absence of complete documentation, Crosse and Fischer refrained from drawing firm conclusions; however, the name they chose suggests that they leaned towards the first hypothesis. The future was to show that they were wrong: super-giant squids were not only a different species, but a different genus, which Steenstrup had first baptized *Architeuthis*.

The spectacular incident of the *Alecton*'s squid had created excitement in the broad public as well as in the scientific world. The press had taken care of presenting an embellished version of the facts.

There appeared an "authentic relation" of the event, supposedly from the mouth of commander Bouyer himself, which was sufficiently incoherent and replete with gross inaccuracies to reveal its apocryphal character. Obviously written by some newspaperman in a hurry on the basis of available documents, it contains expressions taken from Bouyer's official report, spiced up for effect, as well as passages from Mr Sabin Berthelot's letter, and even an emphatic and naive version of the comments of Crosse and Fischer on the possibility of unlimited growth in cephalopods:

"Why shouldn't this beast, which looks like a rough draft, take on gigantic proportions? Nothing stops its growth: neither bones nor shell; one sees no a priori reasons for limits to its development."

What is clearly evident is that although the term "rewriting" had not yet been invented at the time, the practice was already current. I cannot resist the pleasure of reproducing here the conclusion of that "document", which reeks of press room fabrication.. It is truly the kind of piece that one should expect to find in anthologies of the genre:

"Long will this horrible escapee of old Proteus's menagerie haunt my nightmares. Long will I find fixed upon me that fixed and glassy look, with these eight arms grasping me in their serpentine coils. Long will I keep the memory of the monster encountered by the Alecton on November 30, 1861, at two o'clock in the afternoon, 40 leagues from Teneriffe.
Since I have seen that strange animal with my own eyes, I dare not close in my mind the door of credulity to sailors' tales. I suspect that the sea has not said its last word, and that it still holds in reserve some offspring of extinct species, degenerated sons of trilobites, or that it is still developing, in its ever creative crucible, new molds that will frighten sailors and launch mysterious oceanic legends."

The problem with such journalistic exercises, usually written by ignorant scribes without the least concern for accuracy, is that they often confuse well-meaning people. It is this very same "authentic account" of commander Bouyer that the excellent popular science writer Armand Landrin used as documentation in his book "*Les Monstres Marins*", published in 1867. Not the most fortunate selection!

THE INADMISSABLE SQUID OF THE *ALECTON*

In his letter to the Minister of the Navy, the commander of the *Alecton* had said of the giant squid: "It seems to measure 15 to 18 feet up to its head, in the shape of parrot's beak, surrounded by 8 arms 5 to 6 feet long." That was inelegant, but accurate. On the other hand, Mr. Sabin Berthelot had noted that: "This animal measured 5 to 6 metres in length, without including its 8 fearsome arms, covered with suckers, which crowned its head." The unit of measurement was different, but the result the same.

Blending the two versions, but confusing the units of measurement, the press-room hack had sloppily written: "Its body is 5 to 6 feet long; its tentacles, 8 in number, are the same length." He probably figured that 12 feet overall in length was enough to impress his readers, and that, anyway, one should not overdo it... This kind of attitude is typical of many journalists: they blow out of proportion the most trivial incident, but as soon as they are told of a *truly* extraordinary event, they disbelieve it (arguing that they have seen it all and will not be so easily fooled) and tend to minimize it.

Finally, the revised and corrected text of the *Alecton* incident was nevertheless to lead Mr. Landrin to quite a legitimate outburst:

"A 12 foot [3.6 m] long squid! Isn't this the proof of the old fable of the Kraken? That is plenty long enough to terrify fishermen on board small boats as well as to excuse their exaggeration."

How much more lyrical would the author of the *Monstres Marins* have waxed had he known that this 12 foot squid - only 3.60 m - had been summarily shrunken by a journalist, and was actually 20- 24 feet long, i.e. between 6.10 and 7.30 m in length? Would he still have spoken of the Kraken as an "exaggeration" if he had realized that this enormous cephalopodod was nevertheless only of a modest size compared to the 12 m long specimen stranded in Iceland in 1790, or to those which were to be measured in the years to come?

There were more surprises to come. What the reader will probably find most surprising at this point is probably the fact that, even after the *Alecton* incident, there would still be educated people who would simply deny the existence of giant squids, even going as far as demonstrating *a priori* their impossibility.

In this respect, the presentation of the distinguished popular science writer Arthur Mangin in his book *Les Mystères de l'Océan*, is a choice document on the stubborn incredulity and blindness of some so-called scientific minds. An excerpt from this book, often re-edited until 1889, is so damning for this kind of sterile skepticism, which discourages research and paralyses scientists, that it should be read publicly to all students contemplating a scientific carreer. Since this particular ceremony is not yet customary, I will not hold back, and will quote at length from these edifying pages.

One cannot accuse Mr Mangin of being poorly informed, which makes his opinion even more incomprehensible and liable. It is only after quoting the works of Steenstrup and Harting, the testimonies of Rang, Péron, Quoy and Gaimard and of commander Bouyer, and thus with full knowledge of the evidence, that he states his incredulity:

"*...Even when faced with the categorical statements of so many important people, I admit that I cannot help continue doubting. And I believe this doubt to be legitimate, for it pertains to a fact which, if it was demonstrated with absolute certainty, would overthrow all that we currently know about mechanical physiology, all the rules which until now have been considered to guide the organization of living beings. I explain:*

"*As I have already said, there is nothing capricious in creation. Nature is subject to fixed rules, and to believe that all animals may exist with arbitrary dimensions is an opinion which may be held only by people completely unfamiliar with natural philosophy. By all evidence, there exists between the degree of development of various animals and their physiological organisation a necessary correlation by virtue of which it is impossible to believe in the existence of 2 m long infusorians or in a rhinoceros the size of a flea. It is by virtue of that same law that the existence of a squid or an octopus of the size of a whale must appear a priori impossible. For the squid and the octopus are molluscs, and their organisation is incompatible with such an enormous size, which can be found only in animals which have, first, a skeleton, a powerful bony structure capable of holding and supporting their organs, and serving as attachment and support for their muscles; second, a nervous system, a respiratory and circulatory system and a digestive system appropriate to their body motion, and capable of performing the great tasks of nutrition and repair which characterize the life of higher animals; and third, a solid and dense musculature, without which they could neither overcome the resistance of water while swimming nor dive and live in the great depths in which are supposed to live the giant cephalopods described by some voyagers. A mollusc capable of combating cetaceans; a mollusc of the dimensions described by Péron, Quoy and Gaimard, Steenstropp [sic], Harting, S.Berthelot and Bouyer would thus seem in a real, non-figurative sense, to be a true monster, an abnormal, unnatural, fantastic being, one which it would be tempting to rank with the Chimera, the Hydra of Lerne, the Minotaur and other composite animals invented by mythology.*"

Following this apparently rigorous demonstration, there remained to refute the testimony of captain Bouyer and his crew, who had struggled for more than three hours against one of these chimeras. That was child's play for Mangin, who had no trouble discovering some inconsistencies in the reports published about the encounter. Which relation of an exceptional event might not be similarly flawed?

"*The adventure of the Alecton, among others, is most extraordinary, writes Mangin. I believe that it is true in some respects, since it is affirmed by serious and honorable people; however, without casting any doubt on the sincerity and good faith of Mr. Bouyer and his crew, might one not believe that they were mistaken? One might think that such an error, committed by so many people at the same time, is rather unlikely. Perhaps, but one must admit that the animal chased by the Alecton was even less so. Can you*

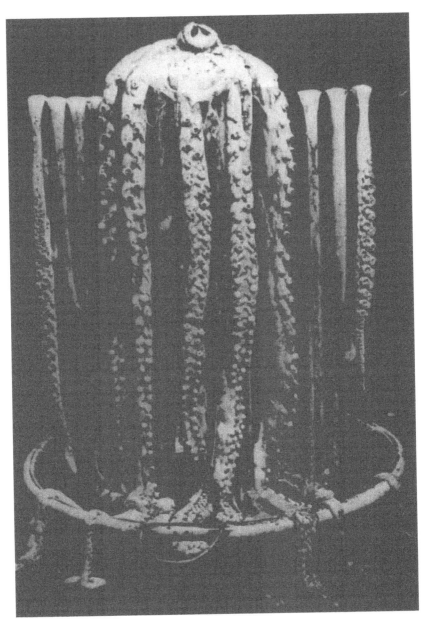

Figure 101. The first photograph of a super-giant squid, hanging over the sponge-bath tub of the Rev. Moses Harvey in St. John's, Newfoundland.

Figure 102. The super-giant squid caught alive on Catalina beach, Trinity Bay, Newfoundland, 1877

Figure 103. The Catalina *Architeuthis*, 1877, after it was killed. (from *La Nature*, 9 June 1883)

Figure 104. Drawing of a giant squid stranded in St. Paul's island, 1874 (after a picture taken by charles Vélain). The shape of the caudal fin strongly suggests that it belongs to a different species

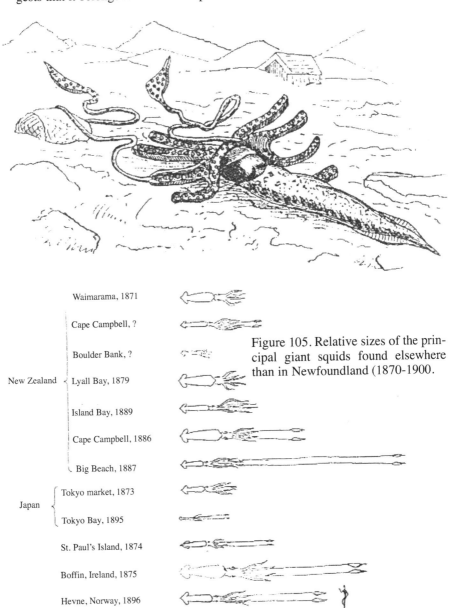

Waimarama, 1871

Cape Campbell, ?

Boulder Bank, ?

New Zealand Lyall Bay, 1879

Island Bay, 1889

Cape Campbell, 1886

Big Beach, 1887

Japan Tokyo market, 1873

Tokyo Bay, 1895

St. Paul's Island, 1874

Boffin, Ireland, 1875

Hevne, Norway, 1896

Figure 105. Relative sizes of the principal giant squids found elsewhere than in Newfoundland (1870-1900.

Figure 106. A full size reconstitution of the Catalina Architeuthis, in the Oceanographic mMuseum, Monaco. (Photo Michel Raynal)

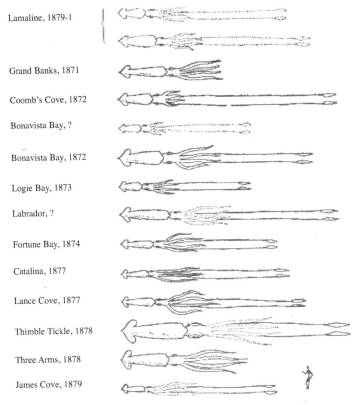

Lamaline, 1879-1

Grand Banks, 1871

Coomb's Cove, 1872

Bonavista Bay, ?

Bonavista Bay, 1872

Logie Bay, 1873

Labrador, ?

Fortune Bay, 1874

Catalina, 1877

Lance Cove, 1877

Thimble Tickle, 1878

Three Arms, 1878

James Cove, 1879

Above: Figure 107 Relative sizes of the main Newfoundland *Architeuthis*, stranded between 1870 and 1880

Below: Figure 108 The first official portrait of *Architeuthis*, after Prof. Addison E. Verrill, 1875.

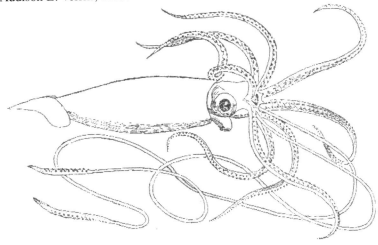

Fig (109) Grey or Risso's dolphin (*Grampus*)

Fig (110) The tail of the scaly squid (*lepidoteuthis grimaldii*): dorsal view on the left, ventral view on the right.

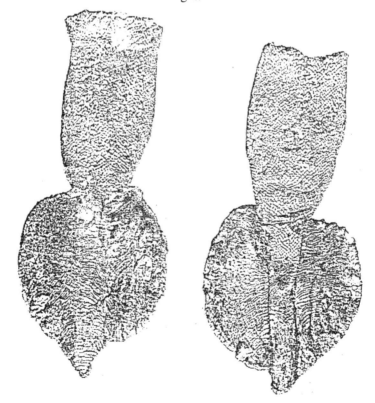

Fig (111). The sea-bishop, after Rondelet, 1554.

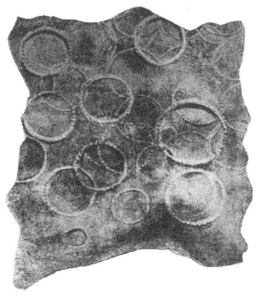

Fig (112) Circular scars left on a piece of sperm whale skin by giant squids, after Murray and Hjort, 1912.

Figure 113. A super-giant squid stranded on the east coast of Scotland stounds the local population. (*Trustees of the National Museums of Scotland*)

imagine this mollusc, whose body alone is 5 - 6 m long and must have weighed up to 2,000 kilos, whose flesh is so such soft that harpoons don't stick to it, and that a slip knot put around it to seize it cuts it in two like a gelatinous mass; this animal which, pursued, shot at and harpooned, remains benevolently for three hours under the blows of its aggressors, rather than fleeing to the abyss, supposedly the usual habitat of its kind, and whose upper part dives and disappears only after having been separated from its lower end!

"The Alecton's commander preserved that lower part on board. Why wasn't it dissected? Why were we not told about its structure and about that of the organs which it contained? It would seem to me that this piece of an unknown monster deserved to be carefully examined. It is also hard to understand how it could have been caught and brought on board, for if the animal was cut in two by the rope, both pieces should have fallen, each on its own side. And that's not all. The report shows us that octopus or squid "vomiting a rather large amount of foam and blood, mixed with sticky matters." It's clear that the reference is to red blood: otherwise, it couldn't have been distinguished from the sticky matters rejected by the cephalopod. However, everyone knows that the blood of cephalopods is colourless[5]. Truly, the more one analyses the report of Mr.S. Berthelot and his description of the giant octopus, the more reason to believe that those from whom he received his information might have been fooled by appearances and by their own imagination..., and the more surprised one is by the fact that the Academy of Science, ordinarily reserved and incredulous, should have so welcomed this report and reproduced it in its Comptes-Rendus."

Of course, in the eye of Mr. Mangin, it would have been preferable for the prestige of the Academy to have refused, with a polite but ironical smile, the communications of commander Bouyer and Mr. Sabin Berthelot, suggesting that they should send their prose to newspapers, or popular magazines, so as to have been in a position, later, to snicker about the frivolous nature of the information presented by the press. That is indeed the usual treatment reserved by scientific journals to revelations of an overly sensational nature, even when member of the editorial board might be convinced of their value.

So, how did the most rational Mr. Mangin explain the *Alecton* adventure, which he held as "true in certain respects"? By a sleight of hand, of course.

About fifteen years earlier, in 1848, Frederic Smith, a passenger aboard the *Peking*, thought that he had seen a sea-serpent near Moulmein, off Burma. When the ship had approached the monster, this worthy gentleman had noticed that his serpent was actually some very large marine algae. Not

---

[5] *To everyone of course, except to those who have examined it and have noticed a bluish tinge. This colour is due to a respiratory pigment based on a copper compound, hemocyanin. (In most vertebrates, the respiratory pigment is red; it is an iron compound, hemoglobin). In any case, what was described there as blood was probably ink.*

only did Mr. Mangin see in this event the explanation for *all* reports about the Great Unknown of the seas, but he also used it to negate the existence of super-giant squids:

"It is quite possible that the fragments picked up by some travellers, which they represented as belonging to giant squids or octopus, were actually debris from a marine plant. The soft consistency of these largish fragments, their viscous surface and their brown or red colour, their strong smell, are all characteristics relevant to many oceanic products, which cannot justifiably be identified as animal rather than as plant material.

"I believe that such considerations, as well as perhaps others which a minimum of reflection will suggest to the reader as well as to ourselves, will suffice to convince the wiser among us, and especially men of science, to accept only for the record stories about extraordinary creatures, such as the sea-serpent or the giant octopus, whose existence would be in a way a negation of the great laws of harmony and equilibrium which rule living as brute and inert matter."

Thus, Mr.Mangin did not hesitate to believe that scientists like Sir Joseph Banks, François Péron, Sander Rang, Quoy and Gaimard, Japetus Steenstrup and many others were incapable of telling the difference between a squid and an alga, even when they had a piece in hand. He preferred to believe that to be possible rather than to believe in the existence of giant squids. A "minimum of reflection" might have perhaps suggested to him that this was equivalent to denying the "great laws of harmony and equilibrium which rule living nature", including the brains of men of science? It would seem that these same laws do not govern the "brute and inert matter" of some popular science writers.

I would like the reader to read and re-read, and even memorize, the astounding demonstration of Mr. Arthur Mangin. When he has become well impregnated with this proof and realizes that it is with dregs from the same barrel that are nearly always refuted reports concerning unknown marine monsters, he will be ready to read my book on the Great Sea Serpent!

After the memorable incident of the *Alecton*, the super-giant squid was to suffer an eclipse of a dozen years or so in scientific circles. The enthusiasm which it had quickly raised among scientists after that famous encounter abated just as quickly. In France, where some of the witnesses of the incident had been available for questioning, there had been the likes of Arthur Mangin to raise doubts as to their lucidity; it is not surprising that in other countries some commentators had not hesitated in qualifying the whole story as "a typical case of collective hallucination." Unfortunately, as no material proof of the incident remained (the tail end soon rotted and was probably discarded), public opinion gradually returned to its former incredulity. And the tentacles and arms conserved in museums all over the world were again also forgotten.

Commander Bouyer's giant squid seemed to have generated more confusion than certainty. The more Jules Verne and his sequel of imitators and disciples were enthusiastically describing colossal octopus at every dramatic corner of their sub-marine fables, the more zoologists withdrew into a careful silence. It seemed that they might have found rather disturbing this sudden irruption of a legendary animal, which had shamelessly intruded into the sacred and orderly Temple of Science. Perhaps by not mentioning it again, one might force it back into the foggy realms of folklore, whence it should never have emerged.

The fact is that, after the memorable session of the Academy of Science of Paris (1861) and the note by Crosse and Fischer (1862), the only favourable comments about giant squids within the following decade, besides a brief passage in Gwyn Jeffrey's *British Conchology* (1862-69), were the popular science book published under a modest nom-de-plume by Alfred Moquin-Tandon (1865), and the even more popular book by Armand Landrin (1867), which owed a lot to the former. That was not much for such extraordinary large animals.

Then, all of a sudden, articles and studies on super-giant squids became so numerous that, for about thirty years, it became difficult to find one's way in this luxuriant literature.

The sudden popularity of colossal squids in scientific media is readily explained. While in the chronicles of the past pertaining to more than two thousand years of History, one detects at most half-a-dozen stranding of such animals, in eight years, there would be more than twenty five: a veritable avalanche! It would have seen that the King of Invertebrates was at last tired of its long anonymity...

Actually, while there was from 1871-1879 a veritable epidemic of "suicides" of super-giant squids on the shores of Newfoundland and Labrador - for which a plausible explanation was soon found - these events also had a catalytic effect. The more strandings there were, the more the skepticism of zoologists crumbled, the more they listened with an open-mind to rumours which in the past merely caused them to shrug, and the more they discovered new testimonies through systematic enquiry, passing from a passive to an active interest.

The time perhaps may have come perhaps to leap from malacology to mammalogy, and to venture a reference to Panurge's sheeps, and the extraordinary breadth of their distribution...

Chapter Eleven

ARCHITEUTHIS GALORE

―――――――

At the end of the XIXth century, in Salem, Massachusetts, it had only been a couple of centuries since they had stopped burning witches. Even then, it was not without regrets that some of the population had abandoned the practice. In 1878, hysterical Mary Baker Eddy, founder of the powerful *Christian Science* sect, was still harassing local tribunals to revive this medieval practice, under the pretext that her personal enemies had caused the death of her third husband, as well as her own discomforts, through some "evil animal magnetism", a new code word for black magic. It was nevertheless in that same puritan city, where the word of the Bible had the force of law, that had been published, since 1867, an excellent scientific magazine, the *American Naturalist*.

It is in the first number of this periodical for the year 1873 that one finds the first traces of the incredible epidemy of strandings which was to afflict *Architeuthis* on the coasts of Newfoundland, for our greatest edification.

After having presented a historical overview of the question of giant squids (unfortunately with that lack of accuracy of which I pointed out the importance earlier), Dr. A.S.Packard Jr., chief-editor of the magazine, reported a number of testimonies which he had himself collected:

"I have been informed by Capt. J.Hammond of Salem, who sailed for forty-one years between that port and the East Indies, that once, while off the Cape of Good Hope, he saw the remains of a squid from eight to ten cubic feet in size (about the size of a 250-280 litre barrel), floating on the surface. The animal had apparently been attacked by whales and dolphins and the arms and head devoured.

"At a late meeting of the Boston Society of Natural History, Hon. N.E.Atwood related the fact that he had seen pieces of squid ten inches [25 cm] in diameter vomited by a sperm whale, and that sperm whales were known to devour giant squids."

All that did not add anything new to the matter, and only confirmed what Dr. Swediaur, Pierre Denys de Montfort, and some other scientists had already put forward. What was most important in the article of Dr. Packard Junior was a reference to an article in the *Cape Ann Advertiser*, which had hitherto escaped the attention of zoologists.

THE KRAKEN AND THE COLOSSAL OCTOPUS

According to that newspaper, in the fall of 1871, the crew of a Gloucester fishing boat which was sailing on the Grand Banks of Newfoundland, had found the floating carcass of a 15 foot [4.60 m] long squid, the arms of which were about 10 feet [3 m] long. The editor of the *Cape Ann Advertiser* had confirmed for Dr. Packard the authenticity of the news item: in fact, a Mr. James G. Tarr, of the firm *Dodd, Tarr & Co* of Gloucester, had personally vouched for its veracity.

Questioned in turn by the curious naturalist, Mr. Tarr had provided additional information about the event. It was around October 20, 1871, that captain Campbell, of the schooner *B.D.Haskins,* had sighted a gelatinous mass floating at the surface of the water, while his ship was anchored on the Grand Bank.

Some of the crew launched a boat to take a closer look at the carcass and found out that it was that of a gigantic squid. The mate measured the body with a ruler and found it to be 15 feet [4.60 m] long, with a circumference of 4 feet, 8 inches [1.42 m]. It was thus rather thin for its length. Of the partly eaten up arms there remained only 9 to 10 feet [2.75 to 3 m] in length; their circumference was estimated at 22 inches [56 cm].

In spite of its thinness, the cephalopod, whose volume was equivalent to the capacity of 8 to 10 barrels, was thought to have weighed about one tonne.

Its beak had been preciously preserved by one of the crew of the *B.D.Haskins*, but the man did not want to part with it, even for its weight in gold. Dr Packard had nevertheless managed to obtain a photograph, which he had sent to his colleague professor Steenstrup in Copenhagen. According to the Danish specialist, this beak came, as far as he could tell, from an *Architeuthis monachus.*

Dr. Packard had also sent his Scandinavian colleague a woodcut sketch of another beak, offered to the *Essex Institute* by captain Atwood, of Provincetown. This particular body part, 4.5 inches [11 cm] long - one of the largest ever found - had been found in the stomach of a sperm whale harpooned, apparently, in the North Atlantic. According to Steenstrup, it belonged to the species *Architeuthis dux.*

Dr. Packard's *American Naturalist* article was not to remain long without echo.

On November 10, 1873, Alexander Murray, a notable of St.John's, Newfoundland, and a member of the Geological Survey of Canada, wrote a letter to professor Jules Marcou, requesting him to forward it to the great Swiss-American scientist Louis Agassiz, the greatest authority of the day in marine biology. The latter hardly had the time to send the letter to the editor of the *American Naturalist* before taking to bed, on December 2, after giving a conference in Fitchburg. He gave up the ghost ten days later, exhausted by overwork and the nervous depressions which accompanied it.

That letter told of the struggle between a fisherman by the name of Théophile Piccot and a really phenomenally large squid, around October 25, near the east point of Bell Island, in Conception Bay, southwestern Newfoundland. This time, it was not possible to dismiss the story, as in the days of Denys de Montfort, as a web of lies and exaggerations, for the brave Piccot had brought back as a trophy one of the monster's tentacles, sliced off with a single blow of his axe. A fragment of this anatomical piece and a few serrated-edged suckers which had fallen off had been preserved in alcohol by Alexander Murray, who would have liked to present them as a gift to professor Agassiz.

According to Piccot, the tentacle had been cut at least 3 m from its base, and when he had brought it to Portugal Cove, another part, 2 m long had been cut off its lower end. Of a tentacle estimated to be 9 to 10 m long, there thus remained after that only a piece about 5.8 metres long, which was to shrink by another 60 cm after marinating in a concentrated brine solution. The end of the arm, about 75 cm long, was in the shape of a narrow and pointed paddle 15 cm wide at its widest. It was covered towards its end with numerous clusters of small suckers with serrated edges. The club-like end of the whip of this super-giant squid was larger than a whole respectable-sized ordinary squid!

Alexander Murray's letter was published in the *American Naturalist* only in February 1874. In the mean time, the incident had been revealed to the public through other periodicals, more concerned with timeliness than a scientific magazine. One could thus read in the December 6, 1873, number of the *American Sportsman* an account of the terrible struggle, penned by Rev. Moses Harvey, presbyterian minister in St. John's. In this very religious country, the fishermen, immediately after their struggle with the monster, had gone to their minister first, and it is to him that they had first shown the mutilated tentacle. The pastor had advised his faithful to bring their trophy to his colleague Harvey, whom he knew to be fond of natural sciences. It is the latter who had finally preserved these remains and entrusted them to the St. John's Museum.

This keen naturalist had immediately perceived the importance of the anatomical fragment. As he still wrote with some emotion a quarter of a century later:

"I had in my possession what all the museums of the world did not contain - a perfect specimen of the gigantic cuttle-fish, commonly named devil-fish, or octopus, of which only some doubtful fragments, widely scattered in various collections, were known to exist. I was thus, by good fortune, the discoverer of a new and remarkable species of fish, the very existence of which had been widely and scornfully denied and had never been absolutely proved."

All this, in fact, because an eleven year old boy had shown extraordinary cool-headedness

Thanks to the numerous accounts of the event, which are all in agreement, we can reconstitute it in its smallest details.

On October 26, 1873, while they were casting their nets off Bell Island, two fishermen, Daniel Squires and Théophile Piccot, the latter accompanied by his young son, suddenly saw a dark and formless mass floating near their boat. Thinking that it was the wreck of some ship, and attracted by the prospect of some valuable loot, they came closer, and one of them struck the object with a gaff in order to get a hold of it. Under the blow, the object suddenly woke up and raised itself in the water, showing an intelligent face, in which two pale and prominent eyes seemed to shine with ferocious intensity. Even more terrifying, the beast opened a parrot's beak as large as a 6-gallon (27 litres) drum, with an appearance of hostility which left no doubt as to its intentions.

Fascinated and petrified by terror, the two fishermen were incapable of the least movement. Before they had time to come back to their senses, the monster, now only a few meters from them, suddenly deployed around its head a number of corpse-coloured arms and threw them towards their boat with the intention of grasping them. Only two of them, a whip and an arm, reached the boat, reached across it, and given their length, wrapped themselves completely around it.

One had to act immediately, otherwise the boat was going to be pulled to the bottom by the mass of the beast. In a desperate effort, Tom, Théophile Piccot's son, seized an axe and with a single blow, cut off the tentacle, and then attacked the arm. With the loss of these precious limbs, the monster retreated backwards in typical squid fashion, and gradually disappeared.

For a moment, the fishermen could still see its pale pink body, and its tail, which emerged above the water surface, cutting it like the bow of a ship. The boy had avoided the catastrophe.

Piccot, who had seen the beast stretched alongside his twenty foot [6.10 m] boat, and could thus estimate its dimensions, claimed that its body was three times as long as the boat. As to its width, it could not be, according to him, less than 5 feet [1.5 m] in the middle and reached as much as 10 ft [3m] at the two lobes of the tail fins.

An 18 m long squid? Scientific commentators of the time were unanimous in declaring that these dimensions must have been exaggerated by terror. Judging from the length, considerable as it may be, of the preserved segment of the whip, that is undoubtedly so. But perhaps Piccot had meant to indicate the total length of the animal, including its long tentacles. On this point, the accounts are ambiguous. If one is to take account of the other body dimensions measured by Piccot, one has to grant that his estimate of 18 metres could not relate only to the body and the short arms, for then this squid would have a rather improbably stretched shape.

With whips nearly 10 m in length, Piccot's squid must have had a body, including head, about 4 metres long, and thus a total length of 14 metres. That is not so far from the 18 metres estimated by the courageous fisherman.

In the letter which he addressed to professor Marcou, Alexander Murray did not only mention the unforgettable adventure of the Piccots and their companion. He went on to describe strandings of two giant squids, which had occured just a few years earlier:

"The Rev. Mr Gabriel now residing at Portugal Cove, but who formerly

*resided at a place called Lamalein on the south coast of the island, states that,
in the winter of 1870 and 1871, two entire cuttle or devil-fish, which
measured respectively forty [12 m] and forty-seven feet [14 m], were stranded
on the beach near that place."*

In a second letter, Murray mentioned, soon afterwards, the capture in
Logie Bay, near St. John's, of a squid smaller than the above specimens, but
still of a considerable size. Its total length was 9.7 m, of which 2.35 m for
the combined body and head and 7.35 m for the whips. The shorter arms
measured 1.8 metres.

The body mass of this cephalopod thus hardly exceeded 4.0 m in
length. Nevertheless, it had been only with difficulty that it had been
overpowered after wrapping itself in a herring net; the fishermen spent half-
an-hour hacking at it with axes and knifes. Thus, the head, the eyes, the
siphon and the forward part of the mantle had been cut to shreds.

The remains, religiously gathered, as it were, by Rev. Harvey, had
been measured and photographed by him artistically stretched above its tub;
this was the very first photographic document of the Kraken. Then the
naturalist-minister had tried to preserve the specimen in brine. After noting
that it still rotted away, he had finally resolved to cut it into smaller pieces so
as to keep it in alcohol-filled jars.

Another passage of Mr Murray's second letter referred to unverifiable,
but much more sensational events:

*"A very respectable person, by the name of Pike, informs me that he has seen
many of these gigantic squids upon the coast of Labrador; and that he
measured the body of one eighty feet [24 m] from beak to tail. He also states
that a certain Mr. Haddon, a school inspector of this place, measured one
ninety feet [27m]. He tells me, moreover, that the monsters are edible."*

That last comment would have appeared absurd or disgusting in
America. That was not the case in Europe, where octopus, squids and
cuttlefish had long been appreciated as food by all Mediterranean people. It is
however not for gastronomic reasons that these latter reports were not taken
seriously by the scientists who were to concern themselves with super-giant
squids, notably professor Addison E.Verrill, of Yale University, their
principal specialist. It was because of their reported dimensions, which were
by far greater than those of specimens already photographed or partly
preserved: four times as large as the monster of Mödrevalle!

Of course, one can understand this prudent mistrust from the purely
scientific point of view, but it completely loses all legitimacy when one sees
the same scientist accept, without batting an eye, other facts, no less nor
more true, but relating to specimens of a size smaller than those already
classified.

In the name of which scientific principle might a witness be more
worthy of credence by the fact that he relates an event for which there is
already a precedent, especially in an area where *everything* is extraordinary and

fantastic? Had it been so quickly forgotten that it was precisely because super-giant squids were not "normal" that their existence had been denied?

Now that there was proof of the existence of gigantic cephalopods, zoologists would soon try to place within their classification those specimens stranded or captured in Newfoundland waters, for which indisputable remains were at hand.

As usual,this did not happen without some clashes.

Already we have seen that after having examined the beak of the squid found dead at sea on the Grand Bank, professor Steenstrup had thought that he recognized in it the species which he had already described under the name *Architeuthis monachus*. As to the enormous beak discovered by captain Atwood in the stomach of a sperm whale, he had placed it in the species *Architeuthis dux*.

These attributions were not to the taste of the English naturalist William Saville-Kent, former assistant-curator of the department of natural history of the British Museum. Regarding the specimens of *Architeuthis monachus* and *Architeuthis dux* which has supported Steenstrup's descriptions, he did not hesitate to write:

"Unfortunately, however, no portions of these animals sufficient for establishing a scientific diagnosis, or for the purposes of positive re-identification, appear to have been preserved."

As we know, that was completely incorrect. The examination of a pharynx "as large as a child's head", armed with a beak more than 11 cm long had perfectly justified the description of *Architeuthis monachus*. Although yet unpublished, that of *Architeuthis dux* was to rely on examination of the essential organs of the specimen found in the Atlantic by Captain Hygom.

Saville-Kent was nevertheless categorical:

"The two fragments preserved in the British and St. John's Museums, in fact, apparently constitute the only substantial material at present available to work upon..."

This statement gives an idea of the degree of ignorance current at the time about the rich collection of giant squids in foreign museums as well as about the works of others on the subject. But, in this case, the ignorance was perhaps not entirely candid.

After all, Saville-Kent was well acquainted with the work of Dr. Harting, of Amsterdam. But, while he quoted him, it was only to support his views that the anatomical fragments which Steenstrup had used to justify his creation of the species *Architeuthis dux* actually belonged to *Ommastrephes todarus*. In the same breadth, the English zoologist contradicted himself, since he recognized that the descriptions of his Danish colleague were indeed based on concrete elements, namely on pieces of squid "sufficient for establishing a scientific diagnosis."

Saville-Kent's behaviour becomes clearer when we notice that he proposes for the giant Newfoundland cephalopod the new generic name *Megaloteuthis*, which would have then assured his own immortal fame.

Saville-Kent had quite judiciously pointed out that the pedonculated arm of Piccot's squid did not carry along its entire length the quadruple row of suckers which characterizes *Ommastrephes todarus:* hence, it had to be a different species, perhaps even a new genus, for which he proposed the name *Megaloteuthis harveyi*. However, this other genus had already been created by professor Steenstrup, because the anatomical pieces which he described could not, in spite of Dr. Harding's opinion, be attributed to an *Ommastrephes*. The study of additional material was soon to confirm this.

Professor Addison Verrill was to be fairer than his British colleague, and recognized the priority of the Danish scientist. As soon as he had heard of the revelations of Dr. Packard Jr., he had taken the first opportunity to examine in person, by paying a visit to its possessive owner, the beak of the large specimen found on the Grand Bank in 1871[1]. He had also studied, in the *St. John's Museum*, the fragment of pedonculated arm severed by young Tom Piccot in 1873. Finally, thanks to the good services of Reverend Harvey, he had secured possession of the various parts of the smaller Logie Bay squid.

On the other hand, professor Verrill had already received from his colleague professor Baird, of the *Smithsonian Institution*, the jaws and two suckers from a squid about ten meters long stranded at some unspecified time in Bonavista Bay, again in Newfoundland. A Rev. Murin had collected these remains. Subsequently, Rev. Harvey also mentioned to him the stranding of an *Architeuthis* at that same place in December 1872. Professor Verrill asked himself whether these two specimens might be one and the same. However, if that were the case, one of his two Reverends must have made some mistake in his estimates. Reverend Harvey had estimated the length of the body at 4.25 m. Further, according to him, the small arms "as big around as a man's thigh" measured 3 metres in length and the whips 9.75 m. All that together was too long for a squid of a total length of 10 metres. The most extraordinary measurement made by the honorable minister was that of the largest suckers of the short arms: more than 6 cm across! Such suckers could only belong to a specimen of at least 14 m in total length.

It is while leafing through the pages of a local newspaper that professor Verrill discovered the longest, but not the greatest specimen, of this horde of giant squids which had come to die on Newfoundland shores. It was mentioned in 1872 in an article by R.T.Bennett, of English Harbour:

---

[1] *The year 1871 was a good one for professor Verrill. Not only was it marked by the stranding of a long series of* Architeuthis, *whose study was to bestow upon him an international reputation, but it also saw the birth of his son Alpheus Hyatt. This latter became the famous naturalist writer Hyatt Verrill, to whom we owe so many fascinating works of popular science, inexhaustible mines of curiosities of the living world.*

THE KRAKEN AND THE COLOSSAL OCTOPUS

"Three days ago, there was quite a large squid run almost ashore at Coomb's Cove, and some of the inhabitants secured it. The body measured 10 feet [3.05 m] in length and was nearly as large round as a hogshead. One arm was about the size of a man's wrist, and measured 42 feet [12.80 m] in length; the other arms were only six feet [1.80 m] in length, but about 9 inches [23 cm] in diameter, very stout and strong. The skin and flesh were 2.25 inches [nearly 6 cm] thick, and reddish inside as well as out. The suction cups were all clustered together, near the extremity of the long arm, and each cup was surrounded by a serrated edge, almost like the teeth of a hand-saw. I presume it made use of this arm for a cable, and the cups for anchors, when it wanted to come to, as well as to secure its prey, for this individual, finding a heavy sea was driving it ashore, tail first, seized hold of a rock and moored itself quite safely until the men pulled it onshore."

After the honorable Mr. Bennett had confirmed all these data by letter to professor Verrill, the latter did not hesitate to include this specimen with extraordinary long tentacles from Coombe's Cove (one of the coves in Fortune Bay) within his list of giant squids from Newfoundland. In 1874, he published a first note on this subject in the *American Naturalist*. In this note, he merely noted that "these remains indicate that there are two distinct kinds of giant squids". In his view, the first type was made up of slender individuals and was probably Steenstrup's species *Architeuthis monachus*. The second, composed of more stocky individuals, might, given the shape of one of the beaks in hand, be Steenstrup's second species, *Architeuthis dux*.

The avalanche of giant squids had however not stopped. Because of the publicity they had received, new reports about them came up from everywhere. Thus, a Halifax, Nova Scotia, geologist, Dr. D. Honeyman, mentioned that an eye-witness had told him of the capture of a 52 foot [15.85 m] long specimen in the Strait of Belle-Isle, on the Labrador coast, at West St. Modest. He did not remember the date at which the monster had been seen:

"It was lying peacefully in the water when it was provoked by the push of an oar. It looked fierce and ejected much water from its funnel."

While the overall length was identical to that of the Coombe's Cove specimen, its body mass was actually much more impressive: its longest tentacle measured only 37 feet [11.27 m], but its body itself was 15 feet [4.58 m] long.

Soon afterwards, there was a report of the stranding, in Harbour Grace, during the winter of 1874-75, of a giant squid which had unfortunately been destroyed without having been measured.

Immediate destruction was the fate of most of these monsters. The poor inhabitants of these regions always found some use for their flesh. For example, in December 1874, a specimen nearly 12 metres in length, including nearly 8 metre long tentacles, stranded in the village of Grand Bank, on Fortune Bay. When the local magistrate, George Sims, had the opportunity to examine it, only a few hours later, the animal was already severely mutilated: the fishermen had carved it up and fed large chunks to their dogs.

In October 1875, according to Captain J.W.Collins of the schooner *Howard*, a large number of giant squids were seen floating on the Grand Bank. Most of them were dead and some of them had already been pecked by fish and sea gulls. There were some of considerable size: Captain Collins measured one whose mantle alone was nearly five metres long. Another had tentacles about 11 metres in length. Fifty to sixty carcasses were collected by fishermen, who transformed them into cod bait.

The squid discovered a year later, on December 20, 1876, in Hammer Cove, Notre Dame Bay, did not have to wait for the arrival of people to be torn to pieces: sea birds and foxes had taken care of it and reduced it to unmeasurable pieces.

The first giant squid to escape destruction completely was stranded alive on the beach at Catalina, Trinity Bay, on September 24, 1877, following a violent storm. It was a medium-sized specimen, of a total length of 12 metres: 2.90 m for the body and 9.15 m for the long tentacles. Its arms were nevertheless 3.3 m long and 43 cm in circumference at their base. Its eyes were 20 cm in diameter: the size of a desert plate! Enterprising people exhibited its corpse for two or three days in St. John's and then took it in brine to New York, where it was bought by *Reiche and Brothers* for the *New York Aquarium*.

Perhaps I have been too categorical in saying that this specimen *completely* escaped destruction... It fell into the hands of an ignorant taxidermist who "misplaced the arms, siphon and other parts, and inserted two large, round, flat red eyes, close together on the top of the head" (A.Verrill). It is this rather distorted monster that New York crowds were invited to contemplate.

The squid which stranded soon afterwards, in October, in the same bay, twenty miles north of its predecessor, was also immediately destroyed, cut up and turned into fertilizer without having been measured.

On November 21 of the same year, a third specimen was cast ashore alive, again in Trinity Bay, at Lance Cove, 5 miles south of the previous stranding. Carried ashore by a spring tide and powerful swells, it desperately tried to escape. Its total length was 13.40 m, of which 3.40 m for the body and 10 m for the tentacles. His remarkably long arms were measured at 4 m in length, thicker at the base than a man's thigh. The strength of the water jet issuing from its siphon was such that it dug a trench in the sand, quite deep and more than 10 metres long. The local people cut up the body of the cephalopod, and unaware of its value, abandoned it on the beach, whence the ebbing tide carried it away.

The following winter, two more truly gigantic squids showed up in Newfoundland, the first on November 2, 1878, at Thimble Tickle, the second, on December 2, at a place called Three Arms. The latter had a body 4.60 m long and an extraordinarily thick, 3.70 m in circumference. Its arms were thicker than a man's thigh and 4.90 m long. There was no mention of tentacles, but one can form some idea of the body mass of the colossus by imagining some kind of rocket, about 10 m long and 1.15 m in diameter.

The first of these two squids was even larger, much larger! It is simply the largest whose existence has been accepted by Science (although it was still slightly smaller than that stranded in 1790 at Modrevalle in northern Iceland). On that account, it really deserves that we should dwell on the circumstances of its capture, communicated by Rev. Harvey to the *Boston Traveller* in a letter published on January 30, 1879:

"On the 2d day of November last, Stephen Sherring, a fisherman residing in Thimble Tickle (Notre Dame Bay) [...] was out in a boat with two other men; not far from the shore they observed some bulky object, and, supposing it might be part of a wreck, they rowed toward it, and, to their horror, found themselves close to a huge fish, having large glassy eyes, which was making desperate efforts to escape, and was churning the water into foam by the motion of its immense arms and tail. It was aground and the tide was ebbing. From the funnel at the back of its head it was ejecting large volumes of water, this being its method of moving backward, the force of the stream, by the reaction of the surrounding medium, driving it in the required direction. At times, the water from the siphon was black as ink.

"Finding the monster partially disabled, the fishermen plucked up their courage and ventured near enough to throw the grapnel of their boat, the sharp flukes of which, having barbed points, sunk into the soft body. To the grapnel they had attached a stout rope, which they had carried ashore and tied to a tree, so as to prevent the fish from going out with the tide. It was a happy thought, for the devil-fish found himself effectually moored to the shore. His struggles were terrific as he flung his ten arms about in dying agony. The fishermen took care to keep a respectful distance from the long tentacles, which ever and anon darted out like great tongues from the central mass. At length it became exhausted, and as the water receded it expired.

"The fishermen, alas! knowing no better, proceeded to convert it into dog's meat. It was a splendid specimen - the largest yet taken - the body measuring 20 feet [6.10 m] from the beak to the extremity of the tail. It was thus exactly double the size of the New York specimen [..] The circumference of the body is not stated, but one of the arms measured 35 feet [10.70 m]. This must have been a tentacle."

From more complete measurements made on other specimens, one can calculate approximately the dimensions of the various organs of this monster. We can imagine a monster with eyes as large as drums (40 cm in diameter), with a beak 20 cm in diameter, with arms as thick as a man's body at their base (85 cm around), and whose largest suckers on the whips must have been 10 cm across!

Couldn't such a titan have been capable of the prowesses attributed to the "colossal octopus" by Pierre Denys de Montfort? It was perhaps time for the doubting Thomases to re-examine their views.

Following the capture of the Thimble Tickle squid, more than 17 m long, and with a body mass reaching a length of 11 m, the last Newfoundland specimens collected over the following year look rather puny.

In October 1879, after a violent storm, two sessile arms 2.45 m in length were found on the beach at Brigus, in Conception Bay.

Finally, at the beginning of November, a "small" specimen, 11,6 m long, with a body 2.71 m long and tentacles 8.85 m long was caught alive at Jame's Cove and ended this strange epidemic of strandings on Newfoundland shores. The beast was moving in the water near shore when a fisherman thought of hitting it with an oar. Infuriated, it dragged itself onto the beach, where fishermen took advantage of this opportunity to slip a rope around it and to drag it ashore. Then, seized by some kind of destructive frenzy, they fell upon it and cut it to pieces.

With more or less complete measurements of about twenty supergiant squids in his possession, as well as a non-negligible quantity of anatomical pieces, professor Addison Verrill studied them carefully in a series of brief notes. Then, in 1879, he finally published the definitive results of his research on the *Architeuthis* of the northeast coast of North America. These results confirmed what a preliminary examination had suggested five years earlier: that two different species of squids were to be distinguished among that sample: one with an elongated body, with arms as long as its body, the other more stocky, also with shorter arms.

Within the first species, he placed the specimens from Grand Bank, Catalina and Lance Cove, as well as that whose beak had been found by Captain Atwood in the belly of a sperm whale; also probably those of Fortune Bay and of Three Arms, and even, in all likelihood, the Thimble Tickle champion. Among them, the particularly great length of the arms of the Lance Cove specimen suggested that it might be a female.

Within the second species, whose type specimen was the "small" robust and stocky individual from Logie Bay, one should also include the Coombe's Cove specimen, as well as those from Bonavista Bay. That from Coombe's Cove, with extraordinarily long tentacles, might have been a female.

This time, after a careful examination of the evidence, professor Addison Verrill no longer related these two types with the species described by Steenstrup, but created for them two new species, *Arthiteuthis princeps* for the elongated types, and *Architeuthis harveyi* for the more stocky types[2].

It is on the strength of this latest baptism that his son, the naturalist, writer and traveller Hyatt Verrill, could say one day, regarding the remains collected by Rev. Harvey and the measurements which he had made, that "Using these documents, my father, professor A.E.Verrill of Yale College, was able to describe the first known giant squid..."

---

[2] *Professor Verrill, well aware of the paucity of the material on which he based his conclusions, recognized nevertheless that his two species might be sexual forms of the same species. This is however in contradiction with his remarks on the presumed sex of some specimens of each species.*

THE KRAKEN AND THE COLOSSAL OCTOPUS

That was of course making short shrift of the meticulous work to which Dr.Steenstrup had devoted a large part of his life. Where would we be if filial piety attempted to upset the rules of taxonomic priority? If Mr.Hyatt Verrill had not then been an untiring and sympathetic octogenarian[3] one would have been tempted to tell him:

" Young man, you exaggerate!"

Before proceeding any further, let us anticipate a little so as not to continue in an atmosphere of ever-deeper mystery, and let us try to answer without further delay the embarrassing question: "Why in the world did all these *Architeuthis* choose this rather limited area of Newfoundland and nearby Labrador to commit these absurd suicides?"

The answer to this question lies in the peculiar situation of the Grand Banks with respect to ocean currents. Why are there in those parts some of the thickest fog concentrations in the world? Because the warm Gulf Stream, coming from Guyana and the Caribbean meets there head-on the cold Labrador Current. "In winter, writes Rachel Carson, the temperature change across the current boundary is so abrupt that as a ship crosses into the Gulf Stream her bow may be momentarily in water 20°C warmer than that at her stern, as though the `cold wall' were a solid barrier separating the two water masses." Near the tail of the Banks, these two water masses are just as easy to differentiate by their colour as two countries on a map: the warm, deep blue waters of the Gulf Stream are in sharp contrast with the bottle-green cold northern current.

According to G.C.Robinson, the eminent British specialist on cephalopods, *Architeuthis* would seem to prefer and optimum temperature of 10°C. This is why it is never found on the Atlantic coast of the United States, bathed by warmer Gulf Stream waters. Off that coast, it must stay below a certain depth to stay in its favorite climate.

What happens when, carried by Gulf Stream waters, the giant squid reaches the latitude of Newfoundland? First of all, as the current gradually cools in its progression towards the northeast, the squid has a tendency to ascend towards the surface to stay at the level of its favorite temperature. If however it wanders by chance off its path, and crosses the cold wall, it finds itself in inextricable difficulties. The only solution would be for it to backtrack, but that would be an unusual reaction. Generally, when it is too cold, it simply rises, since it knows by experience that temperature rises towards the surface. It thus behaves as it is used to do; however, in the Labrador current where it has ventured is so cold that it reaches the surface without finding its favorite isotherm. In these abnormal temperature conditions, its health declines, and soon, moribund, it finds itself at the mercy of waves and tides and ends up on the beach. At least this how professor

---

[3] *Alpheus Hyatt Verrill died in November 1954 in Florida, at the age of 83. He had a new book in the press and other manuscripts in preparation.*

ARCHITEUTHIS GALORE

Robson explained the frequency of these strandings.

The truth was to turn out to be rather more complicated. However one fact was now clear. The great *Architeuthis* does not commit suicide: it faces death because it has driven itself into a corner.

While from 1871 to 1879 there was a veritable avalanche of giant squids on the coasts of Newfoundland, some isolating strandings had also taken place during that period at other parts of the globe, some quite remote.

Thus, in the May 1873 issue of the proceedings of the German Society of Natural History and Ethnography of Eastern Asia, published in Yokohama, Dr. F.Hilgendorf related that he had bought on the fish market of Edo (today Tokyo) a squid sword of an unusual size. He had also had the occasion in that same city to acquire the remains of a squid, offered in public display, whose total length reached 4.28 m: 1.86 m for the body, 45 cm for the head, and 1.97 m for the longest arm preserved. The naturalist had, at first sight, taken this squid for a giant species of *Ommastrephes*, but after reflection, he described it in 1880 as a new genus under the name *Megateuthis martensii*. It was actually an *Architeuthis*, which showed that this kind of squid was not restricted to the Atlantic Ocean, as one might have surmised until then.

Fifteen years later, in 1895, two Japanese zoologists, Mitsukuri and Ikeda, were to confirm the presence of these cephalopods in Japanese waters by announcing the discovery, in Tokyo Bay, of a minuscule *Architeuthis*. That was the smallest ever found: its mantle measured only 72 cm; its arms did not exceed 1.22 m and its tentacles 2.91 m.

The area of distribution of giant and super-giant squids was found year-by-year to be ever more extensive. In the wake of the publicity surrounding the Newfoundland strandings, tongues gradually loosened.

In 1874, reassured by the recent revelations of Dr. Packard Jr., another American naturalist, W.H.Dall, had pointed out that during the winter of 1871-72, a large number of giant cephalopods had been cast ahore on a number of occasions at Iliuliuk, on Unalaska Island, in the Aleutians. Among them there was, among others, a giant clawed squid of the species *Onychoteuthis bergi*, of a total length of three metres and weighing 100 kilos. In addition, W.H.Dall mentioned a variety of testimonies regarding truly gigantic squids in tropical and sub-tropical regions:

"There can be no doubt whatever that some cephalopods in the warmer seas attain an enormous bulk as well as length. Capt.E.E.Smith, an experienced sperm whaler, and a careful and intelligent observer, informs me that he has seen portions of `squid' arms vomited up by the whales in their death agony, as large as a `beef barrel', with suckers on them `as big as a dinner plate".

Dall affirmed that his colleague Henry Hanks, of the *Microscopical Society* of San Francisco, while travelling on a merchant vessel through the islands of the South Pacific, had seen at the water surface a cephalopod "as large as the schooner itself!"[4].

"While this is rather vague, commented the naturalist, still it indicates that specimens much larger than any yet recorded may perhaps exist in those regions. I have also hear rather vague reports, added Dall, of some enormous squid which has been observed in the Gulf of California."

A few years later, these various comments and assertions were to attract the sarcasm of the famous American malacologist George Washington Tryon, in the first volume of his important *Manual of Conchology* (1879-1891). Knowing how large *Architeuthis* can be, we can doubt the legitimacy of his scorn.

Dall's statements confirmed the existence, already noted by Quoy and Gaimard, of enormous squids in tropical seas; even with some exaggeration, Henry Hanks' report supported those of Dom Pernetty and Francois Peron, which had established the existence of a giant species in the southern hemisphere. Indisputable pieces of concrete evidence were soon to confirm the cosmopolitan distribution of *Architeuthis*.

For a long time already, interesting rumors had circulated among the fishermen of Saint Paul, a small volcanic island isolated in the middle of the Indian Ocean, almost halfway between the tip of South Africa and Australia. These good people told to whomever would listen that each year, at about the same time, they saw amidst the schools of fish an enormous cephalopod which raised above them its sucker-studded arms. These stories were of course received with amused skepticism. As luck would have it however, on November 2, 1874, one of these giant squids was cast ashore precisely at the same time that a French scientific mission was visiting the island. Under the direction of captain Mouchez, it had as its mission to observe the latest crossing of the planet Venus across the disk of the sun.

After his return to France, on April 19 of the following year, professor Charles Vélain, of the Sorbonne, the expedition's naturalist, presented himself to the Académie des Sciences and declared:

"In the first days of November, a tsunami cast ashore on the northern jetty a squid of the group Ommastrephes, which was no less than 7.15 m from the tip of its head to the end of its long tentacles. In the expectation of the description which we shall make of it, under the name Architeuthis mouchezi, I have the honour of putting at the Academy's disposal a tentacular arm, the beak and the pharynx of this gigantic cephalopod."

To support his words, Mr.Vélain circulated among the academicians a photo of the whole animal taken by a Mr Cazin. This print, from which was made a rather clumsy sketch, was considered as the first correct

---

[4] *To provide an idea of size, a schooner (a small two masted vessel) measures from 15 to 35 m. Those trading in the south seas at the end of the last century were probably 25 m long or so.*

representation of the appearance of an *Architeuthis*. It is unfortunate that the first portrait published in Europe of the super-giant was that of a rather unusual species...This could only lead to misunderstandings.

Thus professor Gervais, of the Paris Museum, who had already taken an interest in large Mediterranean squids of the genus *Ommastrephes* when he was still teaching in Montpellier about ten years earlier, declared after examining the documents exhibited by Mr Vélain that the *Architeuthis* actually belonged to the group of slender squids[5]. He would certainly not have expressed this opinion if he had seen the picture of an Atlantic *Architeuthis*, whose caudal fin does not have the shape of an ace of diamonds, as in *Ommastrephes,* but that of an ace of hearts. Professor Verrill was soon to point out that the new species *Architeuthis mouchezi* from the Indian Ocean differed from the northern forms by the shape of its fin, narrower and spear shaped. This is why, on a blurry photograph, it could have been interpreted as diamond-shaped.

In fact, the older drawing made of the *Alecton* squid by one of its officers gave a better idea of the form of *Architeuthis* than Mr. Cazin's photograph.

As *Architeuthis'* visits were continuing in Newfoundland, the attention of naturalists from the world over was by now focused on these incredible giants. Each one tried to bring new reports on the Kraken, at least liberated from its cloak of fantasy. After having shown itself in the space of these few years in Newfoundland, Japan and Saint-Paul island, the monster was now to become news in Europe.

First of all, as was proper, it made an appearance in the very home of the *Kraken*, Norway. In 1874, the corpse of an *Architeuthis* floated up Folden fjord, on the 65st parallel. There had been no record of specimens of super-giant squids on that coast since Mgr. Pontopiddan had spoken of the "young and careless" *kraken* stuck in the rocks of Ulwangen Gulf, in 1680. One should not simply conclude from this that not a single stranding had taken place in Norway over those two centuries. The fact is that there is rarely a naturalist, even an amateur, on hand to mention the incident. This particular one also almost went unnoticed. It was merely mentioned in 1878, without the least detail, by G.O.Sars, on the basis of a declaration by the eminent Norwegian mammalogist Robert Collett.

Similarly, between 1877 and the end of the century, there were to occur three more stranding in Norway about which we have very little detail: two in the extreme north, at Kvaenangen and in Tromso Strait, the third at the latitude of the artic circle. Finally, in September 1879, there was a

---

[5] *It is on this occasion that professor Gervais informed his French colleagues of the description of* Architeuthis dux *by his Danish colleague Steenstrup. The latter had sent him the proofs of his definitive work on the question. The publication of this important work was slow to come, since it had already been announced in 1856, twenty years earlier. Impatient naturalists had to wait even longer: Steenstrup's* Spolia Atlantica *were to appear only in 1898, a year after their author's death.*

stranding in Iceland, in Olafsfjord. This time however, an 8 m long arm of the animal was preserved. It is still kept at the Museum of Natural History of the University of Copenhagen, where I examined it in 1961.

To find out more about *Architeuthis*, it had to show itself in a more densely populated country.

In 1877, an Irish zoologist, A.G.More, had the good fortune of announcing the dramatic capture, on the 25th of April of that year, of a truly colossal squid. The event occured off Boffin Island, near the Connemara coast, on the west side of Ireland. Sergeant Thomas O'Connor, of the Royal Irish Constabulary, described it in an exemplary report:

"On Monday last, the crew of a curragh [a large kind of coracle made with wooden ribs, and covered with tarred canvas] consisting of three men, met with a strange adventure north-west of Boffin Island, Connemara...Having shot their spillets (or long lines) in the morning, the crew of the curragh observed to seaward a great floating mass surrounded by gulls; they pulled out, believing to be wreck, but, to their great astonishment, found it to be a cuttle-fish, of enormous proportions, and lying perfectly still, as if basking on the surface of the water. What rarely enough occurs, there was no gaff or spare rope, and a knife was the only weapon aboard. The cuttle is much prized as bait for coarse fish, and, their wonder somewhat over, the crew resolved to secure at least a portion of the prize. Considering the great size of the monster, and knowing the holding and crushing power of the arms, open hostility could not be resorted to, and the fishermen shaped their tactics differently. Paddling up with caution, a single arm was suddenly seized and loped off. The cuttle, hitherto at rest, became dangerously active now, and set out to sea at full speed in a cloud of spray, rushing through the water at a tremendous rate. The canoe immediately gave chase, and was up again with the enemy after three-quarters of a mile [about 1200 m]. Hanging on rear of the fish, a single arm was attacked in turn, while it took all the skill of the men to keep out of the deadly clutch of the suckers.

"The battle thus continued for two hours, and while direct conflict was avoided, the animal was gradually being deprived of its offensive weapons. Five miles [8 km] on the open Atlantic, in their frail canvas craft, the bowman still slashed away, holding on boldly by the stranger, and steadily cutting down his powers. By this time the prize was partially subdued, and the curragh closed in fairly with the monster. The polished sides of the canoe afforded slender means of grasp, and such as remained of the ten great arms slashed round through air and water in most dangerous but unavailing fashion. The trunk of the fish lay alongside, fully as long as the canoe, while, in its extremity, the mutilated animal emitted successive jets of fluid which darkened the sea for fathoms around. The head at last was severed from the body, which was unmanageable from its great weight, and sank like lead to the bottom of the sea... Of the portions of the mollusk taken ashore two of the great arms were intact, and measure eight feet [2.45 m] each in length and fifteen inches [38 cm] round the base. The two tentacles attain a

length of thirty feet [9.15 m]. The mandibles are about four inches [10 cm] across, hooked, just like the beak of an enormous parrot, with a very curious tongue. The head, devoid of all appendages, weighed about 6 stone [a little over 38 kilos], and the eyes were about fifteen inches [38 cm] in diameter."

The remains of this squid, obviously moribund from the very beginning of the encounter, were confiscated by Sergeant O'Connor and sent to the museum of the Royal Society of Dublin. A.G.More thought that it must be an *Architeuthis dux*, but professor Verrill, who had a broader basis for comparison, recognized in it a large specimen of *A. harveyi*. From the size of the remains preserved, it must have had about the same proportions as the 14 metre specimen from Bonavista Bay.

A.G.More took advantage of the incident of the Boffin Island squid to reveal, on the basis of old chronicles, that it was not the first time that a super-giant squid had been caught in Ireland. The world became aware at last of the story of the "two-headed beast" from Dingle-I-cosh, which we have related in detail in an earlier chapter. We recall that that particular cephalopod measured more than 6 metres in length, without the full tentacles.

Proofs of the existence of such titans had already been available in Ireland for two centuries, but nobody had been aware of it...

Somewhat late, More undertook to provide a scientific description of the Dingle-I-cosh squid, but he was rather ill advised in his assessment of the importance of its extendable funnel (the second head), which he thought was sufficient to create a new genus: *Dinoteuthis proboscideus,* meaning the terrible trunked squid.

As Tryon was to point out, quite pertinently, "the extendable funnel is a characteristic common to all enormous cephalopods of the North Atlantic." It is even characteristic of all squids, of whatever size. The Dingle-I-cosh double-headed beast was an *Architeuthis* like all the others.

In October 1880, a third super-giant squid stranded on the west coast of Ireland, at Kilkee, in County Clare. This time again, as in the case of the Dingle squid, it took a while before the incident became known to the scientific world. Not before 1918, forty years later, did the Scottish naturalist J.Ritchie, mention it in a scientific publication.

One should then not be surprised that, in the summer of 1889, the statement of a Cannes fisherman about what must have been an *Architeuthis* probably lost in the Mediterranean was greeted with considerable incredulity. While passing near the Bancales Rocks, not far from St. Honorat Island, across from Cannes, the good man had seen, wriggling in the shallow water, a `fish' at least 5 metres long and of a circumference only slightly smaller at its thickest. It had, said the fisherman, an enormous parrot beak and, on its horrible head, two long horns (most probably its tentacles). This was a rather good description of a super-giant squid, which this uneducated man could not have heard of.

The news of recurring strandings of super-giant squids in Newfoundland was to have repercussions at the other end of the world. In 1879, undoubtedly encouraged by professor Verrill's revelations, T.W.Kirk, a

THE KRAKEN AND THE COLOSSAL OCTOPUS

New-Zealand zoologist and assistant curator at the Colonial Museum, revealed that gigantic cephalopods were also sometimes cast ashore on the coasts of New Zealand.

There were of course long-standing Maori legends about the existence of such monsters, but the white settlers paid no attention to them until their possible authenticity was established on the other side of the world.

Having heard that an enormous cephalopod had stranded, nine years earlier, on the beach at Waimarama (on the east side of the South Island) and that a local settler, F.H. Meinertzhagen by name, was in possession of its beak, Kirk had written to him, asking for more information. On June 27, 1879, his correspondent sent him the beak and told him that the actual stranding had taken place while he was in England. However, one of his trusted friends had examined and measured the monster and kept the beak for him. In a letter dating back to the time of the event, he had provided a non-scientific, but revealing portrait of what Meinertzhagen called the "defunct octopus":

"The beast had eight tentacles, as thick as a man's leg at the roots; horrid suckers on the inside of them, from the size of an ounce bullet to that of a pea at the tip; two horrid goggle eyes; and a powerful beak between the roots of the arms. His head appeared to slip in and out of a sheath. Altogether he was a most repulsive-looking brute."

Some measurements accompanied the letter. From the tip of the tail to the roots of the arms, the beast was nearly 3.20 m ; in the middle of the body, it was more than five metres in circumference; the arms were about 1.70 m long.

There was also a pen sketch of the stranded monster, but only the text allowed to be quite sure that it was not an *Octopus*, but a squid, probably missing its two long whips.

Mr Meinertzhagen also included some interesting details about the attitude of the Maori to the beast:

" All the natives turned out to see him; and the old men say it is a *taniwha* (monster) - a *wheke* (octopus) of that size never having been seen by them. They say that a `taniwha' of this description attacked and swamped a canoe on its way to *Otago*; in fact, they did not hesitate to say that this is the identical animal that did the deed! They also say that these large `whekes' are very apt to seize a man and tear his inside out. No more sea-bathing for me!"

Erudite specialists may well claim *ex cathedra* that no cephalopod has ever attacked a boat. As far as I am concerned, I will not be convinced that identical rumors stemming from Angolese Africans, British Columbian natives and fishermen from Scandinavia and the Indian Ocean, are completely without any basis, or just based on a single exceptional event, exaggerated with the passage of time. To find a common root to legends of people so distant from each other, one would have to go back to paleolithic folklore, at a time when people probably did not even go to sea.

This terrifying giant *wheke* was not an isolated incident. In New Zealand's Colonial Museum, there was the beak of another giant squid, donated by one Mr. A.Hamilton, who had received it in person from a naturalist by the name of C.H.Robson who lived at Cape Campbell. Questioned by T.W.Kirk, Robson had explained on June 19, 1879 the circumstances under which he had acquired the object.

Actually, he had already found on numerous occasions, on the shores of Cape Campbell, the more or less mutilated remains of very large squids. One of them, from which he had extracted the impressive beak as a souvenir, was intact. As far as he remembered, the body of the beast was 2.10 m [7 feet] long, its arms 2.40 m [8 feet], and its tentacles 3.65 m [12 feet]. He was sure that the total length reached 6.10 m [20 feet].

Kirk also found out about a very small specimen, 2.40 m [8 feet] long, caught by fishermen near Boulder Bank, in Nelson, on Cook Strait, as well as about another one of unspecified dimensions found by one Mr. Moore near Flat Point, on the east coast.

What had prompted Kirk to launch an enquiry on giant cephalopods was an incident in which he was personally involved, which had occured a few weeks earlier. On May 23. 1879, the most venerable Archdeacon Stock had notified our naturalist that three boys had discovered on that same morning, in Lyall Bay, the corpse of a "very large" cephalopod.

Aware of the unfortunate tendency that people have to exaggerate, even when they had no preconceived ideas on how large a cephalopod should be, Kirk rushed to the site, fully determined to bring the whole specimen back to the Colonial Museum. Imagine his surprise when he found himself with a squid with a body mass 5 metres long, certainly weighing more than a tonne.

From the tip of the tail to the edge of the mantle, the monster measured 2.80 m, and its short arms were 1.30 m long. Its tentacles, probably mutilated at half length, were at most 1.87 m long.

After Kirk had cut up the beast, he noticed that its pen was taller than a man, measuring 1.90 m .

In 1881, our kiwi naturalist was to describe the Lyall Bay squid under the name *Steenstrupia stockii,* honouring at once professor Steenstrup and Archdeacon Stock. However, this baptism was doubly unfortunate, for the animal was to turn out to be a species of *Architeuthis.* Even if it had belonged to a new genus, the name *Steenstrupia* would not have been valid, for Forbes had already had the idea of giving it to a Coelenterate in 1846. Even in the world of taxonomy, *dura lex sed lex.*

Nobody is immune from this kind of misadventure. That same year, giant squids being in fashion, the eminent British zoologist Sir Richard Owen suddenly decided to baptize the one whose 2.75 m long arm had long been pickling in a jar in the deepest cellars of the British Museum. He invented for it the name *Plectoteuthis grandis* which was also still-born. As you may have guessed, the shriveled arm had also belonged to an *Architeuthis.*

After having been ignored for hundreds of years, the super-giant squid had, over a quarter of a century, been called all sorts of names: *Architeuthis, Megaloteuthis, Dinoteuthis, Megateuthis, Steenstrupia, Plectoteuthis.*

As far as generic names are concerned, the list has not grown since. But the lavish attribution of specific names was to continue for a long time, and has not abated to this day.

In June 1880, an *Architeuthis* nearly 6 metres long, without tentacles, stranded in Island Bay, on the shores of Cook Strait, again in New Zealand. The body and the head were together 2.80 m long; the very long arms reached 3.15 m. Neither the sword nor the beak of that squid were found, probably carried away by fishermen interested in souvenirs. Nevertheless, on the basis of these remains alone, T.W.Kirk thought it appropriate to create a new species: *Architeuthis verrilli.*

A few years later, on June 30, 1886, another *Architeuthis*, much more massive and stocky, was found alive at Cape Campbell. Its body measured 3.15 m in length, its arms 2 m, and its tentacles 5.75 m. This time, the cephalopod was described by C.W.Robson, who had already reported a number of strandings in the region. Robson claimed that hardly a year passed without reports of a stranding of a giant squid during the southern winter (in the months of June and July). Most of the time however, the cephalopod was torn to pieces by sharks, sea lions or toothed whales, and its remains were of little value to a naturalist.

The live specimen received, as was only appropriate, the name of *Architeuthis kirkii.*

After this honour, Kirk found another opportunity to shine on the occasion of a stranding of a bizarre specimen on Big Beach, in Lyall Bay. A fisherman by the name of Smith had brought the beak and the pharynx to the Museum, claiming to have measured the animal and having found its total length to be nearly 19 metres!

Without wasting a moment, Kirk had rushed, excited, towards Big Beach so as to see with his own eyes this record size monster. A double disappointment awaited him on the beach. First of all, he found a cephalopod only 684 inches [17.35 m] long[6]. This of course did not mean that the fisherman had lied to him: long tentacles are very elastic and can easily shrink when drying.

Another disappointment: the mollusc was all tentacles! The longer of the two measured 15 metres. The body itself, head included, did not exceed 2.35 m; with the 2.40 m long arms, this made for a body mass 4.75 m long. This was considerable, but one might have expected larger in a squid more than 17 m long. We recall that in the Thimble Tickle giant, with a total length of 16.80 m, the body part was two and a half times longer. This shows again that in including in the length of a squid that of its elastic longer

---

[6] *In his note, Kirk quotes on the one hand a total length of 684 inches, which gives 57 feet, and on the other 55 ft 2 in, which corresponds only to 662 inches. The first measure must be the correct one, for it is that value which one obtains by adding the length of the mantle (71 inches), that of the head (22 inches) and that of the longest tentacle (591 in).*

tentacles, one does not obtain a reliable estimate of its actual size.

Because these particular features, Kirk gave the Big Beach squid the name *Architeuthis longimanus*.

It is rather astounding that, as soon as the world of science had finally accepted the reality of super-giant squids, there should have happened at nearly every stranding a naturalist who stuck a new label on the specimen examined, the bolder ones claiming the need to create a new genus.

We already know that of all the generic names proposed, only the first one has survived. It gradually became evident that the giant squids are all closely related. As far as specific names are concerned, it is quite possible, and even likely that some may be legitimate, but the least one can say is that most of these descriptions were rather hurried, for the original features of an isolated specimen may be only a matter of differences between individuals, or of some geographical variation which allowed at the best the creation of a distinct sub-species. In fact, a classification by species can only be achieved unequivocally with a large number of specimens in hand. We are still not in that position. One must recognize however that with as elusive an animal as *Architeuthis*, there was little other choice.

Should one insist that zoologists, in order to avoid the proliferation of synonyms, publish only un-named descriptions until a large enough collection of specimens has been assembled? That would clearly be absurd! First of all, even names which have been abandoned are useful reference points. After having read these pages, filled with facts, incidents, and rather unlikely adventures, names like *Loligo bouyeri*, commander Bouyer's squid; *Dinoteuthis proboscideus*, the terrible trunked squid; or *Mouchezis sancti-pauli*, the squid found at Saint-Paul island by Capt. Mouchez's expedition, have a familiar ring and evoque a whole story. Would this be the case if they had all been called *Architeuthis dux*, or *Architeuthis princeps*, for example?

A zoologist can never be absolutely up-to-date on research going on in the whole world on the animal which he is studying[7]. As we have been able to see in these pages, it sometimes takes decades before some works become widely known. In holding his advice or his conclusions, the over-scrupulous zoologist might see a less conscientious, or perhaps merely luckier colleague take advantage of his work and claim precedence and glory. It would be cruel to deprive zoologists - human beings, after all, - of one of the meager satisfactions which they draw from a trade too often obscure and unrewarded: that of immortalizing their name, at least in the eyes of a rather narrow circle of specialists.

---

[7] *The existence of an annual index, the* Zoological Record, *does not completely solve the problem. In spite of the care brought to its compilation, this index may not be absolutely complete; furthermore, unforeseen circumstances (wars, revolutions,...) often slow down its preparation or its publication. Thus, at the end of 1955, the volume devoted to publications having appeared in 1952 had not yet been published; similarly, the 1970 volume only appeared in 1974.*

THE KRAKEN AND THE COLOSSAL OCTOPUS

To return to the story of the super-giant squid, we note that in the magnificent monograph written in 1912 by professor Georg Pfeffer, of the Hamburg Museum, on Oigopsid[8]. Cephalopods, this eminent German specialist pointed out that it was still impossible at that time to distinguish with confidence the various species of *Architeuthis*. That is still the case.

Following his thorough review of the question, the most complete published to date, professor Pfeffer proposed that the various specimens of *Architeuthis* be grouped, at least temporarily, according to the regions where they stranded or were caught. Following a geographical scheme, he thus prudently distinguished between North-Atlantic, North-Pacific and Austral forms. These are very broad areas, but it must perhaps be so for world travellers like *Architeuthis*, which probably cover enormous distances, drifting with large scale currents. For such large animals to feel comfortable, they need an ocean-size aquarium.

---

[8] *D'Orbigny introduced the term Oigopsid to denote decapod cephalopods whose eye lens is not protected by a membrane and is thus directly in contact with the water. In contrast, in Myopsids, the lens is covered by a transparent membrane which is a prolongation of the orbital cartilages. This distinction separates large squids (Ommastrephes, Onychoteuthis...), which are in general Oigopsids, from the common squid (Loligo) and from cuttlefish (Sepia), which are Myopsids. This classification was greeted with considerable reluctance; however, professor Steenstrup was to confirm its relevance on the basis of sexual behaviour. Nevertheless, in modern cephalopod systematics, the cuttlefish and their close relatives, the sepiola and the spirula, are put in a separate sub-order, the Sepioida. The presence of an ocular membrane is only used by some authors to distinguish two groups within the sub-order Teuthoidea, i.e. squids in the broadest sense.*

Part Five

THE SHADOW SIDE

———

"Such things are neither absolutely impossible, nor imposed to us by faith, but belong to the category of what is possible and remarkable"

MAS'OUDI,
The Book of Golden Prairies and of Mines of Precious Stones

Chapter Twelve

FINAL UNCERTAINTIES

"How large these denizens of the deep grow, no one knows; it
is however unlikely that the largest known specimens should
have been fully developed, for they do not come to the surface
except in the exceptional circumstances which I have
described, when their enemies wrest them from their lair."

FRANK BULLEN,
The Cruise of the Cachalot

Only a few months had passed since the American press had announced the
strandings of super-giant squids in Newfoundland when the shipwrecking
"octopus" of Pierre Denys de Montfort was again heard of at the antipodes. It
had taken on a new shape and had adapted to the fashion of the day, since it
was now described as a giant squid. It seemed to have drawn a new and
enhanced strength from this metamorphosis, for now it hadn't merely plucked
a few sailors, but had actually succeeded in wrecking a whole ship.

The news arrived in Europe on July 4, 1874, via the sober *Times*, of
London, not normally given to shallow sensationalism. It referred to a tragedy
which had occured two months earlier and had been forwarded from India by
sea. Here is the text exactly as it appeared:

"The following strange story has been communicated to the Indian papers:

*'We had left Colombo on the steamer Strathowen, had rounded
Galle, and were well in the bay, with our course laid for Madras, steaming
over a calm and tranquil sea. About an hour before sunset on the 10th of May
we saw on our starboard beam and about two miles off a small schooner lying
becalmed. There was nothing in her appearance or position to excite remark,
but as we came up with her I lazily examined her with my binocular, and then
noticed between us, but nearer her, a long, low, swelling lying on the sea,
which, from its colour and shape, I took to be a bank of seaweed. As I*

watched, the mass, hitherto at rest on the quiet sea, was set in motion. It struck the schooner, which visibly reeled, and then righted. Immediately afterwards, the masts swayed sideways, and with my glass I could clearly discern the enormous mass and the hull of the schooner coalescing - I can think of no other term. Judging from their exclamations, the other gazers must have witnessed the same appearance. Almost immediately after the collision and coalescence the schooner's masts swayed towards us, lower and lower; the vessel was on her beam-ends, lay there a few seconds, and disappeared, the masts righting as she sank, and the main exhibiting a reversed ensign struggling towards its peak. A cry of horror rose from the lookers-on, and, as if by instinct, our ship's head was at once turned towards the scene, which was now marked by the forms of those battling for life - the sole survivors of the pretty little schooner which only 20 minutes before floated bravely on the smooth sea. As soon as the poor fellows were able to tell their story they astounded us with the assertion that their vessel had been submerged by a giant cuttlefish or calamary, the animal which, in a smaller form, attracts as much attention in the Brighton Aquarium as the octopus. Each narrator had his version of the story, but in the main all the narratives tallied so remarkably as to leave no doubt of the fact. As soon as he was at leisure, I prevailed upon the skipper to give me his written account of the disaster, and I have now much pleasure in sending you a copy of his narrative:

"*I was lately the skipper of the Pearl schooner, 150 tons, as tight a little craft as ever sailed the seas, with a crew of six men. We were bound from Mauritius for Rangoon in ballast to return with paddy, and had put in at Galle for water. Three days out, we fell becalmed in the bay (lat. 8°50'N, long 85°05'E). On the 10th of May, about 5 p.m., - eight bells I know had gone, - we sighted a two-masted screw on our port quarter, about five or six miles off; very soon after, as we lay motionless, a great mass rose slowly out of the sea about half-a-mile off on our larboard side, and remained spread out, as it were, and stationary; it looked like the back of a huge whale, but it sloped less, and was of a brownish colour; even at that distance it seemed much longer than our craft, and it seemed to be basking in the sun. "What's that?" I sung out to the mate. "Blest if I knows; barring its size, colour, and shape, it might be a whale" replied Tom Scott; "and it ain't the sarpent," said one of the crew. "for he's too round for that ere crittur." I went into the cabin for my rifle, and as I was preparing to fire, Bill Darling, a Newfoundlander, came on deck, and, looking at the monster, exclaimed, putting up his hand, "Have a care, master; that ere is a squid, and will capsize us if you hurt him." Smiling at the idea, I let fly and hit him, and with that he shook; there was a great ripple all round him, and he began to move. "Out with all your axes and knives," shouted Bill, "and cut at any part of him that comes aboard; look alive, and Lord help us!" Not aware of the danger, and never having seen or heard of such a monster, I gave no orders, and it was no use touching the helm or ropes to get out of the way. By this time three of the crew, Bill included, had found axes, and one a rusty cutlass, and all were looking over the ship's side at the advancing monster. We could now see a huge oblong mass moving by jerks just under the surface of the water, and an enormous train following; the oblong body was at least half the size of our vessel in length and just as thick; the wake or train might have been 100 feet [30 m] long. In the time I have taken to write this the brute struck us, and the ship quivered under the thud; in another moment, monstrous arms like trees seized the vessel and she heeled over; in another second the monster was aboard, squeezed in between the two masts, Bill screaming "Slash for your lives," but all our slashing was of no avail, for the brute, holding on by his arms, slipped his vast body overboard, and pulled the vessel down with him on her beam-ends; we were thrown into the water at once, and just as I went over I caught sight of one of the crew, either Bill or Tom Fielding, squashed up between the masts and one of these awful arms; for a few seconds our ship lay on her beam-ends, then filled and went down; another of the crew must have been sucked down, for you only picked up five; the rest you know. I can't tell who ran up the ensign.*

James Floyd, late master, schooner Pearl."

As far as I know, this story was never authenticated; it is probably too late now to hope to find traces of the wreck of such a small vessel.

Was that a hoax? Perhaps, but if so, it was certainly the work of a master, and a most well informed one at that. Everything in the story has the ring of truth, down to the smallest details. The events and the participants are described with precision and the intervention of the Newfoundlander, native of an island which was to turn out to be so rich in *Architeuthis* strandings, is particularly apt. The anatomy and the behaviour of the squid are described accurately and soberly. That the incident took place in tropical waters also speaks in favour of its authenticity, at least in the light of today's knowledge. In fact, while strandings of super-giant squids have never been recorded in the tropics, it is only in those parts that they have been found at the surface offshore. Finally, a becalmed ship would indeed be more likely to be attacked by an Architeuthis. The absence of a load would also make the schooner more likely to lean over and take water, following perhaps a sudden shift in the ballast.

Everything is marvelously consistent.

As far as I am concerned, the only suspicious aspect of the story is that it should have occured precisely at that time when the super-giant squid had figured prominently in the news. Well authenticated strandings in Newfoundland had been disclosed in America in December 1875. There was just enough time, given the means of the day, for the news to reach India and inspire some practical joker to imagine this enormous hoax.

It is perhaps not so important to determine whether this particular attack actually took place. What is really important for our purpose is to show that it *could* have occured. We will thus pay attention to the potential threat posed by Architeuthis and, by the way, also rehabilitate poor Denys de Montfort, guilty of having believed such stories. We need only let the facts speak for themselves.

Since 1880, colossal squids had stopped visiting Newfoundland and not a single specimen was to be cast ashore in the British Isles for a third of a century. The monster was in evidence only in New Zealand, on the other side of the world, in a region from which news were probably rather rare. Although there was an *Architeuthis* stranding at Lökberg, in Norway, near the Arctic Circle, in 1889, the event was not widely known, since it was only in 1946 that Hans Tambs-Lyche pointed out that all specialists had omitted it from their lists.

Thus at that time, the super-giant squid faded from scientific view for about fifteen years. That was enough for it to sink back into oblivion... and to return to the realm of legend. However, *Architeuthis* remained a familiar beast to whalers, who often picked up its remains when moribund sperm whales vomited their last meal.

One of them, Capt. John Ease, of Edgarton, had even talked about it, one day, to professor N.S.Shaler who wrote about him in 1873: "He stoutly maintains that he has seen fragments of squid, where the whales had cut them in two, exposing the cavity of the body, which was as large over as the head of a forty gallon [160 litre] cask. In one case he saw the head of a squid which he believes to have been as large as a sugar hogshead."

That was however a rather unusual revelation, which remained almost unnoticed. In spite of what one might be tempted to believe, sailors are usually completely uninterested in natural history. That is one of the dominant impressions brought back from his famous *Cruise of the Cachalot* by Frank Bullen. If this adventurous young man, as attentive in his descriptions as he was sober in his interpretations, had not been on that whaling ship, we would have missed out on a great number of fascinating revelations.

Thus, after the harpooning of one of the first sperm whales of the expedition, in 1875, his attention had been attracted by

"...*several great masses of white, semi-transparent looking substance floating about, of huge size and irregular shape. But one of these curious lumps came floating by as we lay, tugged at by several fish, and I immediately asked the mate if he could tell me what it was and where it came from. He told me that when dying, the cachalot always ejected the contents of his stomach, which were invariably composed of such masses as we saw before us; that he believed the stuff to be portions of big cuttle-fish, bitten off by the whale for the purpose of swallowing, but he wasn't sure. Anyhow, I could haul this piece alongside now, if I liked, and see. Secretly wondering at the indifference shown by this officer of forty years' whaling experience to such a wonderful fact, as appeared to be here presented, I thanked him, and, sticking the boat-hook into the lump, drew it alongside. It was at once evident that it was a massive fragment of cuttle-fish -tentacle or arm - as thick as a stout man's body, and with six or seven sucking-discs or acetabula on it. These were about as large as a saucer, and on their inner edge were thickly set with hooks or claws all round the rim, sharp as needles, and almost the shape and size of*

Figure 114 The super-giant squid ot Port Simpson, British Columbia, 1922

Figure 115. The "champion octopus" of Baven-on-Sea, Natal, 1924, was clearly a squid. After the sketch of Mr. White

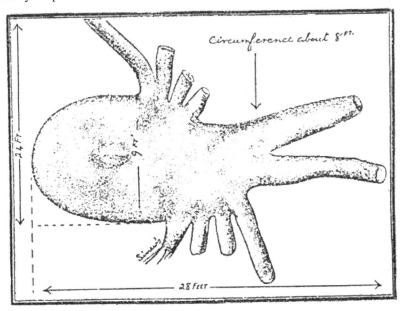

Above Figure 116. The *Architeuthis* stranded in Nigg Bay, Scotland, 1949 (Photo Aberdeen-Bon-Accord)

Below: Figure 117. The 10.5 metre long Architeuthis found in the stomach of a sperm whale harpooned off the island of Fayal, in the Azores, weighed 185 kilos. *(Photo National Institute of Oceanography)*

Above: Figure 118. An *Architeruthis* specimen stranded on a beach at Plum Island, Massachusetts, in 1980 (Photo Carla Skinder, *New England*)

Below: Figure 119. The *Architeuthis* found floating at the surface in Ranheim, Norway, on 2 Oct. 1954, was 9.24 metres long in all (Photo Dr. Erling Siversten)

Top: Figure 120. The attack of the MV Brunswick, mistaken for a sperm whale, by a super-giant squid, according to Capt, Arne Gronningsmeter

Bottom: Figure 121 Relative sizes of the main giant squids described between 1900 and 1955

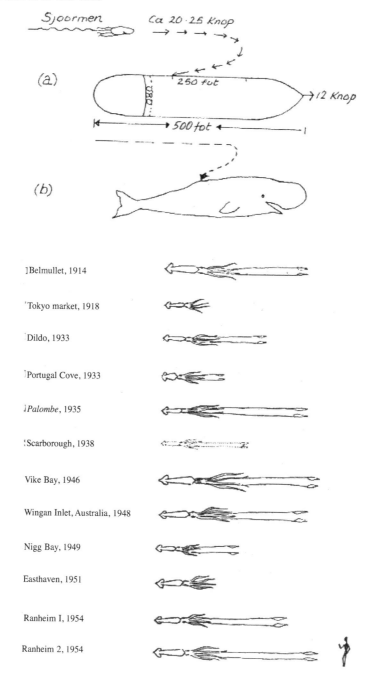

Sjoormen   Ca 20·25 Knop

(a)   250 fot   →12 Knop

500 fot

(b)

]Belmullet, 1914

'Tokyo market, 1918

Dildo, 1933

]Portugal Cove, 1933

]*Palombe*, 1935

{Scarborough, 1938

Vike Bay, 1946

Wingan Inlet, Australia, 1948

Nigg Bay, 1949

Easthaven, 1951

Ranheim I, 1954

Ranheim 2, 1954

Figure 122. Prof. Frederick A. Aldrich examining the Lance Cove *Architeuthis*, 1965 (Photo *L. Moores, Dept. Biology, M.U.N.*)

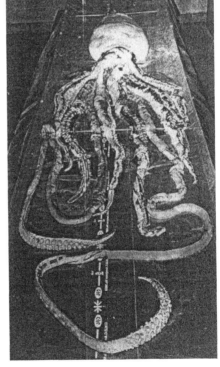

Figure 123. A 6 metre long *Architeuthis* specimen stranded on 8 Oct 1965 in Lance Cove, Trinity bay, Newfoundland (Photo *L. Moores, Dept. Biology, M.U.N.*)

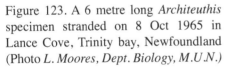

Figure 124. A researcher stradles the Plum Island *Architeuthis* to show its size.
(Photo William B. Colin, *Newburyport Daily News*)

Figure 125. Norwegian fisherman Rune Ystebö with the 10 metre *Architeuthis* specimen which he caught live near Bergen in 1982. It weighed 220 kilos. (Photo Dr. Brix)

Figure 126. Prof. Frederick Aldrich comparing a common squid with the tentacle of a super-giant squid (Photo Greg Loke - First Light)

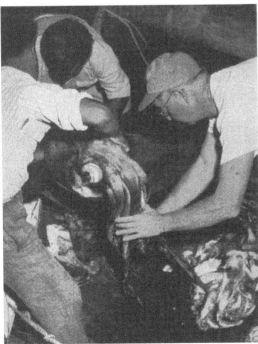

Figure 127. Capture of an *Ommastrephes* in the Humboldt Current off the western coast of South America (Photo Don Ollin-Rapho)

Figure 128. The Thimble Tickle squid, 1878, compared to a skin diver

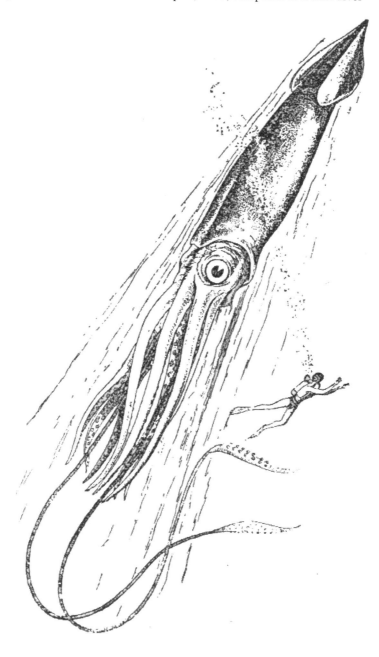

a tiger's.

"*To what manner of awful monster this portion of limb belonged, I could only faintly imagine; but of course I remembered, as any sailor would, that from my earliest sea-going I had been told that the cuttle-fish was the biggest thing in the sea, although I never even began to think it might be true until now. I asked the mate if he had ever seen such creatures as this piece belonged to alive and kicking. He answered, languidly, "Wall, I guess so; but I don't take any stock in fish, 'cept for provisions er oile - en thet's a fact.*"

As we have said, this incident occured in 1875. Judging from the thickness of the arm brought aboard, the squid from which it came must have been larger than the colossus who was to be caught at Thimble Tickle three years later. If that particular catch had not been described in detail and fully authenticated, who would have believed Frank Bullen when he published his book of memoirs in 1899? In any case, those zoologists who had refused to believe in the existence of squids of lengths of 24 and 27 m from beak to tail reported by Mr Pike in Labrador (to this day, Science does not recognize any with a body length appreciably longer than 6 m), would certainly not have believed the statements of the English writer when he wrote in his *Creatures of the Sea* that the squid which he had seen being eaten by a sperm whale was no less than 60 feet [18 m] in length, not including tentacles, and 15 to 20 feet [4.50 to 6.10 m] in circumference. Undoubtedly, sceptics will claim that, given the ferocity of the fight which he had witnessed, Bullen could not have estimated with any precision the dimensions of the combatants. But then, from what sort of giants could those gelatinous masses come that he had seen disgorged by a harpooned sperm whale on another occasion? These floated here and there near the ship and some were estimated by Bullen to be "of the size of our hatch, 8 feet [2.40 m] in length by 6 feet [1.80 m] on a side." On their thinnest plane, these fragments were already 7.20 m around and could only be fragments of some creature of an at least comparable circumference.

Frank Bullen is not the only marine author who witnessed the last moments of the sperm whale and saw it vomit remains of super-giant squids. Herman Melville, the father of the immortal *Moby Dick* - in passing, a paragon of zoological imprecision - spoke of the ejection by one of these agonizing cetaceans of squid arms more than 30 feet [9 m] long. C.W.Ashley also wrote in 1926 in *The Yankee Whaler*: "a sperm whale in his flurry will sometimes vomit pieces of squid half the size of a whaleboat."

Two American naturalist, professor Osmond P. Breland, of the University of Texas, and the writer Willy Ley, independently quote the case of a sperm whale which vomited a fragment of arm 6 feet [1.80 m] long and two feet [0.60 m] in diameter. I have unfortunately not been able to trace the source of this sensational piece of information.

How many such observations of this and other kinds, perhaps even more extraordinary, could some whalers have told us, we shall never know. The discretion of mariners seems as deep as their lack of curiosity. In Frank Bullen's words:

THE KRAKEN AND THE COLOSSAL OCTOPUS

"Sailors are, again taken as a class, the least observant of men. They will talk by the hour of trivialities about which they know nothing; they will spin interminable "cuffers" of debaucheries ashore all over the word; pick to pieces the reputation of all the officers with whom they have ever sailed; but of the glories, marvels and mysteries of the mighty deep you will hear not a word."

This is perhaps too severe or perhaps too general a judgment, as the very story of Frank Bullen himself demonstrates. However, caught between ignorant, indifferent or taciturn sailors on the one hand, and incredulous and mocking scientists on the other, it is not surprising that *Architeuthis* should have so long remained hidden.

It took a providential encounter between scientists and whalers for the giant squid to be in the news again after the epidemic of strandings of 1871-1879. It was first necessary, however, to demonstrate the link that existed between large cephalopods and some cetaceans.

Today, this link seems evident, but this has not always been so. It had long been known that some toothed whales, like the sperm whale, fed almost exclusively on cephalopods, but nobody until the end of the last century imagined that these preys could sometimes hold off their voracious predators.

During their trip on the *Astrolabe*, Quoy and Gaimard had had the opportunity of examining a New-Zealand whale (*Ziphius cavirostris*) with curious markings. "One notices on the lower jaw, they wrote, pores in the shape of small rings and, on the body, small areas of regularly curved white scratches." The two French naturalists, who during a different expedition had gathered the remains of a giant squid, did not draw a connection between the inch-wide rings and the serrated suckers of the cephalopod monster.

Naturalists had long been intrigued by the parallel streaks and the bizarre circular impressions which normally cover the body of Risso's Dolphin (*Grampus griseus*). In 1889, the great Belgian naturalist Henri van Beneden had gone as far as to compare the former to scratches, but without venturing suggestions as to the identity of the scratcher. Capt. Chaves, director of the Museum of Ponta Delgada (Azores) was the first to attribute these various wounds to cephalopods.

On the occasion of the stranding of an *Architeuthis* on the north side of San Miguel island, one of the Azores, Albert Girard, was to divulge that the eminent Portuguese naturalist had observed and photographed "a *Grampus griseus* bearing white on dark marks of the series of suckers of the tentacular arm of an *Architeuthis*, in addition to numerous parallel lines probably created by the action of the horned circlets of these suckers." On one piece of skin of that same cetacean, he had even noted dotted circles which corresponded exactly (by their diameter, the arrangement and the number of points) to the impression that would produce the horned circle of the average sucker of an *Architeuthis*.

A few years later, in 1901, Sir d'Arcy Thompson similarly interpreted the scars found on the New Zealand whale described by Quoy and

Gaimard. He supposed that the circular impressions had been made by the suckers of a clawed squid (*Onychoteuthis*) and the parallel streaks by its claws. In that kind of squid, the smaller arms are still studded with suckers; the longer tentacles are covered with hooks similar to those of other armed squids such as *Cucioteuthis* and *Enoploteuthis*.

This kind of connection may seem strikingly obvious today, but it could be made only when the existence of super-giant squids had finally been founded on sufficient concrete evidence. One might argue that the very discovery on the skin of some cetaceans of the marks made by giant suckers was in itself a strong clue of the presumed existence of gigantic cephalopods. Alas! too often in scientific matters, overwhelmed by naive positivism, one ignores common sense. If, for example, one discovers footsteps in the snow, one doesn't conclude that *someone* walked by: one witholds judgment. It's only when that *someone* is seen walking in the snow that one rushes to examine the prints and exclaims triumphantly: "These prints confirm the passage of the creature which we have seen!"

It is then no wonder that twenty years should pass between the first massive stranding of *Architeuthis* in Newfoundland and the recognition of the origin of some mysterious round scars. Armed and clawed squids had already been known for more than a century when someone figured out that they must leave terrible scars in the flesh of their natural enemies. One might believe that many scientists refuse to believe the truth when it appears too fantastic, or too horrible.

During their submarine hunts, toothed whales have to face various kinds of cephalopods, and it is not surprising that they should bear the scars of numerous fights. The body of *Grampus* sometimes looks like one giant scar, so much so that the natural colour of the skin disappears beneath the overlapping wounds. This brings us to imagine what kind of life these poor beasts have to lead to ensure their sustenance. *The Cruel Sea* is a label which fits reality better than *Pax in Maribus*. And when one thinks of the violent struggles which oppose giant squids and sperm whales, one might readily imagine that smaller, or less well armed cetaceans - perhaps even small sperm whales - might die, suffocated by their prey.

We usually speak of cetaceans and other toothed whales feeding on giant squids. Might the opposite not also be partially true?

On July 18, 1895, a chance encounter of whalers and scientists was to complete the gathering of the various pieces of our puzzle. However, chance, as always happens, was to come to the help of the scientists only because they had already done everything they could to put it on their side.

At that time, Prince Albert I of Monaco was engaged, on board his yacht the *Princesse Alice*, in one of his famous cruises of abyssal exploration which were to lead to great advances in oceanography. His research had taken him to the open Atlantic, near the Azores, when on that day he saw a number of whaling boats pursuing a group of sperm whales. For fear of scaring off the enormous beasts, he kept his ship about a mile and a half from the whalers and witnessed from afar the excitement of the hunt. When he thought that one of the cetaceans had been struck to death, he pulled closer to see the end of the drama.

The scene was atrocious. In a sea tainted red with its blood, the wounded titan moaned, panted and howled mightily. Those on board the Prince's yacht, climbing up into the rigging for a better view, were silently stupefied. Prince Albert, a man of culture and great sensitivity, was shaking with horror at the sight of this homeric butchery:

"For myself, he recalled, struck to the marrow of my bones by the unknown grandeur of this spectacle, I attentively followed its course as that of a vision likely to vanish forever; I was moved by the suffering of this giant, expressed so mightily, and which in the magnitude of all its details seemed more intense than that of less powerful beings [...] All this blood, this great mass killed seemed to me the accomplishment of some great damage, like the fall of a tree, or the wreck of a ship..."

But suddenly, the sperm whale rushes towards the yacht. A brutal fear seizes everyone. Twenty meters from the ship, mad with pain and anger, the beast dives. Will it break the vessel's keel with its broad back? Will it rip out the rudder or the propeller with a single blow of its tail? No: about ten seconds later, it surfaces again on the other side, and this time it hardly moves at all. Emboldened, the whalers come and strike it with their lances with a kind of sadistic frenzy. And the poor beast dies, while the spectators, paralysed by emotion, remain quiet and hold their breath.

"The ship and all the actors of this drama, wrote Prince Albert, were at that time floating on a patch of blood more than an hectare in area, veined with brighter streams, redder than the rest, which still spurted out of the animal and were soon absorbed by the ambient medium as clouds do when they come down from the hills to gradually merge with the fog of the valleys.

"Its enormous head was visible just off our stern, and its lower jaw, loosened by the slackening of the muscles, swung to and fro in the waves, when I saw the mouth, like a yawning cave, vomit in succession a number of cephalopods, squids or octopus, of colossal dimensions. Obviously, this was the result of the whale's last excursion to the abyss, before returning to the surface to be harpooned: a recent mouthful which had not progressed beyond its oesophagus.

"I understood the scientific value of these objects originating from intermediate depths, where there live creatures capable through the strength of their swimming of avoiding all our means of capture and whose existence is sometimes revealed in stories which are often taken as fables."

A small boat was immediately put out to collect these remains with a net; they were to prove a very rich find.

It was a great good fortune, not only for Science, but also for the whalers, that the *Princesse Alice* had been cruising in the area at the time of the capture of the sperm whale. As the animal was too heavy to be brought back ashore by the small whaling boats, Prince Albert offered to help the fishermen and to tow their prize to the nearest bay. This was done. His Grace was rewarded beyond hope for his assistance when his collaborators were allowed on the next day to examine the digestive tube of the cetacean and to gather its contents. He left us an unforgettable description of this operation, which gives a good idea of the horror of one of the more down-to-earth aspects of zoology, little known to the public:

"Already my naturalists had explored the stomach and extracted, from about 100 kilos of nearly digested matters, fragments of giant octopus in a sufficiently good condition that they could later be attributed to completely unknown species. One can guess how captivating this kind of occupation must be for those who carry it out, for they had to rummage through a purple fermenting mush, peppered with eye-balls and beaks resisting the action of gastric juices, and from which rose painful stinks. Towards the end, their own stomach, shaken, manifested externally the results of a storm gradually brewing in its contents, echoes of the event of the previous day which had put in my hands, in the last gasps of the sperm whale, a wealth of scientific treasures.

"Returning to the bay, I found the aspect of the area changed. There were more sea gulls in the air, more fish were leaping out of the water; both were fleeing the invading pestilence; only men struggled in this stinking environment; some for science, some for their own interest."

A few months later, when professor Louis Joubin was to attempt the identification of the treasures gathered in this fashion, he didn't find the olfactory situation much improved:

"Time spent in preserving liquids, such as formaldehyde and alcohol, by these previously well-washed samples could not remove the terrible smell which permeated them; I moved them many times from one recipient and anti-putrefying liquid to another, but each time that I touched them, I could be sure that people whom I approached would question me with anxious looks which left no doubt as to the nature of the smells which I carried with me. Just imagine the work of the Prince's zoologists in that nest of pestilence!"

What would a zoologist not do to collect and study such precious treasures! There were, among other things, an enormous crown of arms of a clawed squid (*Cucioteuthis unguiculata*), each with a hundred claws, as sharp and powerful as those of the largest carnivores. That strange cephalopod had formerly been known only from a few fragments of arms, a few claws and a mouth part. There were three specimens, about one metre long, of another abyssal squid, *Histioteuthis rüppeli*, whose body is covered with luminous organs, as well as two crowns of arms of an unknown related species.

There were also the remains of two man-sized squids, which, incredibly, were entirely covered with scales, like fish. In the honour of the Prince of Monaco, professor Joubin baptized these extraordinary animals *Lepidoteuthis grimaldii*, meaning Grimaldi's scaly squid. And then there was also the fragment of a body, 46 cm long, of a squid which the learned specialist first described with the greatest hesitation as *Dubioteuthis physeteris* (the sperm whale's dubious squid), but later placed in the genus *Architeuthis*. Finally, there were about fifty beaks, all of an impressive size, the largest of which were about 10 cm long. Most of them probably came from some super-giant squid of the same type.

The most surprising specimen in this rich hoard was undoubtedly the scaly squid. I wonder if an accidental stranding of one of these unusual creatures, more or less mutilated, might not be the origin of some Middle Ages and Renaissance fables about Sea Knights. Schele de Vere mentions that in 1305 one of these was captured in the Netherlands; it was entirely covered with a coat of mail. And who knows whether this curious cephalopod might not even be responsible for the fabulous story of the Sea Bishop?

It is particularly appropriate to mention here this episcopal monster, for it has often been confused by some authors with the sea monk, which is as illegitimate as it is disrespectful for his "Excellency". Pierre Belon and Guillaume Rondelet both spoke of this bizarre creature. The latter specified, regarding the portrait which he published of it:

"I saw it in the hands of Gisbert, German physician to whom it had been sent from Amsterdam, with a note affirming that this marine monster in a bishop's habit had been seen in Poland in 1531, and carried to the king of that country, showing by signs his urgent desire to be returned to the sea, wherein he jumped as soon as he was brought back to it."

While it has seemed to me rather unlikely that the sea monk could have been inspired by the remains of a giant squid, such a link does not seem to be excluded for the sea bishop. The metre which extends without break the head and the body closely resembles the end of a squid's body. The floating mantle could very well have been suggested by the fins and the folds of the skin, whose loose appearance is always striking in stranded individuals. In a pinch, the legs could be retracted tentacles, or two remaining arms of a mutilated individual, although this suggestion seems to stretch the interpretation of the available evidence somewhat.

It is possible however that the scales of the monstrous bishop could have been added by an illustrator eager to emphasize the marine nature of the animal: that was current practice at the time. Even if this naive picture does not necessarily relate to *Lepidoteuthis*, there is good chance that it owes its origin to some mutilated squid.

It is even more amusing - edifying, perhaps - to notice what the story of the sea bishop was to turn into, after some distortions, embellishments and unfaithful translations, and particularly after the introduction of a human touch, essential ingredient for the success of news items, following the eternal recipes of perfect American journalism. Here is how Rev. Father Fournier told the story in his 1643 *Hydrography*. In spite of the difference in the date of the incident, the reader will recognize to the smallest details the story told rather more succinctly by Rondelet:

FINAL UNCERTAINTIES

"In the Baltic Sea, around the coasts of Poland and Prussia, there was caught around the year 1433 a merman who had exactly the appearance of a bishop, with mitre on his head and crozier in his fist, with all the ornaments which a bishop normally wears when he celebrates Holy Mass; his vestments could even be readily lifted front and back to his knees, and he allowed people to touch him, particularly bishops from those countries, for whom he expressed his respect in gestures, understanding what people said without speaking. The king wishing to have him locked up in a tower, he communicated that this did not agree with him, and the bishops having beseeched the king to allow him to return to the sea, he thanked them in gestures [...] Having entered the sea to his navel, and having greeted the bishops and the multitude which had gathered, and given his blessing in the sign of the cross, he dived into the sea and was not seen again."

The story was finally inserted in this form by Mgr the Bishop of Sponde into the *Ecclesiastical Annals*. It is surprising that this squid did not end up being canonized.

Let us return to the sperm whale examined by the naturalists of the *Princesse Alice*; we shall find there the answer to other ancient legends.

The squids devoured by the gluttonous cetacean seemed to have presented some serious resistance to their aggressor, for its lips were scarred with enormous circular wounds, caused by serrated suckers. Their examination inspired these reflections to the Prince of Monaco:

"I immediately had the vision of the titanic struggles of which the abysses of the sea are the stage when the terrible mammal dives for its prey {...] Recalling these strange dramas took me back to an incident which occured during my expedition of 1877 on the Hirondelle. I was in the middle of the Atlantic, on my way to the Azores, when, on a clear day, we saw majestic jets of water rising above the calm horizon, and we could easily see that they were caused by the movement of a colossal creature whose head and body sometimes rose like a tower, while the whip of its tail whipped water about in a gigantic spray.

"Soon, the sea closed up on this commotion, but the area where it had taken place kept a white, milky appearance, recognizable more than eight kilometres away, and which might have been a liquid, or simply the froth of whipped water. In spite of my efforts, a contrary wind did not allow the Hirondelle, a modest sailing vessel, to reach that area before the white stain disappeared, although it long stayed visible, and when, many hours later, I reached the area where it had been, I found the freshly severed head of a giant polyp [...]

"Would it be too bold to draw a link between this incident and the new fact reported here, to suppose that in the first instance I had witnessed some particularly dramatic scene where a sperm whale smothered with giant polyps had come to the surface to free itself from their grasp?"

THE KRAKEN AND THE COLOSSAL OCTOPUS

This story of Prince Albert's shed a new light on some old-time descriptions which were difficult to understand or to even believe. The size of the frothy area, "recognizable more than eight kilometres away" accounts for the Chinese *pheg* whale which "beats 3 square sea-miles when it is aroused" . On the other hand, the intensity of the struggle is a strong testimony of the power of cephalopods, capable of defending themselves against a twenty meter long giant. Wouldn't a boat of similar dimensions facing one of those under chance conditions find itself in serious trouble?

The least we can say is that the Prince's report is food for thought. In its wake, professor Louis Joubin, a pioneer in oceanographic research, expressed the following comments, which should have encouraged his colleagues to be less incredulous:

"There are [...] in the zoological literature numerous stories about cephalopods, all seemingly less likely than the other; they have been told by navigators, ancient and modern. Usually, they are merely received with a shrug. From now on, it seems to me necessary to take a second look before rejecting these stories; the descriptions which we have read, arising from the pen of a person who is both a navigator and a naturalist, show that these great displays of nature are absolutely real, in spite of their fictional appearance, and if sometimes the navigators of old, unaware of natural history, have allowed their imagination to amplify what they saw, there nevertheless remains in their accounts some basis for a grandiose truth."

That "basis for a grandiose truth" behind the legend of the *Kraken* had by the beginning of the XXth century gained a firm place in scientific literature: *Architeuthis* had become a star of ocean science. From now on, the creature obstinately ignored for centuries was to receive at each one of its strandings the honours of an obituary in at least one serious journal. That animal which had been held as unlikely, or even absurd, whose existence had even been proven to be impossible, was soon to be mentioned in all textbooks, a sure sign of universal acceptance. That, however, in a stereotyped fashion, which is not without some danger. For thus are born, as we shall see, scientific legends sometimes as remote from the truth as are popular ones.

While human celebrities soon become prisoners of their image, of the simplified personality attributed to them by public opinion, and sometimes emphasize some of its traits into a caricature of themselves, the stars of the animal world are immune to this process. We should not forget it. Impervious to public opinion or to attempts at mythification, the super-giant squid continues its fantastic career in the depths of the ocean, making only rare and variable appearances at the surface. However, another giant squid of more restricted form and dimensions was beginning to swim through the zoological literature.

Towards the end of the last century, at the same time as a minuscule specimen, hardly 2 metres long in its body mass, was fished out of Tokyo

FINAL UNCERTAINTIES

Bay in 1895, two somewhat larger *Architeuthis* were cast ashore in 1896 in Trondheim fjord, a few months and kilometres apart, in Hevne and Kirkesaeteröra. Their total lengths were respectively 10 and 12 metres. With a body length of 5 to 6 metres, they appear to us as the largest specimens measured on the Norwegian coast in the past few centuries. So, where were the "monstrous fish" of Olaüs Magnus, supposed to be three times larger? Their existence was now to reveal itself to us only through some form of coded message.

Not always so, however. In 1973, in their report on observations of large unidentified marine animals on the coast of British Columbia, my Canadian colleagues Paul H. LeBlond and John Sibert of the Institute of Oceanography of the University of British Columbia, mentioned an interesting incident which occured in 1892. One Charles Dudoward, of Port Simpson, had written to them about the following story which he had himself heard from his grand-parents and from other old people when he was a small boy:

"In 1892, the Hudson Bay Company moved their post in that area from the Nass River to Port Simpson and employed many local Indians in the construction of the new post. One hundred and fifty men in fifty canoes were towing in a large log boom (more than 30 across) when all of a sudden their tow stopped and seemed to pull back the 50 canoes against the current. After hours of paddling, the boom moved, but pulled heavy. At last, at dusk, they managed to beach the big raft at high tide. The following morning, at low tide, they found "underneath the boom of logs a crushed octopus larger than the big raft." The arms showed on both sides of the raft. One arm reported to be more than 30 m in length, terminated in a big hook. The suckers on the tentacles were "as big as a basin plate, to saucer-size at the ends." The corpse was later towed out to sea. "

Mention of an "enormous hook" at the end of an arm suggests that the supposed "octopus" was actually a squid. The arms stretching beyond the sides of the log raft could have been the tentacles, such as the thirty metre long one mentioned; since those limbs are highly elastic, it is difficult to draw conclusions on the actual size of the animal. However, it must certainly have been a very large specimen, since the longest tentacles ever measured did not exceed 15 metres. Of greater significance is the size attributed to the suckers. It is hard to imagine a soup plate smaller than 25 cm in diameter. As we shall soon see, such a diameter would suppose a specimen in which the head and body combination would reach about 25 metres!

In 1903, it was the turn of the Norwegian steamer *Michael Sars*, devoted like the *Princesse Alice* to oceanographic investigations, to encounter, north of the Faroe Islands the floating corpse of an *Architeuthis*. Sir John Murray and professor Johan Hjort used the results obtained by the team of researchers on that ship to write a work which will long remain the Bible of Oceanography: *The Depths of the Ocean* (1912). On page 653 of that book one finds a photograph of part of the skin of a sperm whale on which one can distinguish round scars left by the serrated suckers of large squids. Some of them measure up to 2.7 cm in diameter. Because of the classical nature of that book, this latter value is always quoted, in scientific as well as in popular publications, but with unjustified emphasis, as if it were a record.

Actually, these are completely ordinary and quite modest dimensions for the suckers of *Architeuthis*. The largest ones of the swollen ends of the whips, which are always larger than those on the arms, are often larger than 2.7 cm in diameter. For example, in the series of Newfoundland super-giant squids studied by professor Verrill, it is the smallest specimens which have suckers with maximum diameters ranging from 2.3 to 2.8 cm. There is thus no need to get excited about discovering some with a diameter of 2.7 cm.

What dimensions do suckers really attain in the largest squids carefully measured?

Rev. Moses Harvey claimed that the 14 m long Bonavista Bay specimen had suckers 5.7 cm in diameter. This seemed somewhat exaggerated to professor Verrill, and perhaps rightly so. However, the Labrador squid, only slightly larger, had, according to its descriptor, suckers 5.1 cm in diameter. The giant Thimble Tickle squid, at least twice as large as that exhibited at the New York aquarium, must have had proportionately larger suckers, that is twice the size, and thus more than 5.6 cm in diameter.

Using all known data, including those published by professor Verrill, I have managed to derive a rule which expresses, to a good approximation, the relative dimensions of the largest suckers of an *Architeuthis* as a function of its length. The margin of error does not exceed 10%, which is sufficient for our purposes. This rule is expressed as follows:

"If the body, including the head, of an Architeuthis is N meters long, then the diameter of its largest suckers is about N centimetres."

It then follows that for the Thimble Tickle specimen, which measured 6.12 m from the posterior end of its body to its beak, the largest suckers should have been between 6 and 7 cm in diameter. This is more than double, almost triple, the dimensions cited in classical works.

Inversely, a measurement of a sucker, or of the circular wound which it produced, must allow determination of the minimum size of the *Architeuthis* whence it came. I specify *minimum* size, since nothing proves that a particular sucker is the largest found on the animal.

Such calculations lead to surprising results, for the dimensions of the scars discovered on the skin of sperm whales are rather alarming.

After having had the opportunity during a whaling expedition to examine 81 sperm whales (67 males and 14 females) the great British zoologist L.Harrison Matthews wrote in 1938 that:

"Nearly all male sperm whales carry scars caused by the suckers and the claws of large squids, scars caused by suckers up to 10 cm in diameter being common."

FINAL UNCERTAINTIES

If Dr. Harrison Matthews had been a malacologist rather than a mammalogist, he would have been less casual in expressing this remark. At least at first sight, it seems to establish the existence of squids whose body and head would be about 10 m long; 15 m long in adding the arms. These are the dimensions given by Olaus Magnus... The champion of Thimble Tickle is left far behind! Harrison Matthews' observations even suggest that most of the *Architeuthis* examined after strandings are inferior in size to the mean of the species, or at least smaller in size to the ordinary opponents of sperm whales.

What valid objections may one oppose to such conclusions?
One will first ask: are these 10 cm wide marks perhaps ancient scars which may have grown with time, as do, on the bark of trees, the entwined initials of aging lovers? As sperm whales double in length between birth and maturity, their scars would do the same.
This is highly improbable, for the largest scars are usually just as clear and thus just as recent-looking as the smaller ones. It is also unknown whether very young sperm whales are also similarly marked. Harrison Matthews pointed out that *Architeuthis* marks are rare among female sperm whales, which feed mostly on smaller cephalopods. Although he could not provide any similar information about juveniles, whose catching is forbidden, one could rightly suppose that this is *a fortiori* so for them: even more than their mothers do, they must fear attacking very large squids. Of course, as we shall see later, it is not always the sperm whales which attack first. However, could a baby sperm whale survive a battle with a squid capable of leaving 5 cm wide scars on its skin? In other words, can a 10 m long sperm whale hold its own against a multiple-armed monster with a central body mass of a length comparable to its own? It seems to me that the cephalopod, although lighter than its opponent, has over it a distinct advantage: that of never having to come to the surface to catch its breadth. If one finds large round scars only on adult male sperm whales, is it perhaps not because they are the only ones to come out victorious, and thus alive, from battles with such enormous squids?
A more legitimate objection is that which says that 10 cm wide scars do not necessarily come from *Architeuthis* species. That they must have been produced by squids is however indisputable, since they are the only cephalopods which have suckers with serrated edges, capable of scarring the skin of cetaceans. But could they not have been caused by cetaceans with suckers larger, with respect to their body dimension, than those of *Architeuthis*? There are indeed, within the ferocious *Ommastrephes* group, giant squids whose suckers are relatively at least twice as large. *Stenoteuthis caroli*, for example, whose head and body together measure only about 80 cm, have suckers with a diameter reaching about 2 cm.
The various species of that family rarely exceed those dimensions. To suppose the existence of Ommastrephids five times as large to explain the frequent occurence of 10 centimetre suckers would be less justifiable than to accept the existence of Architeuthids only twice as long as the largest

235

specimens known. Let us resign ourselves: that last solution, in spite of its horrible consequences, is by far the most reasonable. Although it is rather difficult to avoid, many zoologists accept it only with poor grace.

It is easy to understand them. For to accept that 10 cm scars must have been produced by squids with a corporeal length of 10 m is to step on a slippery slope. There are reports of *even larger* scars. When faced with such reports, the most incredulous will not only refuse to draw logical conclusions: they refuse to hear the evidence.

Some whalers have mentioned incredible dimensions for the round scars which mark the skin of most male sperm whales. I bring as a witness no other than the naturalist Ivan T. Sanderson, who, while documenting his book *Follow the Whale,* went through just about everything ever written about whales. Here what he points out in his history of our knowledge of marine mammals:

"The largest rings from the largest squids have a diameter of about four inches [10 cm], yet scars left by such suckers on the skin of captured sperm whales have measured over eighteen inches [45 cm] in diameter. It took several centuries for zoologists to accept the existence of the kraken, as the Norse fishermen had always called the Architeuthis, and its size still seems somewhat shocking. The mere suggestion that there might be forms almost five times greater in the depths of the oceans is regarded as so outrageous as to be altogether taboo."

One will recognize that with the most open mind and the best will in the world it is difficult, without feeling like venturing beyond reason, to imagine squids whose head-body combination would measure 45 m in length. The corporeal mass, with short arms, would then reach 60 m and with their tentacles stretched, they would reach 100 m and even perhaps 150 m! Faced with such prospects, one seeks some other explanation, in other to avoid these conclusions and wake up from this nightmare.

In his fascinating work on *The Lungfish, the Dodo and the Unicorn,* Willy Ley reports that "something looking like the mark of a sucking disk more than two feet [60 cm] in diameter" had been observed on the skin of a sperm whale. I don't know where my colleague found that information, but I have always believed that there must have been a typo and that one should read "more than two inches in diameter", i.e. more than 5 centimetres. However, numerous authors have found scars more than 10 centimetres in diameter. Must one reject *all* such reports as based on hypothetical typographical errors?

Although we now temporarily and regretfully close, for lack of sufficient information, the file on these large scars, we should not think the matter settled by any means. Especially since other clues will also militate in favour of the existence truly giant squids.

Whatever may be, novelists and sensational journalists will continue to make hay of the discovery of scars of the size of children's hoops, while on the other hand textbooks will continue to marvel at the existence of circular scars more than an inch [2.5 cm] across. Who would have thought that it would be so difficult to simply tell the truth, that small shred of truth grudgingly revealed by a world so jealous of its secrets.

Of course, the *whole* truth about creatures of the sea is bound to remain for long, and perhaps forever, inaccessible. However, since the beginning of the century, oceanographic expeditions have increased in frequency, and have shed new light on those creatures living in intermediate and deep waters. Following the successful *Plankton Expedition*, organized in the Atlantic by the Humboldt foundation, professor Georg Pfeffer of the Museum of Natural History of Hamburg finally reviewed in 1912 everything that was known at the time on Oigopsid Cephalopods, a group which includes the largest squids. The publication of his remarkable monograph denotes the apex of the ascent of colossal squids in the scientific consciousness[1]. It came a long way: only half-a-century earlier, it was barely emerging from the fog of Scandinavian folklore.

From this time on, reports of strandings of specimens of *Architeuthis*, or the discovery of anatomical debris, become a routine matter. News of that nature flock in from everywhere. In a contest for the greatest number of reports on *Architeuthis* in the XXth century, the competition would be limited to a match between Great Britain and Norway. Newfoundland, the clear winner in the previous century, is not even in the running.

Between 1910 and 1912, two strandings first occured in remote areas of the Norwegian coast, at Grötöy, north of Hinnöy, in the Lofoten Islands, and at Veilholmen, in the island of Smöla, near Trondheim.

Then more than a year went by without any such announcements. In 1914, the annual report of the Irish whaling station on the Belmullet peninsula tells us that in the previous year there was found in the stomach of a gluttonous sperm whale a nearly complete specimen of *Architeuthis*. Its body alone measured 1.80 m and had a circumference of 1.20 m. Its arms were 1.80 m long and its very long tentacles measured more than 6 metres. There was news to reassure the staunchiest defenders of Jonah's story, at least regarding the swallowing capacities of one kind of whale.

On November 21, 1915, in the vicinity of Bergen, a fisherman from Leröy, in Austreim, saw a 7 m long squid floating at the surface in a small bay. Curious, he drew closer, but as soon as he had approached the animal

---

[1] *In professor Pfeffer's remarkable work, which includes the most complete bibliography to date, one finds very few omissions. To me, however, he does not seem justified to deny professor Steenstrup's priority in the first scientific description of the super-giant squid under the pretext that his first diagnosis is based only on popular reports, and that definite work, based on the study of concrete remains, was published only in 1898, after Verrill's works. In fact, the Danish specialist had described in his articles published in 1855 and 1856 his examinations of the remains of two* Architeuthis: *that which stranded in 1853 on the beach at Aalbaek, and that fished out of the Atlantic in 1855 by captain Hygom. It would seem that these two articles, published in little known journals, escaped the usually exceptional thoroughness of the German scientist.*

closely enough, it threw one of its tentacles at him. Luckily, a nearby fisherman witnessed the scene and rushed to his help. Between the two of them they succeed, after a bitter struggle, in killing the mollusc and dragging it ashore.

In 1926, it is in Hellandsjön, near Trondheim, that a new stranding was reported.

The scene then shifts. On November 2, 1917, an *Architeuthis harveyi* was found ashore at Dunbar, on the Firth of Forth, in Scotland. Much further afield, in 1918, a small specimen, less than 3 metres long was caught in Awa province, Japan: one might believe that Japanese waters are perhaps the home of a midget race. The following year a super-giant squid of more normal size is again found in Norway, at Oyvaagen, in Donna Island. In February 1920, it is Great Britain's turn, with news of a stranding at North Uist, in the Hebrides. And in 1911, it is at the northernmost point in Scotland, at Caithness, that another super-squid ends its career.

Another change of scene in 1922. We return to Port Simpson, on the coast of British Columbia, where the informer of oceanographers LeBlond and Sibert, Mr Charles Dudoward, describes another stranding of an enormous cephalopod. It took the combined efforts of twenty men to drag it to Mrs Robertson D.Rudge's Port Simpson Hotel. The animal had eight arms, each 15 m long, and one central tentacle exceeding 30 m; it probably had lost the other one in some battle.

Two years later, there was another spectacular stranding which seems to have escaped the attention of most people interested in this topic. A South African farmer, Mr. W.White, of Fascadale, informed the *Natal Mercury*, in Durban, that on October 25, 1924, he and his friend Frank Strachan had gone to examine a sea monster which, according to local Blacks was stranded on the beach at Baven-on-Sea, near Margate. Having seen the monster, they concluded that it was "a champion octopus".

Their diagnosis is clearly wrong, for on the sketch attached to Mr White's letter one recognizes a mutilated animal where there are still ten stumps clearly visible. It is thus a Decapod and, from its enormous size, probably an *Architeuthis*.

Mr White's error takes nothing away from the interest of his discovery or from his obvious good faith. In his open letter to the Durban newspaper, he invited all interested persons to come and see for themselves the strange flotsam and gave indications on how to reach Baven-on-Sea most effectively.

"I have taken careful measurements in the presence of Mr. Strachan and these are given on the sketch. You will notice the shortness of the tentacles. These would appear to have been bitten off by sharks. The object is lying on its back high and dry when the tide is out, and so far appears quite fresh (no smell). The body is very hard, so much so that Natives have been trying to cut it up, but their large knifes have had no effect. The whole body is white in colour."

FINAL UNCERTAINTIES

From the sketch and the measurements indicated the animal must have been of a really exceptional size, larger that all other *Architeuthis* ever measured, larger even that Olaus Magnus's "monstrous fish"

"I would say, added Mr.White, from the thickness of the two front feelers, that the overall measurement would have been something like 50 ft [15m] had they not been eaten off. You will notice that the tentacle on the left is quite long, and I would say that the body to the tip would have measured 22 ft [6.70 m] or more. The body is 9 ft [2.75 m] in thickness."

Let us recall, for the sake of comparison, that in the Mödrevalle specimen "which a grown man could barely span with his outstretched arms", the body must have had a thickness of about 1.70 m, while the Thimble Tickle squid must have been about 1.30 m thick. If the Baven-on-Sea monster was really an *Architeuthis*, it must have measured, using similar proportions, about 35 m from the tip of its tail to the end of its short arms. That is twice as large as the Thimble Tickle champion! This estimate is confirmed by the thickness of the arms, which, from the sketch, appeared to be about 40 cm thick at their root.

Mr White's extrapolation is thus much too timid, but is readily explained because he thought he was dealing with an octopus, a much less elongated cephalopod than the squid.

Of course, we cannot be absolutely sure that this animal was indeed an *Architeuthis*. The actual shape of the animal, harshly mutilated, probably following an encounter with a sperm whale, is unrecognizable. If it was indeed an *Architeuthis*, all that was left of it was its head.

It is a pity that Messrs White and Strachan could not have turned the wreck over to measure its eyes and its beak and even to preserve the latter for the edification of skeptics. However, on second thought, how could they have lifted that enormous mass, probably weighing many tons?

Following that memorable, though forgotten, incident, two much more modest strandings occured in Norway: one in Hardanger Fjord in 1927, and another in Trondheim fjord in 1928.

At that point, Norway is ahead 7 to 4, but Great Britain will soon catch up. On January 15, 1933, another *Architeuthis* appeared, this time on the east coast of England, near Scarborough, in Yorkshire.

As the specimen stranded in England was somewhat different from all others hitherto examined, G.C.Robson described it under a new specific name, *Architeuthis clarkei,* in honour of the amateur naturalist W.J.Clarke, who had first signaled its presence. It is on that occasion that the specialist from the British Museum offered his explanation of the series of strandings in Newfoundland through their rapid passage from a warm current, the Gulf Stream, to the glacial waters of the Labrador current.

He added that the broader distribution of stranding points on northeast European beaches was "in agreement with that suggestion". In those regions, the transition between the Gulf Stream and the colder northern currents is much more gradual. This would explain that strandings could have occured over a much wider area, including Iceland, Norway, Jutland, Shetland, the Hebrides, Ireland, Scotland and even the east coast of England.

A glance at a chart of the general circulation of surface currents seems to support professor Robson's hypothesis. Where are those oceanic regions where currents of different temperatures converge most sharply? Besides the Grand Banks of Newfoundland and the northwest of Europe, which we have already mentioned, there is the coast of Japan, where the cold Oyashio meets the warm Kuroshio; Tasmania and New Zealand, when the circumpolar current deflects the warm East Australian current; Saint Paul's Island, on the northern edge of the circumpolar current; the subtropical waters of the Canary Islands, near the cold upwelling zone of west Africa; and, finally, the west coast of South America, where the Humboldt current carries cold water towards the equator.

The existence of these critical zones explains the strandings of super-giant squids described by Dr Hilgendorf and a number of Japanese authors in Japan, by Mr. Kirk in New-Zealand, and by Mr. Vélain in Saint Paul's Island. It also helps us understand the appearance at the surface of some of these cephalopods in the Atlantic, near Teneriffe,(Capt. Bouyer) and in the Pacific, off Tasmania (Péron).

In my view, the location of New Zealand strandings is particularly characteristic. Cook Strait, which separates the North and South Islands, is an area of very strong temperature changes associated with large-scale conditions and local stirring of surface and deep waters. It is precisely in the immediate vicinity of that strait that all *Architeuthis* specimens were found.

It is however difficult to believe that a mere change in "climate" should be capable of incapacitating such strong swimmers to the point that they are unable to return to more clement oceanic areas. The geographical distribution of stranding points and of areas of discovery of moribund individuals at the surface may well be related to the convergence of currents of strongly differing temperatures, but the specific manner in which these conditions affect the health of colossal cephalopods must be much more indirect than imagined by professor Robson.

First of all, it is likely that, if Architeuthis usually inhabits a layer of a particular temperature, it is not because they like its lukewarm comfort, but rather because that layer is also the home of its favorite preys. Feeding, for all very large marine animals, raises the same problems. In order to satisfy their enormous appetite, they have to cruise in waters where their prey is highly concentrated.

We have gradually discovered that the oceans have a stratified biological structure. Looking for visual confirmation of the nature of the deep scattering layers, by diving with a bathyscaphe, scientists have found themselves plunged, in Cousteau's words, in a veritable "living soup."

It is of course evident that the phytoplankton, the microscopic plants such as diatoms and flagellates which float in the water column, can only live within favourable conditions of temperature, salinity and light intensity, and is restricted to narrow limits: the photic zone. The animals of all kinds which graze on this floating vegetation, mainly minuscule crustaceans, but also worms, pteropod molluscs with wings like butterflies, small medusas, protozooans and many pelagic larvae, are the animal, or zooplankton, associated by necessity with the phytoplankton. Many fish in turn: herring, sardines, capelin, mackerel, feed on plankton and gather within it in large schools. To feed, giants such as whales, whale-sharks, basking sharks, and manta rays have only to swim with their mouth open through this living bouillabaisse.

Depending on the kind of the phytoplankton at the bottom of the food chain, one finds the rest of the organisms which depend on it at various depths. This might explain the layering of the various scattering layers.

One may presume that the fate of *Architeuthis* is linked to that layer, at a temperature of 10°C or so, where its favorite food is concentrated. If for some reason, a giant squid leaves that rich hunting ground, it soon finds itself in a dramatic situation, for lack of a sufficient food supply. One can imagine also that where the sudden irruption of a cold current disperses its normal food supply, diluting the "living soup", poor *Architeuthis* is faced with imminent starvation. Another kind of danger may also threaten it.

If large animals are rarely struck dead by a sudden change in temperature, this is not the case for the microscopic plants which make up the phytoplankton. Killed by too rapid a temperature change or by different salinities, they not only rot away, consuming some of the water's dissolved oxygen, but they suddenly disappear from the food supply of all those organisms which preyed upon them. These die, and in turn provoque the death of the predators which feasted on their flesh. Once the food chain is broken, a mass hecatomb takes place. Myriads of corpses sink towards the bottom, decomposing in turn and utilizing the water's dissolved oxygen. Sulfate-reducing bacteria enter in action and multiply, and produce sulfur dioxide, a toxic gas. Anyone who has smelled a rotten egg is familiar with its smell. This murderous gas soon poisons the last survivors, be they as large and strong as cetaceans.

Of course, this phenomenon is most disastrous under circumstances - weak currents, stagnant semi-enclosed basins - favourable to prolonged water starvation. Some areas of the oceans are the scene, at regular intervals, of mass strandings of mysteriously killed marine organisms. Thus, nearly each year, between December and January, the sea casts tons of dead fish on the beaches of Walvis Bay, on the Atlantic coast of South Africa.

Strandings of giant or super-giant squids are part of the enigmatic and complex problem of massive mortality of marine creatures. This problem is far from being satisfactorily solved. We know that in some specific cases the epidemic of mortality coincides with the unusual proliferation of some unicellular plants, of the group dinoflagellates. This is the case for the red tide which occasionally occurs on the coast of Mexico: it is estimated that in 1946-47 it killed half a billion fish! This disastrous phenomenon is caused by blooms of a species of dinoflagellates, *Gymnodinium brevis*, which gives the water a red colour. Another dinoflagellate, of the genus *Gonyaulax*, is responsible for mussel contamination, making them likely to cause paralytic shellfish poisoning in people.

Peridinians are organisms which require very little oxygen; as soon as its concentration becomes very low, they proliferate. As they secrete a dangerous toxin, ten times more potent than strychnine - probably a nerve poison - the water itself becomes contaminated. At the source, the phenomenon is probably linked to abnormal variations in temperature and salinity which cause the death of part of the phytoplankton and thus an unusual oxygen demand. It is thus the consequence of perturbations in the pattern of ocean currents, which seems to occur in preference at some times of the year, perhaps due to winds or rainfall.

Nearly a year after the stranding of a specimen in England, *Architeuthis* was again seen in Newfoundland. At the end of December 1933 a small individual was caught in Dildo, in Trinity Bay. It body mass was nearly three meters long; its total length was only six meters. It weighed about a quarter of a ton.

At the same time, the *Evening Telegram* of December 21, 1933, announced the stranding at Portugal Cove of a small individual, with overall length of 3.40 m. In the same issue of the newspaper, old sea wolf Capt. A. Kean, interviewed on the matter, said that he had found at Flowers Cove, more than half a century earlier, a dying squid 72 feet [22 m] long. Needless to say, that declaration, plausible though it may have been, was never taken into serious account.

In the summer of 1935, it was off France, at about the latitude of the Ile d'Yeu, that an *Architeuthis harveyi* was caught in a trawl. Thanks to Mr. J.Cadenat, a technician at the Scientific Office of Maritime Fisheries (La Rochelle Laboratory), we have an account of the circumstances of the capture:

"On July 1, 1935, the La Rochelle trawler Palombe (Capt. Le Bescou) of the fleet of F.Castaing, brought back to the company's warehouse a giant decapod cephalopod. It was more than eight meters long from the posterior end of its body to the tip of its extended long tentacles. It had been caught with an otter trawl, on June 8, 1935, in the Bay of Biscay, by 46°50' North latitude at a depth of about 200 meters. The weather was not particularly good, but nevertheless calm enough to allow normal trawling; this calm period followed a rather long stormy period.

" No one among the trawler's crew, not even the old sea hands who had fished on the banks of Newfoundland, where the presence of giant squids has been mentioned [...] had ever found themselves in the presence of such a large squid. Unfortunately, the arms were cut up and the eyes completely torn off by the crew members. Moreover, before deciding to bring it back to La Rochelle, the body was left exposed to the air on deck for nearly forty eight hours, which accounts for its poor state of preservation when it reached us for examination."

The total length of the animal was 8.18 m, of which the long tentacles made up 6.45 m and the head-body combination only 1.73 m. The longest arm was 1.72 m long. Although the animal seemed enormous to the old sea hands, it was actually rather small for an *Architeuthis*.

In a preliminary note, published in the Proceedings of the French Association for the Advancement of the Sciences, Mr. Cadenat had given a length of 6.15 m for the length of the prehensile arms, and dated that miraculous catch on June 26. Some Arthur Mangin of our times could very well have used these minor discrepancies to claim that the whole thing was the result of the fertile imagination of La Rochelle fishermen and of a gullible commentator. However, since the beginning of the century, such incredulity was no longer fashionable and the reality of the monster was accepted by all without the least contest. When monsters appear often enough, they are no longer monstrous.

Nevertheless, following a rather humorous incident, the capture of this super-giant squid in the Bay of Biscay gave birth to an entirely mythical creature: *Architeuthis nawaji*.

That was the name under which the cephalopod brought ashore by the crew of the *Palombe* was described in Mr. Cadenat's preliminary note. One would in vain dig through the world's libraries to discover the naturalist who described this unknown species, whose name is reminiscent of the Far West. Is it the favorite squid of an Indian tribe? The truth is rather more pedestrian: *nawaji* is obviously an interpretation by a proof-reader unversed in malacology of the word *harveyi*, probably illegibly written by the author of the communication.

So, either because of the myopia of a type-setter, or of the poor calligraphy of a careless correspondent, a mysterious squid has been swimming since 1935 through the zoological literature. I don't believe that there has been so far any effort to kill it once and for all. Many monsters of old were born from such simple mistakes.

In his more definitive study, Mr.Cadenat, after having reviewed previous strandings and captures of *Architeuthis*, pointed out that they always occured in rather cold waters. In his list, he omits the specimens from Norway, Ireland, Scotland, the Shetlands, the Hebrides, England, Japan, Saint Paul Island, and New Zealand, but his conclusions nevertheless remain valid. No super-giant squid has ever been stranded or caught equatorward of 30°North or South.

Mr Cadenat asked himself whether there was some link between that distribution and the fact that the temperature, that year, was particularly low in the Bay of Biscay. His query is not without pertinence: it is possible that in order to catch a giant squid off the coast of Vendée in the summer the water had to be colder than usual. However, at a depth of 200 m, it is more likely that the colossal squid would have found its favorite temperature, even in normal times. I believe that one should rather see in the capture of an Architeuthis by the *Palombe* an example of a lucky catch.

After this miraculous catch, the Norway-Great Britain competition resumes in earnest. The Scandinavians are still in the lead, 7 (*Architeuthis*) to 5, but not for long. On November 7, 1937, a giant squid is caught off Bell Rock (Angus), in eastern Scotland. In October 1938, another individual, nearly 5 metres long, stranded on a beach near Scarborough, in England, about a mile north of Ravenscar. A fisherman by the name of Shipley, who discovered it, declared that its arms were as thick as those of a man. Of course, he cut up the whole mollusc into cod bait, but the beak was preserved. A lucky stroke.

After these two events, the score was tied. Norway would again take the lead on the following year, following a stranding in the Lofoten Islands. The Second World War was declared, raged and ended without any futher news of giant squids. In 1946, Norway scored again in this polite contest with Great Britain, when two *Architeuthis* were found, one at Lyngholmen, in Sunnhordland, the other in the bay of Vike. Both of these specimens had bodies shorter than two meters long; with their tentacles stretched out, the second reached 9.35 m overall.

In September 1948, there was a diversion: a giant squid with a total length of 8.50 m was found ashore in Wingan Inlet (Victoria). It was the first time that a giant squid had been seen in Australia proper.

A year later, Norway took a comfortable lead, thanks to a stranding which took place on an island near the mouth of Hardanger fjord. But at the same time, Great-Britain made a spectacular recovery. On October 2, 1949, an *Architeuthis* was found for the second time on a Shetland beach, at the point of Whalefirth Voe. One hardly need to mention that the animal was immediately cut up for bait by the local fishermen. However, its beak, collected by Dr. Peterson, from Camb, was offered to the Royal Scottish Museum, together with a description and a coarse sketch of the animal.

On November 30 of the same year, another squid of the same type stranded one morning near a salmon fishermen's shack in Nigg Bay (Aberdeen county) in Scotland. When the fishermen first saw it, it was still alive and convulsively clinging to the rocks of the shoreline. The same afternoon, it was found dead, although it had apparently not suffered any damage. Its total length was only 5.87 m, of which 1.45 m for the head and body; it was, however, in perfect condition, and -O miracle! -this time the fishermen immediately notified the Marine Laboratory in Aberdeen. This allowed Mr. Bennet B.Rae to publish excellent photographs and to study this cephalopod with great care. In spite of its small size, it weighed more than 100 kilos. Finally, on December 14, 1951, the already mutilated corpse of an *Architeuthis harveyi* was thrown onto the rocks east of Easthaven (Angus) also in Scotland. As it was in a rather inaccessible area, it was not discovered for some time, and when reached was in a sorry state. Its head and body could still be measured, together about 2 meters long, and its jaw, its rough tongue and a piece of one of its arms, with a few suckers, were preciously collected.

As we have noticed from these short accounts, even today, in spite of the speed of means of communication and transport, the stranding of marine monsters, even in the most civilized areas of the world, does not always offer scientists the opportunity for a careful examination. Sometimes no material proof remains of such incidents. One can be almost assured that when they occur in more savage regions there is little chance to collect about them more than vague rumours or stories loaded with exaggerated or fabulous details. Lucky if there is even someone on the spot to witness the event; all

Figure 129. Navy hard-hat diver Ledu, attacked by a giant octopus in the Missiéssy dock, Toulon (after *le Petit Journal Illustré*, 15 juin 1912)

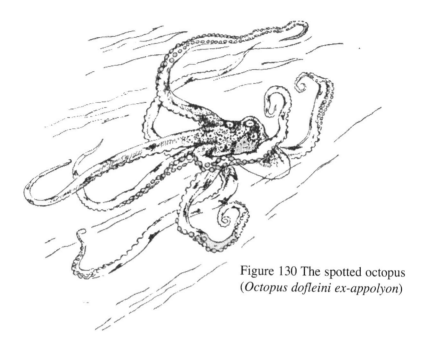

Figure 130 The spotted octopus
(*Octopus dofleini ex-appolyon*)

Figure 131. An octopus with eyes as large as saucers, attacking a Japanese fisherman, after a frawng bu Hokusai (1760-1849)

Figure 132. Diver struggling against a spotted octopus in PUget Sound, Washington; the animal has torn off his mask (Photo Gene Daniels-Rapho)

Figure 133. A 8 kilo spotted octopus out of the water (Photo Gene Daniels-Ropho)

Figure 134. Giant octopus attacking a Japanese fishing boat and the subsequent sale of its tentacles at the market, drawn from a book by Ki Konc, Products of the land ad the sea (from Lat Nature, mar 1874)

ABOVE: Figure 135. Two fishermen struggling against a giant octopus

BELOW: Figure 136. Fight between a lone fisherman and a giant octopus

Figure 137 Dr. De Witt Web (1840-1927), who revealed to science the colossal octopus stranded on Anastatia Island, Florida, in December 1896. (Archives of *Saint-Augustine Historical Society*)

Figure 138. Gary S. Mangiacopra, who pieced together the story of the Octopus giganteus of Anastasia Island, Florida

Figure 139. This spotted octopus has a span of 4.8 metres; some have been measured with twice the span (Photo Gene Daniels-Rapho)

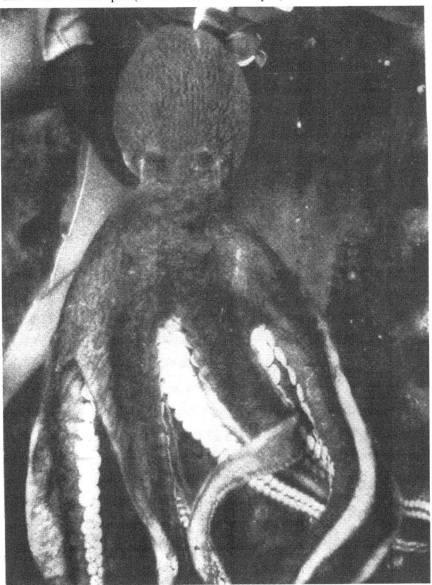

Figure 140. The Anastasia Island carcass, showing the lateral extemsions of the body (drawing by Alpheus Hyatt Verrill after a photograph taken on 7 Dec 1896)

Figure 141. Moving the 5-ton carcass required a major effort. Dr. DeWitt may be seen directing operations (Photo taken during the second week of January 1897)

the shores of the world are not as densely settled as are those of the North Atlantic. It is quite likely that it is not only because of their privileged geographical position that most *Architeuthis* strandings have been recorded in countries like Newfoundland, Norway and Great Britain.

On this point, one might have thought a few years ago that the British Isles would end up with as many records of *Architeuthis* on their coast as Norway. However, the latter is still in the lead.

In 1953, there was yet another stranding in the area of Hardanger Fjord. In 1954, two *Architeuthis* were discovered one after the other in Trondheim Fjord, near Ranheim, exactly where one of their congeners had been found a quarter of a century earlier.

The first of this pair of monsters appeared on July 30 of that year to astounded fisherman Fritz Lauritsen as he was picking up his salmon nets at the end of the day. The fearsome mollusc, 7.50 m long overall, of which the body mass made up 3.20 m, was dead, tangled in his net. Perhaps he had just succumbed to the polluted effluent of the Ranheim paper mill. Alas, that squid was already severely damaged when scientists were notified of its capture.

It's on October 2, of the same year that the second squid was seen in the same area, floating at the water surface. Because of the proximity of the paper mill, it was first supposed to be a large piece of drifting brown paper. However, when it stranded on the shingles, mill quartermaster Alvin Anderson recognized it for what it was. However, it was four young boys who took the initiative of calling the Trondheim Museum. As it was evening, they could not come on the same day to take delivery of this ambassador from the Kingdom of Neptune. However, on the next, Dr. Erling Sivertsen and his assistants could verify with satisfaction that the animal was perfectly preserved. It was a rather nice specimen, with a total length of 9.24 m: 2.14 m for the body, head included, 1.90 m for the longest arms, and 7.10 m for the longest of the two tentacles.

The announcement of this stranding in the newspapers caused quite a sensation. Such a crowd showed up at the museum that the administration decided to set up a special exhibition to allow the people to contemplate the marine marvel. In one of the rooms open to the public, the squid, first injected with formaldehyde to saturation, was laid out on a large canvas. A strong fan was used to disperse the objectionable vapours.

For two weeks, astounded visitors filed past the monster, which finally attracted more than 10,000 people. Each day, the curators of the natural history section had to work in shifts until 8 pm to provide the public with the necessary explanations.

What of course was of great concern to most of the visitors was the repeated appearance of the monsters off Ranheim beach, where the local population usually went swimming. Although the commentators pointed out that the *Architeuthis* usually reach the beach in a moribund state, it is likely that, from that day, many of the visitors gave up swimming in the sea.

Here is the most surprising of all these events: it is neither on the beach nor on the rocks, but in the stomach of a sperm whale, that the largest

THE KRAKEN AND THE COLOSSAL OCTOPUS

Architeuthis of this century have been collected. Rees and Maul have reported that on June 12, 1952, one of these large cetaceans, harpooned near Sao Lourenço, in Madeira, vomited a giant squid 10.35 m long, weighing 150 kilos. It had been swallowed whole, and still gave signs of life!

This record of gluttony was quickly broken. On July 4, 1955, Robert Clarke witnessed, on the island of Fayal, in the Azores, the flensing of a 14 metre sperm whale. In its stomach, there was found a squid nearly as long: 10.50 m in overall length. A 185 kilo mouthful. That beats Gargantua!

One striking feature of the group of super-giant squids collected over the world since the beginning of this century is their relatively small size. Except for the poorly identified specimen from Natal, and those found within sperm whales, none reaches an overall length of 10 metres. A body mass of length over 5 m is exceptional: the average is closer to 3.50 m. The length of the body alone is only 1.5 to 2 m. If we stick to properly measured specimens, there is no comparison with the famous Newfoundland crop, where specimens shorter than 10 metres in overall length were the exception. The recent crop is a far cry from those titans, whose body mass often exceeded seven meters, or of the champions like that of Mödrevalle and Thimble Tickle, the latter reaching near 12 m.

While today the ex-*Kraken* has become a rather ordinary animal, whose existence is no longer in doubt, the relatively small size of the specimens collected in the past half-century nevertheless leads naturalists to underestimate its size. Already one is talking of the giant squids of Newfoundland as exceptional, nearly fabulous phenomena. Some have gone as far as to suggest that they were measured with "elastic rulers". However, a number of clues suggest that they are not true champions. Treatises and textbooks may well state that "*Architeuthis* may reach a length of 17 m", nothing allows one to suggest that this is an upper limit.

Let's not even bring up the "monstrous fish", more than 15 m in length, evoked by Olaüs Magnus, whose authenticity is debatable, or the 18 m squid (without tentacles) which Frank Bullen saw struggling with the sperm whale, or those even more gigantic individuals supposed to have been measured on the Labrador coast. Let us grant that the description of such monsters may have been affected by myopia or exaggeration. Let us even forget for the moment the serious clue provided by the presence of very large circular scars on the skin of sperm whales. There still remain more convincing pieces of evidence: to which titanic monsters belong those fragment of enormous arms collected by whalers from the jaws of agonizing sperm whales?

We recall that Dr Schwediawer spoke of an arm 8.20 m long, "as thick as a ship's mast"; Capt. Benjohnson, of a fragment of arm 10.65 m long and as wide as a mast at its root; Capt. Reynolds, of an arm 13.70 m long and 75 cm in diameter; Capt. Dens, of the end of an arm, 7.6o m long and "as thick at its base as a foreyard"; Capt. Andersen also of an arm 7.60 m long, "so thick that it could hardly be embraced"; Capt. E.E.Smith, of portions of squids' arms "as thick as a meat barrel", with suckers as wide as plates; Frank Bullen, of a segment of an arm as thick as a stout man; Herman

Melville, of arms more than 9 m long; Osmond Breland and Willy Ley of an piece of arm more than 60 cm in diameter...

Must all these various testimonies and reports be rejected out of hand? That would be absurd. There is no reason to believe them less than the Newfoundland fishermen and pastors on whose word rely most of Verrill's descriptions. Let's not forget that out of nineteen specimens of *Architeuthis* collected in Newfoundland and Labrador, there remained not the smallest debris from twelve of them. For those, one had to rely entirely on the word of those who saw them.

Since that is the case, one should in all fairness grant equal interest to the above testimonies, especially since they all support each other.

Let's specify immediately that in most of the cases described, one is not talking about tentacles, whose length is known often to exceed 10 metres. That of the Big Beach, New Zealand, specimen reached 15 m. Those whips are of uniform thickness along their whole length (except for their easily recognized paddle shaped end) and, even in the largest specimens studies, their thickness did not exceed that of a man's arm. The above descriptions clearly almost always refer to a short, or sessile arm.

On the basis of the dimensions of these arms, carefully measured on a number of specimens, we can calculate by extrapolation how big they would have been in the Thimble Tickle champion, whose head and body together measured 6 metres. If, as supposed professor Verrill, it was an *Architeuthis princeps*, it would have had arms 6 m long and 25 to 30 cm thick at their root. Might it not be rather imprudent to state that none longer have ever been found?

If we suppose, quite arbitrarily, that the enormous arms vomited by sperm whales are always those of the genus *Architeuthis*, what dimensions should we ascribe to those from which come the longest fragments described?

In *A.harveyi*, the most stocky species, the arms are always at at most as long as the body, while in *A. princeps*, the more slender species, they are not much longer than the sum of the body plus the head. So, if to find the length of the body mass of an *Architeuthis* one multiplies the length of one of its arms by two, one is certain of obtaining an underestimate.

Thus, the 8 m long arm mentioned by Dr Schwediawer must have belonged to a squid at least 16 metres long, without including tentacles.

That collected by captain Benjohnson most probably came from a specimen of dimensions twice as large as the titan of Thimble Tickle: more than 21 m, just for the length of its body mass.

The gigantic arm fished out by the crew of Captain Reynolds, and mistaken for a sea-serpent, was three times longer than those of the Three Arm specimen; it logically must have belonged to an individual at least three times as large: a squid 27 or 28 metres long, without tentacles.

The severed arm found by the crew of Captain Dens could not have come from a squid smaller than 11 metres, again without the whips. That is also the case for that of an identical size which Captain Andersen could barely embrace and which was thus at least 50 cm in diameter. From its thickness, that limb must have belonged to a specimen at least 20 metres long.

THE KRAKEN AND THE COLOSSAL OCTOPUS

To judge from the size of its suckers, probably three times as large as those of the Thimble Tickle squid, the arm "as large as a meat barrel" examined by Captain Smith would have been that of a 36 m long monster.

Frank Bullen told us that the arm whose capture he describes in his *Cruise of the Cachalot* was of the thickness of a stout man's body. That is of course rather vague, but it suggests that it belonged to a specimen at least as large as that of Thimble Tickle.

The "arms" of which Melville speaks of, more than 9 m in length, could of course have been tentacles.

As for the fragment mentioned by Breland and Ley, its thickness, at least twice that of the arms of the official champion which we are using as a reference point, suggests that it must have belonged to an individual at least 24 m long, again without tentacles.

And let's not forget the mutilated specimen of Baven-on-Sea (Natal), whose arms had a diameter of 40 cm, and which must have measured 35 m without tentacles.

If all those remains are really those of super-giant squids, there would thus exist evidence for individuals 15 m, 16 m, 20 m, 21 m, 24 m, 28 m, 35 m, 36 m and maybe more, without whips. As the latter, when stretched, always extend beyond the arms by at least as much as the length of the body mass, there would thus exist squids of a total length of 30 m, 32 m, 40 m, 42 m, 48 m, 56 m, 70 m and 72 m! And that's not all. Among females, the whips are much longer (they stretched beyond the arms by twice the length of the body mass in the Coombs' Cove *Architeuthis*.) One would then have to assume the existence of even longer individuals, perhaps even reaching 100 m!

Even if one takes into account the natural propensity that people have to exaggerate, and takes away 50% from all the reported dimensions, we are still well above the official record size of 17 metres.

We are thus led to ask ourselves, once again, by which right the testimony of " a respectable person, by the name of Pike" was rejected, when he claimed having measured in Labrador a stranded squid, 24 m long from tail to beak? Similarly, why would the report of Captain Kean have been rejected *a priori*, when he found at Flowers Cove (Newfoundland) a specimen of a total length of 22 metres, and also the statement of Mr Haddon, school inspector for Labrador, who claimed to have measured a specimen 27 m long.

The round-number dimensions (80 and 90 feet) given by Messrs Pike and Haddon were obviously estimated by eye, and probably exaggerated. Even if we cut them in half, we still have monsters two or three times as large as the Thimble Tickle champion, monsters such as those to whom belonged the arms examined by Captains Reynolds and Smith, or the enormous lumps of gelatinous flesh vomited by a sperm whale under Frank Bullen's very eyes.

This is the time to recall that, according to the study of round scars left by squid suckers on the skin of sperm whales, individuals twice as long as that of Thimble Tickle would not be that rare. And if the dimensions put forward for some of these marks are correct, the wildest claims would be justified.

FINAL UNCERTAINTIES

Among the various facts reported on the existence of the kraken which are stranger than fiction, there are certainly some which are exaggerated; even some which are false. But there is no way that one could reject this entire body of converging proof of visual observations, large sucker prints and gigantic arms.

Anyway, the existence of squids reaching a body mass length of twenty meters can no longer be denied in all good faith. Wouldn't then such titans have been capable of the prowesses attributed to the "colossal octopus" by Pierre Denys de Montfort? Could they not have tipped the *Pearl*, which displaced 150 tonnes.

To answer these questions, we must first of all agree on a maximum weight an *Architeuthis* squid can reach.

Let's first recall that the density of living creatures is only slightly higher than that of water, so that we can as a good approximation consider that a cubic decimetre of living flesh weighs about as much as a litre of water, i.e. one kilo, and a cubic metre about one tonne. The approximate weight of a squid is rather easy to calculate since, except for its long whips, the animal is in the shape of a spindle whose length and circumference are usually given.

There is no problem then in checking that the *Alecton* squid, 6 to 7 metres long, must have weighed about three tonnes, as had guessed the sailors who had tried to pull it out of the water. As for the Thimble Tickle "champion", 11 to 12 metres in length, and thus of twice the size, we can estimate its weight as eight times (two cubed) as much. It must thus probably have weighed near 24 tonnes. According to the more precise calculations of the American specialists G.E. and Nettie MacGinitie, it would have weighed 29.25 long tons without its tentacles and 30 (30,450 kilos) in all. Our estimate is thus rather conservative[2].

Frank Lane, who wrote an excellent popular book on cephalopods, *Kingdom of the Octopus*, cast in doubt on the validity of all these calculations under the pretext, among others, that "of all giant squids which had been measured, none of them weighed more than 2 tons."

This is a rather lame excuse, since the difficulty in weighing an animal becomes nearly insuperable beyond a certain size. A specialist from the British Museum for Natural History whom Frank Lane had queried on that question had nevertheless also supported this view and declared that "a weight

---

[2] *According to the MacGinities: "Two tentacles of* Architeuthis *have been found which measured 42 feet [12.80 m] in length. If one were to reconstitute the body of a squid using proportional measurements from other squids, the specimen to which these tentacles belonged would have had 4 ft 6 in [1.37 m] in diameter, a body length of 24 feet [7.32 m] and a total length of 66 feet [20.11 m]. It would have weighed 42.5 tons [43, 130 kilos]."*

Such an extrapolation is however inappropriate, since the tentacles are very elastic, and their length varies considerably. Let's remember for example the Big Beach Architeuthis*, which had a 15 metre tentacle, while its head and body combination did no exceed 2.35 m.*

of 30 tons for *Architeuthis* is quite ridiculous".

In this case, the laws of physics and mathematics must then also be "quite ridiculous" for a rather simple calculation justifies such estimates.

If we forget about the squid's very long tentacles and gather the arms so that they come to a point together, we have a kind of spindle, or cigar, if one prefers. While the volume of a cylinder of length l and radius r is $r^2 l$, that of two cones base to base is $r^2 l/3$. The volume of a spindle is intermediate between the two, approximately $r^2 l/2$.

Let's apply this formula to the second Ranheim specimen (4 m long, 20 cm in radius in cross section). We find $1/2 \times 3.14 \times 0.2^2 \times 4 = 0.217$ m^3, i.e. 271 kilos. Professor Sivertsen had calculated that its weight was between 200 and 300 kilos.

The *Alecton* was not quite as precisely measured, but it was between 6.10 and 7.30 m long and, from the original sketch, about 1 m in diameter. Our formula now yields a weight between 2.39 and 2.88 tonnes. According to the estimate made at the time by those who had lifted it, it weighed between 2 and 3 tonnes.

Let's now consider a large specimen, accurately measured but not weighed. The Three Arms squid (1878) was 10.65 m long, and 1.15 m in diameter in cross-section. This gives us a calculated weight of 5,325 kilos.

The case of the Thimble Tickle squid is rather more delicate, for all we know about it is that it measured 6.10 m from the tip of the tail to its beak. Verrill thought that it was an *Architeuthis princeps*. According to the usual proportions of this species, it would have measured a dozen metres to the end of its short arms. There was also no mention of its thickness, but its radius could have been 35 cm for a very slender specimen, 65 cm for an average thickness, and 1.20 m for a very stocky one. For these various values, our by-now well-established formula, gives the weights of 2,304, 8,508 and 27,126 kilos respectively. Dr. MacGinitie was not very far from the truth with his 29.25 tons, if we think of his estimate as a maximum possible value.

Since then, the MacGinities have been subjected to such sharp criticism from their colleagues about the enormous weight which they claimed for the animal that in 1968 they had to issue a prudent retraction. Admitting that they had made an error in their calculations, they re-evaluated the weight at 7,627 kilos: less than 8 tonnes. What happened is that in the mean time a first specimen of *Architeuthis* had duly been weighed, that of Conche, which had stranded in Newfoundland in October 1964. Its body mass measured 5.70 m and its diameter was 35 cm, but it weighed only 154 kilos.

However, according to my formula, he should have weighed 274 kilos.

That is only a factor of 1.82, less than double, in a calculation which is very approximate. One could point out that the Conche specimen might have been particularly dehydrated and might have lost much of its weight while waiting for two weeks on the beach before being picked up by experts and put on the scales. I will not go that far. I have always said that one should take the thickness of the specimens into account, and this is why, for example, I have calculated that according to its thickness, the Thimble Tickle specimen might have weighed 2.3, 8.5 or 27 tonnes. These are even greater differences than a factor of two.

Nevertheless, we have established by strict calculation that there must be *Architeuthis* weighing more than 5 tonnes, and some even larger ones which must weigh between 2 and 27 tonnes, the normal weight being around 8 tonnes.

There are good reasons to believe that there may even exist specimens twice as long as that of Thimble Tickle, which depending on their girth, might have weighed between 16 and 216 tonnes, but more likely around 64 tonnes.

The weight of the largest cetaceans is estimated at 150 tonnes. Sailors were perhaps right to say that the squid is the largest fish in the sea and that the *kraken* was the largest animal ever seen on earth.

Since there undoubtedly exist super-giant squids whose weight is measured in tonnes, perhaps even tens of tonnes, one should perhaps look more kindly upon the "fable of the colossal octopus", spread by ancient navigators and of which Denys de Montfort became the bold defender. Wouldn't such monsters, as heavy as they are powerful, be able to threaten the crew of ships displacing 50 tonnes by grabbing on to the yards. Such was the tonnage of the larger Norwegian vessels of old: a *hav-skip* (high sea vessel) or a *kaup-skip* (merchant vessel) generally displaced about 50 tonnes and could carry only about 15 tons of freight.

I am not asking my readers to believe blindly the stories of seamen prone to exaggerating: I merely invite them to consider a simple problem of mechanics.

Furthermore, let's recall that, except in the case of the *Pearl*, no one has ever said that a cephalopod had tipped over a ship of moderate tonnage, or even a smaller boat; all that was ever claimed was that these marine giants were *capable* of doing it. Regarding the wreck of the 150 ton *Pearl*, if it really happened, it must have been caused mainly by the sudden unlatching of the ballast, triggered by the inopportune boarding.

In such incidents, the worst always seemed to be avoided through immediate action. Whether they were the donators of the Saint Malo ex-voto, or the sailors of Captain Jean-Magnus Dens, the oriental crew mentioned by the Chimsgau Indians, Newfoundland fishermen, or those from the Irish isle of Boffin, men always managed to dispose of their aggressor by cutting its tentacles with their axes, knifes or boarding sabres. Only in the incident of Captain Dens' ship were there three deaths, because of the rather exceptional circumstances: caulking of the ship's hull at sea in a calm ocean.

Overall, then, from the mouth of those involved in those incidents, an attack by a super-giant squid is certainly one of the most terrifying spectacles in the world, but it is exceptional, and can be readily handled, except for the effect of surprise.

This is the time to ask ourselves why would a large cephalopod suddenly attack a ship.

THE KRAKEN AND THE COLOSSAL OCTOPUS

In many cases it seems that there was no provocation at all on the part of the crew, although one could not possibly swear to it. What do we know about the attack of the oriental two-master, reported by the Chimsgau? Further, nothing proves that one of Captain Dens' sailors, busy with cleaning the ship's hull, might not have struck an exploratory blow with a mop or a brush on some intriguing floating mass, with the result that this kind of harassment usually brings. In the cases cited by Denys de Montfort, which are the only ones, along with the *Pearl*, about which we know some details, the attack appears to have been violent and unprovoked.

It would be absurd to deny the aggressivity and fierceness of active predators such as squids. Those rare fishermen which have faced squids even of moderate size will confirm that they are terrifying enemies, of a voracity exceeding that of starving sharks. In 1940, the members of the oceanographic expedition led by Michaël Lerner decided to fish for giant Ommastrephes (*O.gigas*) in the Humboldt current, off Peru. These are good size cephalopods, reaching up to 4 m in total length and weighing up to 150 kilos. The largest caught by Lerner and his colleagues were no longer than 2.40 m and weighed only about 50 kilos. This was however far from a casual fishing operation, of the kind that one might engage in for relaxation over the weekend. David D.Duncan reports that these demons managed to cut with their sharp beak the thickest steel wires used as fishing lines, and to take large chunks off the wooden gaffs used to pull them on board. Night and day they swarmed around the bait, but would not be satisfied with it; if one of them was wounded, the others would rush it and tear it to pieces.

These *Ommastrephes* are midgets compared to *Architeuthis*. Although everything seems to suggest that they are fiercer and faster than the latter, *Architeuthis'* appetite must be in proportion to its size. From the point of view of the prey, one trait replaces the other.

Following the terrifying adventure of those few saved from the sinking of the troup transport *Britannia,* during the Second World War, we can no longer underestimate the danger that super-giant squids represent for people. The tragedy was reported, although somewhat inaccurately, in the *London Illustrated News* of November 1, 1941, and also in the *News Chronicle* of October 21, 1941.

The *Britannia* was sunk on March 25, 1941 by the German raider *Santa Cruz* in the middle of the tropical Atlantic, about half-way between Guinea and Brazil. Twelve men, of whom Lieutenant R.E.G. Cox, escaped immediate death thanks to a small raft, no larger than a throw-rug. Given their numbers only a few of them could sit on it at any one time, and they took turns. The others could then hold on to the side and keep their head above water.

One night, a large squid rushed at the shipwrecked, threw its whips around a sailor and in a sudden jerk pulled him off the raft. The unfortunate man, famished and exhausted, immediately went under and did not reappear.

One can hardly imagine the fear which then took hold of the other passengers. It only grew when, soon after the first attack, Lieutenant Cox was also attacked. He felt an enormous tentacle wrap itself around his leg, which immediately caused an agonizing pain. For some unknown reason, the monster immediately released its grasp. However, brief as it had been, this contact had been enough to inflict on the officer terrible wounds. With its serrated suckers, the squid had torn from his leg, as though with a hole cutter, disks of skin and flesh the size of a penny.

From the size of those wounds, one can deduce that of the squid which produced them, probably with the largest suckers on its tentacles. It must have been an *Architeuthis* with a body about as long as that of a man, and, whips included, no longer than 7 m. Some Ommastrephes, about three times smaller, could also have caused such wounds, but I doubt that they would have had the strength to pull off a man.

After five days in the water, during which time most of the shipwrecked died one after the other, the last three were rescued by a Spanish vessel. Among the survivors was Lieutenant Cox, whose wounds required extended care. Two years after the event, they were examined by the English biologist Dr. John L. Cloudsley-Thompson. Even today, circular scars on the leg of the British officer bear witness to the horrifying attack of which he was victim.

I wonder if our arm-chair experts would have the courage of going to the hero of this atrocious adventure and, in their usual know-it-all manner, claim that stories of agression of people by squids are only old wives' tales!

Attacks by giant squids have always taken place in tropical regions, that is in regions where strandings have never been reported. In warm seas, *Architeuthis* which find themselves in a cold current at depth are not in any special trouble. Simply by rising, they will find their optimal temperature of 10°C before reaching the surface, and will not be in peril of stranding.

It is of course where they are not handicapped by ocean conditions that they can demonstrate the greatest vitality. We note in this respect that appearances of super-giant squids in good condition usually occur in tropical regions, where a cold current upwells from below, for example in the Benguela current. This fits with what we know about the favorite temperature of *Architeuthis*. In cooled tropical waters, they may find their favorite 10°C isotherm near the surface.

The adventurers of the *Kon Tiki* have strikingly confirmed this fact, verifying with their own eyes the abundance of giant squids in the cold Humboldt current, in the tropical pacific:

"More than once, during such nights, says Thor Heyerdahl, we would be startled by the sight of two large round and shining eyes suddenly emerging from the sea near the raft, which looked at us without moving. Our visitors were often giant octopus [read: giant squids] which came to float near the surface, with their great green diabolical eyes shining in the dark like phosphorus."

If the companions of the *Kon-Tiki* got to see, on numerous occasions, a sight that very few people have had the occasion to contemplate, it is of course because they adopted a most quiet and discrete means of tranport. With our modern propeller-driven steamers, we can expect to be approached ever less frequently by marine monsters such as super-giant squids, which in the words of prof G.C.Robson must be, like their cousin the octopus, "prudently agressive" and of a "furtive nature".

We note that the two sailing ships which were assailed at the end of the XVIIIth century, that from Saint-Malo and that of Captain Dens, were immobile in the water when they were attacked, the first being at anchor, the other becalmed. The *Pearl* was also becalmed. The quiet and innocuous impression given by the two ships might probably explain the sudden attacks from an animal which is perhaps quite, but not very bold. In one case at least, two sailors working very near the sea surface on boards lowered alongside the ship were essentially bait for the voracious monster. That was also the case for the shipwrecked of the *Britannia*.

Regarding the attacks against the ships themselves, one might perhaps consider the hypothesis of an accident.

The Japanese zoologist Katsuya Tago reported in 1937 that one day "an enormous squid, 6 m long, leaped into a fishing boat near Kinkazan."

It is in the habit of squids to make great leaps above the sea surface, and they sometimes land by accident on the deck of ships. The fall of such a large squid, weighing hundreds of kilos, must have caused some damage. And if it had been an *Architeuthis*, weighing many tonnes? Even a large vessel would have been seriously affected.

It would seem that these are also the arguments put forward by Miss Joyce Allan, curator of the mollusc section of the Australian Museum, in Sydney, when she rather bravely attempts to find some basis for the legends involving "octopus" attacking ships:

"We must realise, however, how much smaller ships were in earlier days, how relatively slow their passage and nearer they would be to the waterline than now. Therefore, if a giant squid, such as we recognise today, shot to surface waters with its jet-propelled speed, grasped part of a small craft with its tremendous arms, and with its added weight behind it, brought any pressure to bear, it is not unreasonable to assume the craft would turn turtle."

It is not always necessary to invoke blind chance to explain attacks by giant squids on some ships. The declarations published a few years ago by captain Arne Gronningsaeter, of the Royal Norwegian Navy, in the scientific magazine *Naturen*[3] , will perhaps give us a clue to this enigma. This old navigator related, in rich detail, how his own ship, a 15,000 ton tanker, was attacked during the period 1930-33 in the tropical Pacific. As in previous cases of agression, these incidents took place in warm waters.

"I was steaming between Hawaii and Samoa, told Captain Gronningsaeter,

---

[3] *He had the good grace of repeating these comments to me personally for a film which I made in 1962 for French Television:* The Phantom of the Colossal Octopus.

FINAL UNCERTAINTIES

when my ship, the M/V Brunswick, was attacked three times by a giant squid, each time in exactly the same manner."

From the bridge, the captain had been able to analyse at his leisure the technique used by his curious agressor. Swimming parallel to the ship at a speed of 32 to 40 knots, the animal rapidly caught up with it, as it travelled at only 12 knots (19 km/h). When he had passed it by about 30 m on the port side and found itself about 50 metres off the port bow, it would turn around, describing a circle, and would rush directly to a point about one third down from the bow. The cephalopod had finally managed to get a grip on the hull, about 50 metres from the bow and at a depth of 4 to 5 metres. However, unable to hold on to the smooth metal plating, it had gradually slipped towards the stern and finally ended chewed into pieces by the propeller.

Captain Gronningsaeter explained these attacks against a ship in terms of the hostility which reigns between squids and sperm whales. According to him, *Architeuthis* mistakes the ship for its hereditary enemy and attacks it about one third of the length from the bow, because it can get the best grip on the large cetacean approximately at mid-body. It is at that place that the latter has a prominent hump.

One might object that the size of the largest sperm whale -25 metres - is far less than that of a 150 m long tanker. However, we should note that the part of the hull which is underwater is somewhat shorter than the whole ship.

Furthermore, among marine animals, it is rarely simply the appearance of the prey which triggers the aggressive reflex. The appetite of sharks, for example, is stimulated mostly by odours, or rather tastes: that of blood turns them into killers as surely as pushing a button. Among barracudas, those sea-pikes of the Caribbean, the attack is provoked by a precise speed of 29 km/h which is that of the fish on which they feed. A barracuda can be caught with a lure of any size, shape, form or colour provided it is dragged at a speed of 29 km/h.

It is possible that such a stimulus may be at play in the giant squid: perhaps a speed near 19 km/h may be the cruising speed of a sperm whale (although we know that a fleeing sperm whale, or one that is furiously charging, may reach a speed of 32 km/h). It is then not surprising that the noise of the *Brunswick*'s propeller did not deter the attacks of the *Architeuthis*. Only the existence of a reflex mechanism similar to that which governs the behaviour of sharks and barracudas could explain its indifference to such noisome vibrations.

From all these attacks of ships by cephalopods that we have heard of, emerges a lesson. Super-giant squids only rush onto their enemy (or what they mistake for it) when they find it immobile and suppose it to be asleep, or when they see it cruising in a leisurely fashion. What is certain is that some irresistible compulsion sometimes throws them onto objects many times their size, sometimes with fatal consequences. For sperm whales have sharp teeth, and tankers have propellers.

THE KRAKEN AND THE COLOSSAL OCTOPUS

Let us take stock. It seems most probable that *Architeuthis* normally live within an oceanic layer of a narrow temperature range, which is that of their favorite prey. However, in order to satisfy their insatiable appetites, they have to be constantly on the move. In this, their great speed serves them well. So as not to wander too far from their favorite medium, their *biotope*, as a zoologist would say, it is to their advantage to follow the path of major ocean currents, to drift with them, adjusting their depth to gradual cooling or warming of the waters. (This would make them rather cosmopolitan, but would restrict to some limited areas the oceanic regions where they travel near the surface.) As long as they remain in the closed loop of a warm current, the Gulf Stream, for example, they are assured of being able to stay within a zone of favorable temperature. However, if in temperate latitudes they should suddenly, by accident, enter into a glacial current, or even if they should be carried by a branch of the current which gradually meanders towards polar regions, cooling all the while, they may find themselves in a critical situation. Used by instinct to ascend when they wish to find a warmer temperature, they will now come all the way to the surface without finding the layer where live their prey. They are then subject to famine. Where a sudden convergence of currents of greatly different temperatures occurs, their existence is threatened even more directly, because of the accumulation of dead organisms which makes the waters unhealthy. This explains at the same time the relative frequency of their strandings and their occurence in some well-determined temperate regions.

On the other hand, their enormous size, sometimes much greater it would seem than that of duly examined and catalogued specimens, and the voracity which it implies in such exclusive meat eaters, makes them dangerous adversaries for anyone, man included. The fact that they can hold their own against the sperm whale, the world's top predator, is proof enough of their power. However, it follows from what we saw earlier that the regions where man and the super-giant squids are likely to meet are extremely restricted. *Architeuthis* are pelagic animals, and approach the shore only when they are moribund; in the open sea, under tropical latitudes, they rarely meet with isolated swimmers, except for poor shipwrecked sailors, holding on to a raft or some other flotsam.

True, they sometimes happen under some conditions to rush against a ship, which they probably mistake for their greatest enemy, the sperm whale. However, such attacks, not only may go unnoticed in large, high-decked vessels, but are also unlikely to be harmful to anyone but the attacker. It is unlikely that squids would think of fishing preys located so high above the water. All this make it quite unlikely, these days, that a sailor should be pulled off by an *Architeuthis*. However the exceptional nature of such an incident makes it even more likely to fuel the imperishable flame of marine superstitions.

Thus, if it wasn't for its sensitivity to unexpected temperature changes, and for the presence of dangerous oceanic zones, the colossal squid would never have emerged from the depths, nor the *Kraken* from the mists of legend. And we would probably still be reading today with the same ironical smile, the legend of Saint Brendan, or some passages, like the following one, from Thor Heyerdhal:

"Around 2 o'clock, when the helmsman had trouble distinguishing the black sea from the black sky, he saw in the water a weak light which gradually took on the form of a large animal. It was impossible to tell if it was plankton phosphorescing on its body or if the animal itself had a phosphorescent surface, but the ripples of the black water gace this ghostly creature vague and wavy contours. Sometimes it seemed round, sometimes oval, or triangular, sometimes it even seemed to split into two parts, which came and went separately under the raft.

"Finally there were three of these large shiny ghosts slowly circling around us. There were real monsters, their visible parts exceeding 6 to 8 metres in length."

In a letter to Rachel Carson, Dr. Heyerdhal added that:

"Their dimensions seemed to exceed those of the raft (Kon Tiki measured about 15 m by 6 m).

"We hastily gathered on the open deck to follow their fantastic dance, which lasted hours, continued the Norwegian ethnologist in his book. Mysterious and silent, our shining companions stayed a fair depth below the sea surface, especially on the starboard side, where hung the lantern, but they also sometimes stayed under the raft or appeared on the port side. The light reflected off their back suggested that they were larger than elephants, but they were not whales, for they never came up to breathe.

"Were they giant rays, whose shape appeared to change as they turned sideways? They were not disturbed by our putting the light right at the water surface so as to be able to tell what sort of creatures they were. And like all ghosts and leprechauns, they disappeared into the depths at the first rays of dawn."

Perhaps those were super-giant squids. It is on that vision of spectral entities with vague contours and imprecise dimensions that I would like to close the story of these titanic princes of the sea. For they are somewhat similar to our knowledge of *Architeuthis*. We know they exist, we know the shape and dimensions of some of them, and we believe that we can distinguish among them, on the basis of rather meager samples, a series of different species, although we are not sure on how to recognize differences between sexes. However, we still don't know how large they can become, and we don't really know if they are the world's largest cephalopods. About their behaviour, their swimming speed, their migrations, their hunting tactics, their means of camouflaging, their usual diet, their sex life, their development and longevity, their gregariousness, and other aspects of their life, we are at best at the level of hypotheses and more or less bold extrapolations; at worst, in complete ignorance.

THE KRAKEN AND THE COLOSSAL OCTOPUS

Phantoms, gargoyles, ghouls and freaks is what they have been for us, and this is probably what they shall always remain, in spite of truncated arms, shriveled and discoulored which we keep in jars, which can hardly be recognized as animal remains, and the dissected jaws sleeping in dusty boxes, carefully labeled and numbered. Perhaps even because of these very relics. For in our nightmares, these immense tentacles come to life, crawl, undulate and come to gradually smother us; we drink of the pain to the dregs in these suckers similar to cups with serrated edges; these crooked beaks, as large as a hand, gnaw on our liver, poor Prometheuses that we are, who have dared penetrate the secrets of oceanic gods.

Chapter Thirteen

THE INCOMPLETE STORY OF
THE COLOSSAL OCTOPUS

———

"To be ignorant of one's ignorance
is the malady of the ignorant."

A. B.ALCOTT
Table Talk

And what about the giant octopus, will you ask? In the preceding chapters, each time there was some gigantic polyp or some titanic octopus, it was always upon closer look, a squid. Does this mean that there is no gigantic octopus?

This is certainly the conclusion that many naturalists have reached. If you open recent zoology text-books, encyclopedias of natural history, and popular science works, you will always find the same refrain: squids can actually grow to a gigantic size, but giant octopus are found only in adventure novels.

I need only to quote as an example a characteristic passage, from the excellent book by the American naturalist N.J.Berrill, *The Living Tide:*

"Octopi do not grow so big, reaching a maximum spread of only seven or eight feet [2.20 -2.40 m], which means they have arms rarely longer than three or four feet [0.90 to 1.20 m]. Squids are another matter. Some are tiny and cute, but other are the only animals that fight the big whales, though they only do it to save themselves from being the whale's dinner. The largest found have measured fifty feet [15 m] from tail tip to end of tentacles, and sucker marks found on whales indicate that some may be three times as large."

Thus, untypically for a professional zoologist, Mr. Berrill does not hesitate to recognize the existence, still rather speculative, of squids up to 45 m in length. On the other hand, as far as octopus are concerned, his estimate remains well below actual, verified measurements.

The popular legend of the *Kraken* has been replaced nowadays by what Maurice Burton rightly call a scientist's legend. The man in the street is not the only one to succumb to specious generalisations, to rely on authority, or to reach a hasty conclusion...

It is on purpose, so as not to confuse what was already a complicated enough business, that while tracing the history of giant cephalopods, I have put aside everything which clearly related to the octopus.

Actually, the whole problem begins at that point where I cast doubt on the identity of the octopus feeding on salt fish, at Carteia, and determined that it must have been a squid. It is in that text of Pliny that the confusion between squids and octopus begins, although the Roman encyclopedist is not responsible for it. All he spoke about was a *polypus monstrosus,* meaning a monstrous creature "with many feet", a term which applies equally well to all cephalopods. In the days of Pliny, the rules of zoological taxonomy were still far from being well established, and one could hardly blame him for his lack of precision.

We are thus brought back, in the business of giant octopus, to our starting point, the Homeric legend of the pernicious Skylla. This creature was without doubt an octopus. It lived in a submarine cave, which is proper for a benthic cephalopod; on this point the poet leaves no doubt.

Of course, we can hardly reach, simply on the basis of Homer's text, conclusions regarding the undoubtedly considerable size of a cephalopod capable of picking a man off the deck of a boat. However, isn't it significant that it is precisely there, in the Strait of Messina, the home of Skylla, that a bold diver will again, two millennia after Homer, mention the presence of enormous octopus?

Our informer is himself part of a legend. He is a Sicilian diver of the XIIth century, Nicolas by name, whose extraordinary skill at swimming had earned him the nickname Nicolas the Fish[1]. His fame must have been widespread, for numerous authors of the Middle Ages, and later of the Renaissance, have mentioned him, in a number of variants.

According to the Englishman Walter Mapes, who lived for a long while in Italy at the end of the XIIth century, the man was so at home in the sea that he had penetrated all its secrets and could even predict storms. He died because he had been taken away from the sea to be shown to the King of Sicily, William the First (who reigned from 1154 to 1166).

Of course, with the passage of time, more imaginative commentators began to attribute to this undoubtedly exceptional diver achievements which would be completely impossible without a diving suit, such as being able to remain under water for three quarters of an hour. In the XVIIth century, the jesuit Father Athanasius Kirch er went so far as to write that Pescecola sometimes lived four or five days in the sea and that he could stay a whole day under water without breathing... At the same time, rumours began to the effect that he had grown webbing between his fingers (Kircher) and even fish scales (Jovianus Pontanus, in the XVth century).

---

[1] *Also under other forms:* Nicolas Pesce *(Walter Mapes),* Pescecola *(Father Athanasius Kircher),* Colas Poisson *(P.G.Fournier) and even the* fish Nicolas *(Cervantes)*

THE INCOMPLETE STORY OF THE COLOSSAL OCTOPUS

One of Nicolas' greatest feats, according to Gervais of Tilbury, had been the exploration of Skylla's cave, undertaken at the order of King Roger of Sicily (probably Roger II, who reigned from 1101 to 1154). Athanasius Kircher on the other hand says that it was King Frederic II (who reigned from 1355 to 1377) who enticed our hero into diving into Charybdis' lair by throwing into it a golden cup. That is obviously an error. Even if the exploit of the famous diver has been placed at different times, it is nevertheless likely that a spectacular sporting accomplishment is at the origin of all these legends.

In any case, the Fish-Man came back from his memorable descent into the Strait of Messina with a description of a hostile and terrifying word; in Father Kircher's words:

"...I saw herds of terrifying octopus holding on by their arms to the rocky underwater cliffs; they wave their other arms away from the wall, and that is the most terrifying thing that I saw in that deep. I saw some whose body was as long as that of a man, and whose arms had more than ten feet in length [3 m]; I have no doubt that if they had seized me, they would have killed me by their embrace alone."

In a rather surprising statement for a generally incredulous zoologist, Mr. Moquin-Tandon wrote in 1865 that "this report has not received enough attention by the naturalists of our day".

On the look-out for everything pertaining to the colossal octopus, Pierre Denys de Montfort could not fail to be interested in the story of Nicolas the Fish. He also related a very similar contemporary testimony:

"The respectable Fortis told me that he knew a fisherman in Venice who was a very skilled diver, who, terrified by the marine monsters which he had seen below a certain depth, had decided no longer to dive from the fear they had caused him, and gave up that dangerous occupation."

There again, it was the sight of very large octopus which had terrified the diver. I have also heard an identical anecdote from the very mouth of one of the witnesses of a similar incident.

At that time, a fisherman of my acquaintance - one of the best in the Department of Var - well known in the Hyères Islands as Joseph le Bicot (locally, as Zé le Bic), was engaged, with one of his cousins, Napoléon by name and a hard hat diver, in harvesting urchins at a depth of 15 m. This very profitable fishery is also forbidden, for at that depth, they can be scooped by the shovel full, to the point that an area can be quickly fished out.

One day, our two "rapers of the sea" decided to operate just off the sewers of the town of Cassis. Zé was working the pump and Napoléon had put on the diving suit. The latter had been under water for some time when he indicated by repeated pulls on the rope that he wanted to come back up quickly. When le Bic had unscrewed his brass helmet, he discovered his colleague Napoléon pale and stuttering. He had suddenly found himself face to face with an unexpected fisheries patrol officer: an enormous octopus "with eyes as large as those of a cow."

The mollusc had not made the least gesture of aggression towards the diver. Nevertheless, Napoléon did not dare dive again in that part of the coast, and from this day Joseph le Bicot also said a definite good-bye to his diving suit, which he sometimes donned on to replace his colleague. Still, these Mediterranean fishermen were quite familiar with octopus, and did not usually fear them, even when they were quite large: they regularly capture with their bare hands specimens weighing 10 kilos, with a span of nearly three metres. The octopus which they had encountered on that day must have been of a particularly impressive size!

Readers accustomed to adventure novels where octopus always have "eyes the size of saucers" will probably not be over-impressed by Napoleon's description. However, novelists who use such expressions have probably never seen an octopus. Its eyes are rather small, as are a frog's... and, yes, also a cow's. Here again the error comes from the confusion with the squid, which has enormous eyes. What is striking in the octopus is not the size of its eyes, but the intensity of its stare. One is astounded to discover under water a glance which is not fixed, cold and inexpressive, as that of a fish, but moving, attentive, and intelligent.

So, in order to have "eyes as large as a cow's", the octopus of the Cassis sewers must have been of an extraordinary size.

How large, it is difficult to say. One should not forget that under water everything seems about a third larger than in air. Fear leads one to forget about the necessary correction, even to exaggerate in the opposite direction...

Since the publication in 1851 of Jean-Baptiste Vérany's book on *Mediterranean Molluscs* we have some rather precise ideas about the size that octopus may reach in that sea:

"The largest octopus that I have seen, relates the director of the Cabinet of Natural History of the City of Nice, was about 3 metres long and weighed 25 kilos; a very skillful and clever old fisherman came upon it against the breakwater of the port of Nice, grabbed it with his own hands by leaning out from his boat, turned its mantle over and finally mastered it, not without considerable effort.

" Fifteen kilo octopus are not rare in Nice, and those weighing 10 kilos are quite common."

As we are talking of the *Octopus vulgaris*, whose head is about one sixth of the total length, Vérany's champion must have had tentacles about 2.50 m long and a total span near 5 metres.

As we know, the indisputable report of the Nice naturalist did not keep Dr.Chenu from stating, seven years later, in his *Encyclopédie Zoologique* that there were no very large octopus, except in the Pacific Ocean, where they reached "a span of nearly two metres"! One is never a prophet in one's own parish, even in the kingdom of the octopus.

THE INCOMPLETE STORY OF THE COLOSSAL OCTOPUS

Might not Vérany's champion be bested? Is it really to be thought as the ultimate size limit for *Octopus vulgaris*? Given the persistent rumours which circulate on the existence of really enormous octopus, I took upon myself to verify if there did not exist indisputable traces of encounters with or, even better, captures of even larger individuals, especially in that Mediterranean Sea, where still hovers the ghost of the terrible Skylla. This enquiry took me into research areas of unsuspected scope, but also as demanding and they were disappointing. The scientific literature, through which one readily finds one way if one has been initiated early, has been silent on this matter since Vérany's report. It seems to confirm the record which he reported. Its silence however seemed to me almost a conspiracy, designed to keep tightly shut the doors of fantasy. I was thus led to consult various colleagues, specialists in the study of cephalopods and marine biology, as well as participants in oceanographic expeditions, who might be aware of unpublished information on the subject. I also carried out a personal enquiry among fishermen and scuba divers in the Mediterranean, and undertook an active correspondence with representatives of the French navy; I went through the archives of various harbours and read through works of recollections of hard hat and scuba divers; I methodically sifted through collections of magazines and newspapers. Overall, it took me a number of years to accumulate the information presented here. If I insist on this point, it is to show that it is sometimes difficult to obtain precise and reliable information on events which are nevertheless of undisputed reality.

My best clue at the start, was a paragraph in an article entitled *Le monde extraordinaire des pieuvres (The extraordinary world of the octopus)*, published in 1952 in a popular science magazine:

"In the last century, J.-B.Vérany, of Nice, whose large tome on cephalopods is still authoritative, saw, fished out of the harbour in Nice, an Octopus vulgaris with a span of 6 metres. In Villefranche, Mr. Trégouboff, director of the Marine Biology Laboratory of the University of Paris, caught himself during the war an octopus with a span of 7 metres. In Toulon's Museum there has long been preserved a specimen of similar dimensions, found in the wreck of the Liberté when it was refloated."

The article was signed by Jean Dragesco, but I soon learned that this excellent nature photographer was not the source of all the information contained in the article: following this execrable journalistic practice, the article had been "rewritten". Apparently, to the great dismay of its author, the original article had been re-arranged, truncated and extended by a third. The final product contained rather fantastic details on the mating of octopus - this certainly did not inspire much confidence! Even the reference to Verany's champion was distorted: a 3 m long octopus cannot have a span of 6 m! It is as if a 1.60 m tall dancer spread over 3.20 m when she did the split...

All this was rather suspicious. Nevertheless the reference to Dr.Trégouboff seemed to me to provide enough credibility to justify further investigation. Not wishing to disturb a colleague undoubtedly absorbed by his own research, I looked first through his own publications for the source of the information which was attributed to him, and then through those of researchers of the Zoological Station in Villefranche. Having found nothing, I finally resolved to contact Dr.Trégouboff himself, who answered with as much speed as cordiality. Here is the actual information which he had released:

"After the war of 1914-1918, I believe in 1920, there came from Toulon a barge of the French Navy (I cannot remember its name, and never knew that of its commander) to lift and repaint the mooring blocks anchored in the bay (of Villefranche) for use by the squadron which periodically cruised through, once or twice a year [...] The hard hat diver who had gone down to the bottom to detach the block from its principal anchor (this mooring block for battleships was anchored in a depth of 42 metres) having mentioned the presence of a large octopus under one of the struts, had been provided with a harpoon and had managed to bring the ocotpus back on board. The animal was then killed and measured. Its length, arms outstretched, was 3.50 m, and its weight around 18 kilos. I was not present for this capture; the information which I am now providing you was given to me by the sailors, who had cut up the octopus with the intention of eating it or using it as bait.

"...As you can see, aptly concluded Dr. Trégouboff, we are far from the 7 m octopus, figment of a journalist's imagination."

One should say, in defense of the journalist, that the expression "its length, arms outstretched" is ambiguous, when it is the span which is meant. Nevertheless, the circumstances of the capture were reported with rare inaccuracy , probably on the basis of hasty notes. Furthermore, the animal's span was incorrectly calculated, simply doubling its presumed length. Finally, it was unthinkable that an 18 kilo octopus should have a span of 7 m, when a specimen with a 3 m span already weighed about 10 kilos.

This particular journalist also seemed to have some strange ideas about the weight of cephalopods. In the same article, he wrote of the giant squid that "its length can exceed 16 m, for a weight of more than 500 kilos." In fact, according to the MacGinities' meticulous calculations, it should be around 30,000 kilos, which is a lot more than 500 kilos!

My research on the octopus discovered in the wreck of the *Liberté* were less diappointing, but more laborious.

That particular battleship had sunk in Toulon harbour in 1911 following an explosion in the hold. This catastrophe had claimed hundreds of victims and widowed all of France; many still kept of the tragedy a horrified memory. As the lifting of the hulk had taken place only a few years before the war of 1940, I searched through the press of the time, as well as in maritime archives for mention of a news item as sensational as the discovery of an authentic marine "monster".

Nothing. The gigantic octopus seemed to have evaporated; it slipped through my fingers as easily as its smaller cousins which I had often tried to catch underwater. I had nearly given up, when a friendly correspondent told me that one of his friends remembered having read that a giant octopus had attacked a diver in Toulon harbour between 1900 and 1903. That was vague, but I became interested. This was well before the time of the explosion of the *Liberté* and of course also of its salvage. However, that large and agressive octopus might have been that same one who had later set its home in the wreck. I had to find out, and went back to my research starting at an earlier date.

While my investigations in scientific libraries, museums, and maritime archives had come to naught, it was, by irony of fate, in the illustrated supplement of the *Petit Journal* where I finally found "my" giant octopus. In the issue of June 16, 1912, the following information accompanied one of these dramatic drawings of which that astoundingly trashy sheet made a specialty:

DIVER ATTACKED BY OCTOPUS

"It is in Toulon harbour that took place this attack of a diver by an enormous octopus. Sailor-diver Ledu had descended in the Missiéssy dock to recuperate objects fallen in the water, when all of a sudden, issuing from a crack, a giant octopus leaped upon him." "The diver having forgotten his knife could not defend himself, and pulled on the alarm bell, but the marine monster had already had the time to embrace his body with its tentacles, 8 metres long. When poor Ledu was brought back up on the pump boat, he had completely lost consciousness. Energetic attention was required to revive him. The cephalopod mollusc which was attached to the diving suit was killed with knife blows. It weighed no less than 60 kilos and its suckers were as large as a 5 franc coin."

I later obtained the confirmation that this was indeed the *Liberté* octopus. My correspondent's friend had been mistaken in placing the incident at the beginning of the century rather than on the year that followed the catastrophe. It was also by mistake that the rewriter of the popular science magazine had associated the discovery of the monster with the lifting of the wreck. In fact, it was in order to search for objects blown out of the *Liberté* by the explosion that the sailor-diver Ledu had descended into the Missiéssy dock. These were the fine threads which linked the enormous Toulon octopus to the sunken battleship.

Following this explanation, the dimensions attributed to the monster require some discussion. Judging from its weight, this octopus couldn't have had 8 metre long tentacles. Probably, the reporter had meant that its tentacles, spread out, had a span of 8 metres. Indeed, the draftsman charged with illustrating the incident had shown an animal of this latter size, in spite of his director's taste for the horrible, monstrous and spectacular.

To be convinced, in a more scientific way, of the legitimacy of such a correction, it suffices to draw a curve of the weight of *Octopus vulgaris* as a function of its length. We can do so, roughly, using only those specimens already mentioned in this book. An octopus 30 cm long weighs a little more

than a kilo and a half; it reaches about 5 kilos when it is twice that length. This could be confirmed by any fisherman or fish merchant. The 9 kilo octopus of my friend Cestino was slightly shorter than a man (1.60 m?). That of the bay of Villefranche (3.50 m in span) must have been around 2.10 m long with a weight of 18 kilos. Finally, that of Nice, mentioned by Vérany, was 3 metres long and weighed 25 kilos. If we extend the curve which approximately (let's not forget that all these lengths have been measured on an extremely elastic animal) goes through the series of representative points of octopus known to us, we find that a 60 kilo *Octopus vulgaris* should measure between 4.50 and 6 m in length. An octopus with a span of 8 metres, i.e. 4 m long tentacles, would be about 4.80 m long. It would then be normal for it to weight about 60 kilos, Q.E.D. Continuing along the same curve, we discover that an octopus with 8 m
long tentacles would most certainly weigh hundreds of kilos!

Are there such enormous octopus in the Mediterranean? Some reports suggest it. Although, as far as I know, the *Octopus vulgaris* of the Missiéssy dock is the largest specimen ever captured and duly measured, nothing proves of course that it be the largest existing. However the information which I have been able to gather on supposedly larger specimen encountered in our inland sea is extremely vague.

Following an enquiry on the perils of skin-diving, the *National Geographic Society* assessed that the least danger of this sport are the monsters of the deep, so feared by the ancients. On this subject, it stated that even the largest octopus will flee from a man who moves determinedly above the bottom:

> *"Perhaps the largest octopus yet met underwater was one with eyes "the size of saucers" seen by a reputed diver off Piraeus, Greece. The man, as courageous as are all inveterate deep divers, admitted he was frightened, but said the animal beat him to the retreat.*[2]

So we encounter again the octopus with eyes "the size of saucers", familiar to novels of submarine adventure, but this time, it is not in a work of imagination (we hope). Without pronouncing judgment on the report of the "reputed diver", I would like to bring into sharper focus the size of an octopus with such impressive eyes.

After having measured a series of saucers and having found their mean diameter to be 15 cm, I calculated, to the best possible approximations the relative size of the eyes of *Octopus vulgaris* so as to be able to determine the dimensions of a specimen whose eyes would be that large.

In an octopus with a span of 1.20 m, the eye, with its horizontal pupil, is nearly 1 cm wide, If the same proportions hold, an individual 3.50 m in span, as that of the bay of Villefranche, would have an eye nearly 3 cm in diameter. An individual 5 m in span, similar to Vérany's, would probably have an eye more than 4 cm in diameter. The 8 m span Toulon octopus

---

[2] *Quoted from a report in* Science Digest, *November 1953.*

probably had eyes with a diameter nearly 7 cm across. To have eyes "the size of saucers", an ordinary octopus would have to be more than twice as large, with a span of about 18 m!

Let's suppose that our Piraeus diver might have been fooled by refraction in water, and that the eyes might only have *appeared* as wide as saucers to him; this would mean that they would have in fact been smaller by about a third, i.e. about 10 cm in diameter. This dimension still implies an octopus with a span of 12 metres.

Are there really in the Mediterranean octopus with eyes the size of saucers, with a span in excess of 10 metres, and with a weight of hundreds of kilos? No concrete information supports this affirmation. At best, some witnesses livid with terror have come forward to report them, stuttering. In this story of the colossal octopus, the last word has not been said yet. Much still lies in the shadow.

We are nevertheless convinced by now that, contrary to the opinion of many naturalists, there really exist, even in the Mediterranean, large octopus, specimens with a span largely exceeding the 2.40 m grudgingly conceded these days by N.J.Berrill, and *a fortiori* the 2.0 m which Dr. Chenu and his colleague Desmarest conceded, not without some reservation, only to those species of the remote Pacific.

On this point, what about those exotic octopus, inevitable bogeymen of tropical adventures on sunny islands? We have so far spoken only of Mediterranean octopus; perhaps they grow much larger elsewhere. It has often been observed that in warmer waters the same species of molluscs enjoy a more considerable development. Do perhaps also octopus?

One thing for sure is that in the last century, the discovery of a 3 metre long octopus was as big news far from the Mediterranean as it is on its shores.

In 1872, the *American Naturalist* of Salem published under the title *A colossal Octopus* a letter from one Mr. J.S. George, of Nassau, in the Bahamas. It was reported that an octopus slightly longer than 3 m long had been found dead on the beach. The scientific correspondent added:

" *This is the first specimen I have seen during twenty-seven years of residence in Bahamas, but tradition has it that there are here specimens of immense size.* "

Mr George went on to say that the weight of the mollusc had been estimated at 90 to 140 kilos, but this estimate appears to have been made quite casually, since a Mediterranean octopus of the same size, mentioned by Vérany, weighed only 25 kilos. It is true however, that there exist octopus which are more or less stocky, according to the genus, or the species, and that no indication was given on either in the letter.

In any case, Mr. George's report provoked intense interest in scientific circles; it was soon reproduced by numerous cephalopod specialists. Dr. Packard Jr. was the first to quote it, in the following issue of the

THE KRAKEN AND THE COLOSSAL OCTOPUS

American Naturalist, of which he was the chief editor, on the occasion of the strandings of super-giant squids in Newfoundland. He also reported, on his own account, that professor Brewer, from Yale College, had told him having measured at the market in San Francisco octopus of 4.25 m in span. Packard Jr. then concluded with:

"Accounts of colossal species of octopus are not uncommon. They occur in the mid-Indian, Atlantic and Pacific oceans, and seem to be as large and much more common than the ten-armed squids."

In that same article, Dr.Packard nevertheless reported the discovery, on the Grand Bank of Newfoundland, of a slender squid about 7.60 m long, without tentacles. Its weight had been probably underestimated at about one tonne... To reach a similar mass, a common octopus would have to be more than 9 metres long, with a span of 15 metres.

We can see that in those days, those who believed in the existence of gigantic cephalopods still believed, against all evidence, that the true champions in size were still to be found among the octopus. Times have changed!

Professor Brewer's report on the presence of octopus with a span of more than 4 m on the fish markets of the Pacific coast of the United States was to be confirmed in 1879 by professor Addison Verrill:

"Dr. W.O.Ayres tells me that he has often seen this species exposed for sale in the markets of San Francisco (where it is eaten chiefly by the French), and that specimens with arms 6 or 7 ft [1.80 -2.10 m] long are common."

What is "this species" to which professor Verrill is referring? It is *Octopus punctatus*, described by Gabb in 1862, which is recognized by the large ring-shaped spots spread around its body. It continues to this day to be considered by specialists to be the largest of the whole sub-family Octopodinae, but it has often changed names since. In 1912 Berry described it again under the name of *Polypus apollyon,* and in 1923, Naef created for it the new genus *Paroctopus*, considering that it differs from all other known octopus, especially in its overall shape. As the species name *punctatus* was already in use, before Gabb's work, to describe another cephalopod, it could not be kept for reasons of priority, and the specific name *apollyon* won out. The scientific name of the animal would then be *Paroctopus apollyon* (or *Octopus apollyon,* if one refuses to recognize its classification in a different genus). Fortunately, Grace E. Pickford discovered that the species had already been described in 1910 by Gerhard Wülker under the name *Polypus dofleini*. It is now most commonly known as *Octopus dofleini.*

In 1873, the American naturalist William H.Dall, to whom we owe the first description of the natural riches of Alaska, revealed the surprising dimensions often reached by this octopus, but also emphasized the slender nature of this delicate giant, and its low mass in relation to that of large squids:

THE INCOMPLETE STORY OF THE COLOSSAL OCTOPUS

"The Octopus punctatus Gabb, which occurs at Sitka abundantly, reaches a length of sixteen feet [4.90 m], or a radial spread of nearly twenty-eight feet [8.50 m], but the whole mass is much smaller than that of the decapodous cephalopods of lesser length. In the Octopus above mentioned, the body would not exceed six inches [15 cm] to a foot [30 cm] in length, and the arms attain an extreme tenuity toward their tips."

The relative lightness of this kind of octopus confirms this slenderness. MacGinitie informed us of the weight of a specimen captured in Monterey, California, which "could have touched without difficulty the sides of a circle of 10 foot radius." This animal, with a span of 6.10 m, weighed only 50 kilos.

These monsters, which are most common along the whole Pacific coast of North America, are probably responsible for the aggressive incidents mentioned by many authors since the last century.

Thus, in 1821, as he was returning from his voyage around the world, the German navigator Otto von Kotzebue reported what the Aleuts had told him about the gigantic octopus:

"It sometimes happens that an octopus of this kind will throw its long arms, twice as thick as those of a strong man, around the boat of an Aleut. It would carry him down into the depths if the man did not have the presence of mind to cut with his knife the fleshy arm of the octopus, which is covered with large suckers. The octopus usually sits with his body attached to the bottom of the water, and generally chooses a spot where it can reach the surface with its arms. The latest incident of this kind occured in the pass between the northern most point of Oomnack Island and the small island near it. No ship can venture there on account of it being so shallow."

It would seemed that similar incidents also occured further south. We have from Japan various rumours regarding the existence of gigantic octopus: for example, a drawing by the famous Hokusai (1760-1849), depicting an attack on a man by an octopus with eyes at least as large as saucers. We also know, thanks to the 25 April 1874 of *La Nature*, of the existence of a work entitled *Products of the Soil and the Sea*, attributed to Ki Konc, wherein is shown the struggle between a similar octopus and a fisherman, and the subsequent sale at the fish market of the enormous arms which the latter cut off. They certainly measure more than 2 metres in length, and are as thick as a man! One simply cannot believe that these illustrations, published as they are in a work of documentation, could refer to purely imaginary events.

Towards the end of the last century, the Canadian naturalist George Mercer Dawson, rector of McGill University, also referred to similar rumours, coming from the other side of the ocean:

" Indian women are reported to have been drowned by being clasped by huge Octopods whilst bathing in the Pacific, on the coasts of British America, and

THE KRAKEN AND THE COLOSSAL OCTOPUS

among the Indians are traditions of narrow escapes."

These giant octopus indeed continue to be in the news in that region. Shortly before the last world war, John D.Craig, a hard-hat diver specialized in underwater photography for Hollywood productions, was almost a victim of one of these octopus. Off San Benino Island, in Baja California, he had seen while diving two enormous octopus which seemed to occupy the entire bottom of a hole in the rocks with a diameter of about 12 metres. Knowing that abrupt movement could provoke these cephalopods to attack, he remained immobile. As his Japanese colleagues had recommended, he allowed one of the beasts to feel his leg, without moving. When the curious animal had completed its examination, it withdrew. Craig thought that the moment had arrived to slip away. However, he had not waited long enough. The mollusc caught up with him immediately and grabbed his ankle. Fortunately, they were on a gravel bottom and the animal could not hold its victim down. This is what saved Craig, who returned to the surface, completely enveloped by the monster. His assistants immediately attacked it with axes and cut most of its arms off. One, which Craig kept as a souvenir was 2.50 m long. The animal must have had a span of more than 5 meters.

There are larger ones still, we know for sure, and they are no less agressive. The Canadian diver A.E.Hook, from Vancouver, in British Columbia, reported that in 1928, he had to fight, at a depth of 25 metres, twice as large a monster, in Queen Charlotte Strait. His story is however full of implausible statements.

Fortunately, we have more coherent and precise details on another of these aggressive Pacific octopus, which indulged in a similar attack sometime before the first World War. This information has come to us through a dispatch from San Francisco, which was reproduced in numerous newspapers at the beginning of February 1912:

"An epic struggle, reminiscent of that of Gilliatt in the Toilers of the Sea, took place, according to the Franco-Californian, off Point Lobos [a headland west of the city of San Francisco], between a gigantic octopus and a Japanese fisherman, T. Xamaguchi. The latter, wearing a diving suit, was picking up shells from the sea bottom when he suddenly felt enveloped by the monster's tentacles. He struggled under this terrible embrace before being able to send to his companions in the boat the agreed-upon signal for being pulled back up to the surface. Once there, it was not without difficulty that they managed to cut away with an axe, one by one, the arms of the beast, which was trying to get a hold of the boat to tip it.

"The octopus weighed at least 125 kg; its tentacles measured 4.90 m in length and its head was the size of a pumpkin."

Knowing that an *Octopus punctatus* with a span of 6.10 m weighs 50 kilos, we can easily calculate that the weight of a specimen with a span of 9.75 m should indeed be 125 kilos. This calculation speaks in favour of the authenticity of the information and of the precision of the data it contains.

THE INCOMPLETE STORY OF THE COLOSSAL OCTOPUS

Soon afterwards, a personal experience would allow W.H. Dall to confirm such numbers and to revise his earlier, somewhat timid, estimates:

"In 1874, I speared an octopus in the harbor of Iliuliuk, Unalashka, which was afterward hung, by a cord tied around the body immediately behind the arms, to one of the stern davits of the coast survey vessel under my command. As soon as the animal died and the muscles relaxed, I noticed that the tips of the longer tentacles just touched the water. On measuring the distance with a cord, I found it to be sixteen feet [4.90 m] , giving the creature a spread from tip to tip of the longest pair of arms, of not less than thirty-two feet [9.80 m]. The arms toward the tips were all exceedingly slender, but rather stout toward the body, which was somewhat over a foot long. The largest suckers were two and a half inches [6.3 cm] in diameter; the whole creature nearly filled a whole washtub. Parts of this specimen are now in the U.S. National Museum."

According to E.G.Boulenger (*A Natural History of the Seas*), the span of the spotted octopus would reach up to 12 metres in Australian waters. This would imply a weight of about 200 kilos. I have not been able to find the specific evidence on which the ex-director of the Aquarium of the London Zoo has based this assertion, and I suspect that it might be a mistake.

In any case, it is probably the spotted octopus, the most common in the Pacific, which should be held responsible for innumerable attacks on skin and suited divers, regularly mentioned in the Australian press. Prof. Osmond P.Breland, of the University of Texas, wrote in 1952 that:

There have probably been more reports of octopus attacks from Australia than from anywhere else. Only recently two such encounters were described in the Australian newspapers. In both instances, the men escaped after terrific struggles, and only after they had reduced the relatively small octopuses to mincemeat with the knives carried by all divers in those regions."

The readers will recall that in the case of the hard-hat diver of the Moyne River, attack in 1879, the aggressive octopus was also "relatively small", with a span of less than 3 metres. However, if Mr. Smale had been surprised by a specimen with a span of 9 metres, three times as large, and probably twenty-seven (three cubed) times as heavy, he might have not been able to return and tell us of his adventure. The monster would have quickly torn his suit, cut his air tube and safety line, and dragged him to his lair. His companions would have merely pulled back to the surface the clues of his tragic death. We would have of the drama only a vague version, enough for another one of those news items in response to which those zoologists busy in their laboratories merely shrug. By the nature of their habitat, octopus attacks are quite unlikely to take place in front of witnesses, and even less so in the presence of what specialists call "qualified witnesses". The only witnesses are usually the *dramatis personae* themselves, most of the time humble pearl or sponge fishermen, aborigines from the islands of Pacific.

Thus, even if it were true, as claims professor Breland, that "no authenticated case of death caused by an octopus attack could be discovered", might that not simply be due to the very nature of such an aggression?

An octopus does not kill a man in a few seconds, as would a shark under the eyes of horrified and powerless witnesses. In slow motion, the octopus gradually smothers its prey before beginning to eat it. Having lassooed its prey with the tip of one of its tentacles, it can only begin its feast once it has passed it from sucker to sucker all the way to its beak. The more energetically the prey struggles, the longer this process takes. When hard-hat diver Martin Lund, in the hire of the Coast Wrecking Company, was attacked in 1909 by a large octopus while he was examining the wreck of the *Pomona*, down at a depth of 10 m in the vicinity of Fort Ross, it was only after an hour of a determined struggle that the beast let him out of its viscous grasp. (A whole hour. Sixty minutes of which only the genius of Victor Hugo could describe the full horror.)

Thus, if an octopus attacks a man in the presence of a witness, the latter will have plenty of time to assist his companion before anything drastic happens; with a knife or an axe, even only with a sharp object, it is possible to overwhelm even a large specimen since their movements are slow and it is easy to anticipate them. This kind of story would add up to the thick file of narrow escapes. But, in that case where an aggression takes place in the absence of a third party, the larger the octopus, the less likely will the man attacked be able to untangle himself from its embrace. It is possible that he might - I have already given a number of examples - and a new story will be added to the survival file. However, if, immobilized by an oversized ennemy, he does not manage to free himself and is carried away to some dark corner, nobody will ever hear of him again. His case - the only one of an actual death - will be classified under "Disappeared for unknown causes."

How is it possible then to ever find out about victims of octopus attacks?

In order to fully measure the danger that octopus may represent for people, we always come back to the question of the size which they can attain.

In these last years, the opinion of specialists was that the weight of giant octopus was only in the tens of kilos, while that of giant squids could be counted in tons.

Formerly, there was no hesitation in attributing a very large size to some octopus of the Indian Ocean. In 1777, the British zoologist Thomas Pennant related, on the basis of a report by one of his friends, resident in India, that an octopus 3.65 m wide in its central part had been seen in the local seas; its arms were 16.50 m long. With what we know today, even the most open minds are likely to believe that such dimensions are more likely to indicate a squid of the genus *Architeuthis*.

And if there *really* existed octopus with a size comparable to that of their ten-armed cousins? The sea has already had so many surprises for us! Who knows whether those monstrous arms vomited by sperm whales, and which can't come from giant squids of accepted sizes, might not come from such monsters?

So, even if there existed, at a depth of 100 metres or more, real colossal octopus, we would have little chance of establishing their existence according to the demanding requirements of science. Until we have at our disposal extremely flexible machines to systematically explore the depths of the ocean, we must continue to examine with the greatest care the contents of the stomach of those poor sperm whales which we are so cruel as to hunt and harpoon.

At the end of the last century, it was thought for a moment that the existence of the colossal octopus had been established on the basis of irrefutable material proof. The news had first been published in a local paper, the *Florida Times-Union*, of Jacksonville, on December 1, 1896, under the title *Big Octopus on the Beach*:

"*St.Augustine, Fla. November 30.*

- A great fish. supposed to be a whale, 22 feet [6.70 m] long, 8 feet wide [2.40 m] and about 6 feet [1.80 m] high, was found partly buried in the sands of Anastasia Beach yesterday evening by Herbert Colee and Dunham Coxetter who were making a cycle run to Matanzas inlet. The immense carcass lies about four miles [6.5 km] south of Dr. Grant's hotel and has probably been beached several days, as it is well embedded in the sands: it is where a whale was caught in the Matanzas river two years ago. President Webb of the scientific society examined the monster this evening and says it is an octopus. The tentacles are gone, either by the action of the sands or through sharks eating them off."

The stranding had thus occured on Anastasia Island, in front of St. Augustine, about fifty km south of Jacksonville. Soon the *New York Herald* would join the local press. On December 2, it published a dispatch which had been sent by telegraph, with the result that the names of the witnesses were distorted and some of the wreck's dimensions were misquoted: "Its body, which is estimated to weigh about five tons, has sunk into the sand to a considerable depth, but that portion above the surface measures twenty-three feet [7 m] in length, four feet [1.20 m] in height and more than eighteen feet [5.50 m] across the widest part of the back."

There was however an additional point mentioned in that strangely titled "*Last of this Sea Serpent*: "The hide is of a light pink color, nearly white, and in the sunshine has a distinct silvery appearance. It is very tough and cannot be penetrated, even with a sharp penknife."

The rest of the story has been painfully reconstructed, mainly thanks to the American researcher Gary S. Mangiacopra, from letters sent by its principal protagonist, Dr. DeWitt Webb, a physician obsessed with natural history and president of the *St. Augustine Historical Society and Institute of Science*. Here is what he reported to Mr. J.A. Allen in a letter dated from St.Augustine on December 8, 1896:

THE KRAKEN AND THE COLOSSAL OCTOPUS

"You may be interested to know of the body of an immense Octopus thrown ashore some miles south of this city. Nothing but the stump of tentacles remain, as it had evidently been dead for some time before being washed ashore. As it is, however the body measures 18 feet [5.50 m] in length by 10 feet [3 m] in breadth. Its immense size and condition will prevent all attempts at preservation. I thought its size might interest you, as I do not know of the record of one so large."

Dr. Webb's letter finally ended up in the hands of professor Addison Verrill, already famous at that time for his brilliant studies of the super-giant squids. He immediately published a scientific note on this topic in the *American Journal of Science* of January 1897. He indicated however that, from the proportions given, it was more likely to be a squid than an octopus, a squid weighing 4 to 5 tonnes, and clearly larger than the specimens duly authenticated in Newfoundland.

Professor Verrill soon received from Dr. Webb additional information on the amazing stranding. The St. Augustine physician told him, among other things, that excavations had led to the discovery in the sand of a stump of an arm 36 feet long [11 metres] and with a diameter which still reached 10 inches [25 cm] where it had lost its extremity. Faced with all the details available and after examining photographs of the carcass, professor Verrill had concluded that one should agree with Dr. Webb's diagnostic: it was really an octopus.

He formally recognized it in the Sunday supplement of the *New York Herald* on January 3, 1897:

"Dr Webb has sent me photographs, four different pictures of the animal. They were taken on the same day he examined it. They show that the body is flattened, pear shaped, largest near the back end, which is broadly rounded and without fins. This form of the body and its proportions show that it is an eight-arm cuttlefish, or octopus, and not a ten-armed squid like the devil fish of other regions. No such gigantic octopus has been heretofore discovered."

In the American *Journal of Science* of February 1897, professor Verrill proceeded with the formal christening of the wreck:

"This species is evidently distinct from all known forms, and I therefore propose to name it Octopus giganteus. It is possible that it may be related to Cirroteuthis, and in that case the two posterior stumps, looking like arms, may be the remains of the lateral fins, for they seem too far back for the arms, unless pulled out of position."

The triumphant Dr. Webb also communicated to Verrill a valuable report from one John L. Wilson, who had been there when the carcass had been first thrown on the beach:

Figure 142. The Anastasia Island carcass and the three tams (four horses and six men altogether) which managed to drag what was left of it onto planks to carry it away from the beach (Photo taken during the second week of January 1897)

Figure 143. Michael Raynal baptized the Florida octopus *Otoctopus giganteus*, the giant eared octopus (Photo Raynal)

Figure 144. Forrest G. Wood initiated modern research on the Anastasia Island carcass (Photo B. Heuvelmans)

Figure 145. Prof. Joseph F. Gennaro Jr. applied modern analysis techniques to identify preserved tissues of the Anastasia AIsland carcass (Photo B. Heuvelmans

Figure 146. Ciliated octopus, 2.50 m long, photographed in 1984 from the pocket submersible Cyana at a dept of 2,600 m (Photo IFREMER, Biocyarise, 1984)

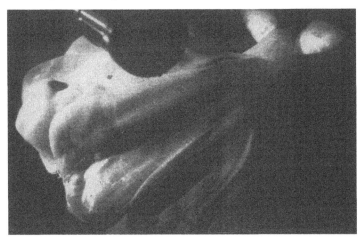

Figure 147. Tissue samples seen in polarized light; left, squid; middle, octopus; right, the Anastasia Island, Florida, carcass. Upon examination, the structure of the carcass tissue more closely resembles that of the octopus than that of the squid (From *Natural History*, Mar 1971)

Figure 148. Prof. Franco Tassi's comparison of the dimensions of the Florida colossal octopus with some specimens of the super-giant squid (L'Orsa, Milano, Dec. 1989)

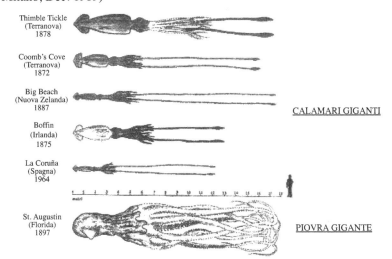

Figure 149. Tentative sketch of the Florida colossal octopus, represented as a ciliated octopus, by Stefano Maugeri (L'Orsa, Milano, Dec. 1989)

THE INCOMPLETE STORY OF THE COLOSSAL OCTOPUS

"One arm was lying west of body, 23 feet [7 m] long; one stump of arm, west of body, about 4 feet [120 cm]; three arms lying south of body and from appearances attached to same (although I did not dig quite to the body, as it laid well down in the sand, and I was very tired), longest one measured over 32 feet [9.15 m], the other arms 3 to 5 feet [90 cm to 1.50 m] shorter."

Soon after Mr. Wilson's examination, the corpse had been carried away by a violent storm and had ended up further south on Anastasia Island, at Crescent Beach, which, according to Dr. Webb, explained how it had lost its arms. Following this, on January 16, 1897, the conscientious president of the *St. Augustine Historical Society and Institute of Science* gathered together three teams, altogether four horses and six men, armed with pulleys, to move the well-bundled carcass, and he managed to pull it out of its hole and to put it on boards, about 12 metres higher on the beach. He then undertook to cut out samples 25 cm by 7.5 cm, which he sent, preserved in formaldehyde, to Dr. Verrill at Yale University and to professor William H. Dall the mollusc specialist of the Smithsonian Institution. Later on, he had the stinking mass pulled by horses to its final position, on South Beach, near the railway terminus.

The story ends there, for Verrill's giant octopus did not pan out. In April, the champion of *Architeuthis* published another article on this matter in the *American Journal of Science*, stating this time that: "Additional facts have been ascertained and specimens received that render it quite certain that this remarkable structure is not the body of a Cephalopod." Concerning the arms which had been examined, he declared that: "Subsequently, when it was excavated and moved, this statement proved to be erroneous. Apparently nothing that can be called stumps of arms, or any other appendages, were present. Folds of the integument and mutilated and partly detached portions may have been mistaken for such structures."

Professor Verrill's new opinion was essentially based on his examination of the skin samples taken from the carcass: "These masses are from 3 to 10 inches [7.5 to 25 cm] thick, and instead of being muscular, as had been thought, they have a structure similar to the hard, elastic variety of blubber-like integument found on the head of certain cetaceans, such as the sperm whale." He then concluded, after presenting a detailed description of this structure: "Although such an integument might, perhaps, be supposed compatible with the structure of some unknown fish or reptile, it is certain that it is more like the integument found upon the upper part of the head of a sperm whale than anything else I know."

This prudent suggestion was soon transformed into a categorical affirmation by Dr. Frederic A. Lucas, curator of the *U.S. National Museum*:

"The substance looks like blubber, and smells like blubber, and it is blubber, nothing more nor less." The chronicler of the *London Natural Science* snickered: " *The moral of this is that one should not attempt to describe specimens stranded on the coast of Florida, while sitting in one's study in*

THE KRAKEN AND THE COLOSSAL OCTOPUS

Connecticut."

Such is the epitaph of *Octopus giganteus*. For my part, I had so much confidence in the authority of professor Verrill, that I did not even think of mentioning the incident in the first edition of this book. It seemed to me that it would merely burden an already heavy document. I was wrong. In Science, one should never trust anyone.

Nevertheless, the numerous photographs which we have of the wreck clearly show "stumps of arms", or at least extensions of the body. Further, there are statements, difficult to dispute, by Dr. Webb and John Wilson, who spoke of pieces of arms found on the beach. And everyone has forgotten the description of the carcass given by one of the first people to see it, Dr. George W. Grant, the owner of the hotel near which it had stranded. His report appeared on December 13, 1896, in the *Pennsylvania Grit*, of Williamsport:

"The head is as large as an ordinary flour barrel and has the shape of a sea lion head. The neck, if the creature may be said to have a neck, is of the same diameter as the head. The mouth is on the other side of the head and is protected by two tenacle [sic] tubes about eight inches [20 cm] in diameter and 30 feet [9 m] long. These tubes resemble an elephant's trunk, and obviously were used to clutch in a sucker-like fashion any object within their reach.

"Another tube or tenacle [sic] of the same dimensions stands out on the top of the head. Two others, one on each side, protrude from behind the monster's neck and extend fully 25 feet [4.50 m] along the body and beyond the tail, which is separated and jagged with cutting points for several feet, is flanked with two more tenacles [sic] of the same dimensions as the others and 30 feet [9 m] long. The eyes are under the back of the mouth, instead of above it.

"This specimen is so badly cut up by sharks and sawfish that only the stumps of the tentacles remain, but pieces of them were found strewn for some distance on the beach, showing that the animal had a fierce battle with its foes before it was disabled and beached by the surf."

Professor Verrill's about-face is impossible to understand, and one can't believe that a rather summary examination of integuments could have justified it. It would seem to have been suggested to him by his colleague Dall, who was also an indisputable authority. One should not forget that Verrill occupied an enviable official position within the scientific establishment, and that he had probably been criticized for his christening of the octopus, found premature. The comment which appeared in *Natural Science* was representative of the prevailing opinion of scientists, which, one must admit, is always of an excessive prudence, bordering on cowardice. This time, there was for Verrill no question of back-tracking, and he expressed his definitive advice on the matter in the April number of the *American Scientist*:

THE INCOMPLETE STORY OF THE COLOSSAL OCTOPUS

" ...my present opinion, that it came from the head of a creature like a sperm whale in structure, is the only one that seems plausible from the facts now ascertained."

The oracle had spoken. The great cephalopod specialist had stated. One had to concede. Nevertheless, what Dr.Webb and his colleagues had seen of the remains of the animal did not agree with this pronouncement. He mentioned this in a letter of March 17, 1897, addressed to professor Dall:

"I do not see how it can be any part of a cetacean, as Professor Verrill says you suggest. It is simply a great bag and I do not see how it could have been any part of a whale."

Verrill did not pursue the polemic. As for Dr. Webb, he continued to preside with dignity his "provincial" scientific society, for a total of 34 years. But we can be sure that he must have continued to have the incident in mind.

It took nearly 75 years for the matter to be reconsidered. In 1957, a researcher at the oceanographic department of the *Naval Undersea Research and Development Laboratory* of San Diego, Forrest G. Wood, happened to discover some old newspaper clippings about the St. Augustine marine monster. Uneasy about the way the case had been summarily dismissed, he began to go back to sources and to exhume the correspondence exchanged between the protagonists. After about a decade of real detective work he discovered that fragments of Verrill's *Octopus giganteus* had been preserved in a jar at the *Smithsonian Institution,* in Washington.

Wood interested one of his friends from the University of Florida, Joseph F.Gennaro Jr. in this search; Gennaro, who was soon to become professor of biology at New York University agreed to carry out a histological study of the preserved fragments.

A meticulous microscopic examination of thin cuts made in the sample was to settle the question: "After 75 years, the moment of truth was at hand. Viewing section after section of the St. Augustine samples, we decided at once, and beyond any doubt, that the sample was no whale blubber."

Was it then the remains of some very large *Architeuthis*? A comparison of the specific structure of the connective tissues seen on the sample with cuts from octopus and squids showed that it was much closer to what was found in the former than in the latter. Professor Gennaro was forced to conclude:

"The evidence appears unmistakable that the St.Augustine sea monster was in fact an octopus, but the implications are fantastic. Even though the sea presents us from time to time with strange and astonishing phenomena, the idea of a gigantic octopus, with arms 75 to 100 feet [25 to 30 m] in length and 18 inches [45 cm] in diameter at the base - a total spread of some 200 feet [60 m] - is difficult to comprehend."

The results of the amazing research of Wood and Gennaro were published in March 1971 in *Natural History,* the magazine of the *American Museum of Natural History of New York.* It was a scandal! The revelation was so astounding that, the issue of the magazine having been offered for sale at the beginning of the previous month, many people imagined that it must be an April fool's joke. Museum officials had to issue a press statement insisting that it was definitely not a hoax.

Where could these gigantic octopus live, whose representative had been found stranded on a Florida beach?

It was Forrest G.Wood who first suggested as a hypothesis that these incredible monsters might inhabit the region of the Bahamas, across Florida Straits. Years before, in 1956, his fishing guide, Duke, had told him about the giant "scuttles" with 75 foot [23 m] long arms, with lived at great depths. *Scuttle* is a local distortion of *cuttle,* and is also probably assimilated by local folk with the verb *to scuttle,* perhaps based on the ship-wrecking reputation of large cephalopods. Further, the Island Commissioner of the archipelago had told Wood that, when he was a child, his father and his friends had pulled out, under his very eyes, from a depth of 80 metres, a very large octopus, attached to their line. Luckily, the animal had let go of the line and had gotten a hold of the keel of their boat before disappearing, to the greatest relief of all witnesses, frozen with terror.

Gary S. Mangiacopra followed up on Wood's idea and confirmed that the Florida current could in fact carry the corpse of a giant cephalopod - Denys de Montfort's colossal octopus finally rehabilitated - from the Bahamas to Anastasia Island. His French colleague Michel Raynal undertook the task of digging up all old reports on observations of octopus of alarming dimensions in that region.

The first one of these is rather vague, but nevertheless significant. It emanates from Pietro Martire, of Anghiera, the biographer of Christopher Columbus, who in his book *De Orbe Novo,* written around 1500, spoke of the Spaniards who had just arrived in the Carribean islands: "One night that one of them had gone ashore and was sleeping in the sand, a monster came out of the sea and grabbed him by the middle of the body, and took him away, in spite of the presence of his companions, and jumped back in the sea with his prey." It must have been an octopus, the only large animal capable of performing such a feat, and the place where this happened must have been near the Bahamas, the first islands visited by Columbus.

Raynal then quoted the story, which I cited earlier, mentioned by the French journalist Revoil, of the attack on a ship by a gigantic octopus "in the shoals of the Lucayes islands", meaning the Bahamas. We recall that the helmsman had cut with his axe an arm which was still more than 3.50 m in length and was the thickness of a man.

Our astute   cryptozoologist then recalled the observation by J.S.George, of Nassau, in the Bahamas, who found an octopus more than three meters in length, dead on the beach.

He then quoted the observation, in 1941, by J.C.Martin, a retired researcher from the *U.S.Naval Undersea Research and Development Center,* in

San Diego, who, off Fort Lauderdale, in Florida, saw, floating in the water a creature about 9 m long, whose arms, about the same length, were more than 80 cm in circumference.

Raynal finally reviewed a whole series of rumours which showed that there exists in the Bahamas a genuine tradition about gigantic octopus, exemplified by a number of specific incidents. They are thought to inhabit the Blue Holes, immense holes like those found in Swiss cheese, found at the bottom of the ocean and ideal refuges for benthic cephalopods, like octopus. One should not be surprised that a marine monster, locally called the *lusca*, is supposed to lurk in these, in wait for fishing boats, which it has the reputation of pulling under. It is sometimes nicknamed *Him of the Hands*, or *Him of the Hairy Hands*.

This is one of the elements which suggested to Michel Raynal to try to find more precisely the identity of this gigantic cepahlopod.

Let's not forget that Gennaro's histological analysis was subject to a certain degree of uncertainty. The structure of the tissues of the sample taken indicated that its owner was closer in nature to octopus than to squids, but to what extent? Further biochemical analyses had also confirmed this point: there was no question of affinity to cetaceans, but a clear similarity to what is found in cephalopods. However, as the identity of the monster remained unknown, only hypotheses could be formulated.

Raynal then reminded us that Verrill himself had first suggested a possible kinship with the Cirrates, or ciliated octopus, because of the position of the "two posterior stumps", which seemed more like "the lateral fins" found in the latter. This is what accounts for their popular name of "eared-octopus".

This hypothesis was rather bold at the time, because the Cirrates then known were all of rather small dimensions. About thirty centimeters at most, except for the type specimen of *Cirroteuthis magna*, captured by the *Challenger* in Antarctica which reached the exceptional length of 1.15 m. But in March 1984, the French submersible *Cyana* filmed in the eastern Pacific a ciliated octopus of a length of 2.50 m. A large Cirrate was now in the realm of possibilities... The giant Florida octopus might have been one, and its nickname *Him of the Hairy Hands* might have been refer to the ciliae which entirely cover the arms of Cirrates.

This is why Michel Raynald suggested to change the name *Octopus giganteus* to *Otoctopus giganteus:* the giant eared- octopus. A new name certainly seemed necessary, for it seems highly improbable that the Florida octopus, whose weight must have reached 3 or 4 tonnes - just remember the arrangements necessary to move it! - should be of the same genus as the common octopus.

In fact, there remains the question as to which, of *Architeuthis* or *Otoctopus* is the heaviest. Keeping in mind that the size of living creatures in the sea, as well as on land, is limited by their physiology, we might gamble that the extreme weight of a sedentary benthic cephalopod must be greater than that of a pelagic cephalopod, built for speed.

After many detours and by-ways, we have finally managed to answer some of the various questions which faced us in the first pages of this book. No, there is no doubt that the millennial terror of divers and navigators of those marine monsters known as *Kraken* or Colossal Octopus are fully justified. In all the seas of the world, a man can find himself facing octopus of a weight nearly equal to his, but with so many arms, like some Indian vengeful deity, and with a method of immobilization worthy of the god of Sodom and Gomorah, that the struggle is uneven from the start. An encounter even by a large boat of a squid as large as a bomber, or of an octopus with a span of 60 metres, must be an experience more horrific than the most morbid imagination could invent. Even the story of the fight of Laocoon and his sons against the two snakes sent from Tenedos to strangle them pales in contrast with this terrifying image.

We have heard of some such stories, which usually ended without damage for the sailors, either because their boat was a large one, or because the beast was a smallish one, or half-dead. I doubt however that the occupants of a whaleboat grasped by some tentacular monster, weighing tens of tonnes, and still in energetic health, would have had the opportunity to come and tell us of their misadventure or of their prowesses. And if this kind of encounter "which would not have pleased even a god" happened to an ordinary bather? Recall the shipwrecked sailor from the *Britannia* pulled of in a wink by what was probably a medium-sized squid.

The *National Geographic Society*, following the enquiry previously referred to, declared that no skin or hard-hat diver had ever found himself face-to-face with an *Architeuthis*[3]. This assertion seems to me absolutely gratuitous, for if such a thing had happened, it is more than likely that we would have never found out about it. The learned assembly may well point out that this carnivorous beast normally lives at a depth in excess of 160 m, and thus greater than that ever reached by divers; we have however had the opportunity to find out that in some cases the super-giant squid comes up to the surface; sometimes it comes so close to the shore that it ends up being thrown ashore. It is quite possible that on such occasions one of these monsters might have accidentally run into some isolated swimmer, a shipwrecked sailor, for example, or the occupant of a frail canoe, kayak, or raft. What news would have reached us of such an event except: *Disappeared at Sea*, or *Mysterious drowning*? How many unsuspected dramas, horrifying pains, or unspeakable agonies may hide under these terse formulas?

Of course, such tragedies, of which mankind keeps the memory thanks to the testimony of those who miraculously escaped, can only be exceptional, since they are based on a set of very rare circumstances. But we know that it is precisely upon exceptional facts that legends are built. It would then be absurd to reject it under the pretext that it is unlikely: it's very essence is to be unlikely.

---

[3] *Captain Harry Rieseberg, great salvager of sunken treasures and smiter of colossal octopus in front of the Almighty, claims that this happened to him and that he is indeed the only one to whom it ever happened. We shall leave to him the entire responsibility of his declaration.*

THE INCOMPLETE STORY OF THE COLOSSAL OCTOPUS

One rightly criticizes hasty generalizations, but little is said of those who automatically reject everything exceptional, unusual, abnormal, extraordinary and fantastic. However, these two attitudes are not without some similarities: they are really antithetical. The first consists in believing that an event is usual, normal, or general because one witnessed it; the second consist in not believing it because, it being rare, unusual, and not general, one does not have the opportunity to witness it. In both cases the conclusion is based on limited experience. From a mathematical point of view, both attitudes contradict the laws of probability. One corresponds to recognizing in a bell-shaped Gaussian curve only one of its extremities; the other to deny the existence of these extremities.

An old joke is that of the stranger who, having just landed at Le Havre, wrote to his family that in France all women had red hair, because the first woman that he had seen when setting foot ashore had hair of that colour. Would this brave man be any less ridiculous if, after having walked all day in town without meeting a single redhead, he was to inform his relatives that there was not in France a single woman with fiery hair?

Paris, Ile du Levant, 1954-1957,
1974,
1994.

BIBLIOGRAPHY

"In this kind of study of Histories one must consult without difference all kinds of authors, old as well as new, foreign and French, to learn those things of which they diversely discourse."

MICHEL de MONTAIGNE

PART 1
HERE, ANYTHING IS POSSIBLE

ALBERTUS MAGNUS, *De Natura Animalium* in *Opera Omnia* (t.VI, Lyon, 1651).

--- *Man and the Beasts* (de Animalibus, trans. J.Scanlan, Bainghampton, N.Y., 1987)

ARISTOTLE, *Histoire des Animaux* (bk.IV, chap.VII, § 12).

--- *The Complete Works* (J.Barner, ed.,Princeton, Princeton Univ.Press, 1984)

BANCROFT (Edward M.),`On the fish known in Jamaica as the Sea-devil' (*Zoolog.Journal of London*, **4**, pp. 444-457, 1829).

BATTEN (Roger),`The need to classify' [about Neopilina] (*Natural History*, New York, **67**, pp. 148-155, Mar 1958).

BOSCHMA (H.), `Quelques baleines de petite taille' (*Endeavour*, London, **10**, pp. 131-135, 1951).

BRASIL (L.), `Sur le Mésoplodon bidens échoué au Havre en 1825'(*C.R.Séances Soc.de Biol.*, Paris, **67**, p. 479,

1909).

BRELAND (Osmond P.), `Which are the biggest?' (*Natural History,* New York, **62,** pp. 67-96, Feb 1953).

2

BULLEN (Frank), *The Cruise of the Cachalot (*London, John Murray, 1898).

--- *La Coisière du Cachalot* (Paris, Silliver, 1950).

CARSON (Rachel), *The Sea around us* (New York, Oxford University Press, 1950).

--- *Cette mer qui nous entoure* (Paris, Stock,1952)

CHEVEY (P.), `Capture d'un Requin-Baleine, *Rhineodon typus* A.Smith, en Cochinchine'. (*Institut océanographique de l'Indochine,* Station Maritime de Cauda, 30 avril 1936).

GINSBURG (Isaac), `Captured "Living Fossil" fish' (*Science News Letter,* Washington D.C. January 17, 1953).

HARRY (capitaine), `La Manta peut atteindre deux tonnes' (*Science et Voyages*, Paris 18 novembre 1920).

LANDRIN (Armand), *Les Monstres Marins* (Paris, Hachette, 1867).

McCANN (Charles), `The Whale Shark *Rhineodon Typus* (Smith)', (*J.Bombay Nat.Hist.Soc.,***52**, (Nos 2 and 3), pp. 326-334, 1954).

MUNSTER (Sebastian) *Cosmographia Universalis* (Basel, Henrich Petri, 1489-1552).

--- *La Cosmographie universelle* (Paris, Sennius, 1575).

NORMAN (J.R) and FRASER (F.C.), *Giant Fishes, Whales and Dolphins* (London, Putnam, 1937; New York, Norton C°, 1938).

ONFFROY DE THORON (Vicomte Enrique), *Amérique équatoriale, son histoire pittoresque, sa géographie et ses richesses naturelles, son état présent et son avenir*

BIBLIOGRAPHY

(Paris, Vve. J.Renouard, 1866).

OPPIAN *Les Halieutiques* (Paris, Lebegue, 1817)

--- *Halieuticks: of the nature of fishes* (Oxford, Printed at
the Theatre, 1722)

PARÉ(Ambroise), `Des monstres marins' (Appendice au *Livre
des Monstres*, chap.1) in *Oeuvres Complètes* (Paris,
Baillière, 1840-1841).

3

REVOIL (Benedict-Henry) *Shooting and Fishing in the Rivers,
Prairies and Backwoods of North America* (London,
Tinsley Bros., 1865).

RONDELET (Guillaume), *Libri de piscibus marinis in quibus
verae piscium effigies expressae sunt* (Lyon, 1554)

--- *L'Histoire entière des poissons composée premièrement en
latin [..] maintenant traduite en français sans avoir
rien omis étant nécessaire à l'intelligence
d'ixelles avec leurs pourtraits au naïf* (Lyon,
M.Bonhome, 1558).

SANDERSON (Ivan T.), *Living mammals of the world* (pp 209
-220, New York, Hanover House; London, Hamish Hamilton,
1955).

--- *Follow the Whale* (Boston and Toronto, Little,Brown and
C°,1956).

SMITH (J.L.B.), *Old Fourlegs:the story of the Coelacanth*
(London, Longmans, Green and C°).

TRYON (George Washington), *Manual of Conchology, structural
and systematic* (vol.1, pp. 70-74, Philadelphia, 1879).

ULLOA (Antonio de) and Jorge JUAN, *Relacion histórica del
viaje a la America Meridional* (vol 1, Madrid, 1758).

--- *A Voyage to South America* (vol 1, p. 130, London, 1758).

VERRILL (A.Hyatt), *Strange prehistoric animals* (Boston,
L.C.Page & C°, 1948)

--- *Les Etranges animaux préhistoriques* (Paris, Payot, 1953).

2

ON THE AGRESSIVITY OF THE OCTOPUS, IN
LITERATURE AND IN THE OCEAN.

***.: `A Marine Diver Attacked by an Octopus' (*Ill.
Australian News*, no 183, p. 187, Nov 28,1879).

4

*** : `A Fight with an Octopus' (*New Haven Journal and
Courier*, Aug 3, 1894).

*** :`Octopuses Kill Fishermen' (*New Zealand Herald*,
Auckland, Sep 10, 1984).

*** : `Search for the Blue Ringed Octopus' (*Science News*,
Washington.D.C., **93**, Mar 9, 1968).

*** : `Les brigands de la mer' (*Yachting Gazette*, Paris,
16 décembre 1899).

*** : `Kraken zogen zwei Fischer in die Tiefe' (*Welt*, 11
September 1984).

AELIANUS (Claudius), *De Natura animalium*. (Basel,
A.Gronovius, 1750)

ALDROVANDI (Ulisse),*De Reliquis animalibus exanguibus ut
pote de mollibus, crustaceis, testaceis et zoophytis
libri quatuor post mortem ejus editi* (Lib. I, chap.II,
Francoforti, imp. J.Treudelii, 1618).

ARNAULT (Rene) `Le charmeur de pieuvres' (*Sciences et
Voyages,* Paris, n. sér. no 17, pp. 16-20, aout 1969).

ATHENAEUS, *Deipnosophistarum sive Caenae sapientium libri XV*
(Venetiis, apud Andream Arrivabenum, 1556).

--- *Deipnosophists* (London, Will.Heinemann Ltd., 1927-1941).

BARTLETT (Norman), *The Pearl Seekers* (London, Melrose,1954).

BARTSCH (Paul), `Pirates of the Deep: Stories of the Squid

and Octopus.'(*Ann.Rep.Smithson.Inst.*for 1916, Washington, D.C., pp. 347-377, 1917).

BERGE (Victor) and Henry Wysham LANIER, *Pearl Diver* (London, Heineman, 1930).

BERRILL (Norman John), *The Living Tide* (pp. 206-207, New York, Dodd, Mead & C°, 1951).

BERRY (S.Stillman and Bruce W. HALSTEAD, `Octopus bites: a second report' (*Leaflets in Malacology,* Redlands, Calif. **1,** pp. 59-66, 1954).

BOULENGER (E.G.), *Searchlight on animals* (p.171, London, Robert Hale & C°, 1956).

5

--- *The Aquarium Book* (p.87, New York, D.Appleton-Century

C°, 1934)˙

--- *A Natural History of the Seas* (New York, D.Appleton -Century C°, 1936).

--- *La Faune des oceans: histoire naturelle de la mer.* (p.110, Paris, Payot, 1939).

BRELAND (Osmond P.), `Harmless or deadly?' (*Natural History*, New York, **62**, pp. 402-407, November 1953).

BRUCE (Henry I.),*Twenty Years under the Ocean* (London, Stanley Paul, 1939).

BUCKLAND (Frank), *Log-book of a Fisherman and Zoologist.* (London, Chapman and Hall, pp. 170-179, 1883).

BURFORD (Virgil), *North to Danger* (New York, John Day, p.173, 1954).

BURTON (Maurice), *Animal Legends* (London, Frederick Muller, 1955).

CHEESMAN (Evelyn), *Six-legged Snakes in New Guinea* (London, G. Harrap, pp.78-80, 1949).

COLL (Pieter), *Das Meer* (Wurzburg, Arena, p.55, 1968)

COUSTEAU (Jacques-Yves) *Le Monde du Silence*
(Paris, Editions de Paris, 1953).

--- *The Silent World* (New York, Harper, 1953)

CRAIG (John D.) *Danger is my Business* (London, Arthur Baker,
1938).

CROSSE (Henri), `Un mollusque bien maltraité' (*Journal de
Conchyliologie*, Paris, pp. 177-181, 1866).

CUVIER (Georges), *Le Règne animal distribué d'après son organisation*
(t.III, pp. II et ff., Paris, Déterville, 1817)

--- *The Animal Kingdom* (London, Whittaker & Co., 1835).

DAETZ (Gary), `Meet and eat the Octopus' (*Natural History*,
New York, **64**, pp. 210-213, 1955).

DANIEL (R.J.), *Animal Life in the Sea* (London, Hodder &
Stroughton, 1925).

DENYS DE MONTFORT (Pierre), *Histoire naturelle des
mollusques*, in BUFFON, *Histoire naturelle générale et
particulière*, nouvelle édit. par C.S.Sonnini (vol.
LIII, t.II, Paris, Duffart, an X [1801 or 1802]).

DEPPING (Guillaume), `Rencontres avec les poulpes' (*Journal
des voyages*, Paris, no 664, pp. 195-196, 30 mars 1890).

DOUKAN (Gilbert), *La Chasse sous-marine* (Paris, Chanteney,
2 edit. 1948).

DRAGESCO (Jean) `Le Monde extraordinaire des pieuvres'
(*Sciences et Avenir*, Paris, no 69, novembre 1952).

DUNCAN (P.Martin), editor, *Cassell's Natural History* (vol
V., London, 1876-1883).

EVANS (Bergen), *The Natural History of Nonsense* (New York,
Alfred Knopf, pp. 81-82, 1947).

FLECKER (H.) and COTTON (B.) `Fatal bite from octopus' (*Med.
Journ. Austr.*, **2**, pp.429-431, 1955).

BIBLIOGRAPHY

GIBSON (John), *Monsters of the Sea* (London, Thomas Nelson, pp. 106-113, 1894).

GILL (Rev. William W.), *Life in the Southern Isles* (London, Religious Tract Society, 1876).

GILPATRIC (Guy), *The Compleat Goggler* (London, Bodley Head, 1939).

GORSKY (Bernard), *Dix mètres sous la mer* (Paris, Durel, 1946; Plon, 1953).

GOSS (Michael), `Giant Squids on the Attack' (*Fate*, Highland Park, Ill, **38** (7), pp. 34-41, July 1985; **38** (8), pp.78-82, Aug. 1985).

GRIMBLE (sir Arthur), *A Pattern of Islands* (London, Murray, 1952).

HALSTEAD (Bruce W.) `Octopus bites in human beings' (*Leaflets in Malacology*, Redlands, Calif. **5**, pp. 17 -22, 1949).

HASS (Hans), *Welt und Wasser* (München, Molden, p. 164, 1973).

HUGO (Victor), *The Toilers of the Sea* (transl. Mary W.Artois, New York, G.H.Richmond, 1896)

7

KIPPEN (Ken), *Buried Treasure* (Garden City, N.Y., Garden City Publ.Co., pp.97-111, 1950).

KLINGEL (Gilbert), *Inagua* (New York, Dodd, Mead & Co.,1940).

LANE (Frank), *Nature Parade* (3rd ed., p 52, London, Jarrolds, 1946).

MABBETT (Hugh), `Death of a skindiver' (*Skin Diving and Spear Fishing Digest*, Sydney, pp. 13-17, Dec 1954).
MANGIACOPRA (Gary S.), `Fishermen Fables' (*Of Sea and Shore*, Port Gamble, Wash. D.C. **9**, (no 4), pp.207-208, Winter 1978-79).

MENARD (Wilmon), `Never Again' (*Natural History*, New York,

57, p.324, Sep.1948).

MICHELET (Jules), *La Mer* (Paris, Hachette, 1861).

MILLOT (Gilles), *Les Pieds-Lourds: Histoire illustrée des scaphandriers à casque français de 1850 à nos jours* (Douarnenez, Editions de l'Estran, pp.244-245, 1987).

OPPIAN, *op.cit.* (chap. iv).

PARKER (G.H.), `The power of adhesion in the suckers of *Octopus bimaculatus*' (*Jour. Experim. Zool.*, **33**, pp. 391-394, 1921).

PLINIUS SECUNDUS (Caius), *Historia naturalis (lib. ix).* (Lugduni Batavorum, Elzevier, 1635)

--- *Histoire naturelle des Animaux* (trad. P.C.B.Guéroult, Paris, Delance et Lesueur, 1802).

PYCRAFT (W.P.), `The Tale of an Octopus' (*Illustrated London News*, London, **170**, p.1,000, Jun 18, 1927).

RIESEBERG (Harry E.), `Octopus, Terror of the Deep' (*Mechanix Illustrated*, **21**, (no 4) ,pp 42-44, Feb 1939).

--- *I Dive for Treasure* (New York, Dodd and Mead, 1942).

--- *Treasure Hunters* (New York, R.M.McBride, 1945).

--- *The Sea of Treasure* (New York, Frederick Fell, 1966).

ROBSON (G.C), *A Monograph of the recent Cephalopoda based on the collections of the British Museum* (London, British Museum, 1929-1932).

8

RONDELET (Guillaume) (*op. cit*).

RUSSELL (Eric F.), *Great World Mysteries* (London, Dennis Dobson, pp. 143-144, 1975).

SCHILLING (Tom), *Bêtes sauvages et tendres* (Paris, Calmann-Levy, pp. 173-175, 1957).

BIBLIOGRAPHY

SEBILLOT (Paul), `Victor Hugo: les Travailleurs de la mer'
(*Revue des Traditions populaires*, Paris, **5**, pp.567-569,
1890).

--- *Le Folklore de la France* (Paris, Maisonneuve et Larose,
Vol.III, p.359, 1906).

SUNTER (George H.), *Adventures of a Trepang Fisher* (London,
Hurst & Blackett, 1937).

TEMPLE (major Sir Grenville T.), *Excursions in the
Mediterranean, Algiers and Tunis* (London, Saunders &
Otley, 1835).

TRYON (G.W.) (*op.cit.*)

VERANY   (Jean-Baptiste),   *Mollusques   méditerranéens*   (Gênes,
Imp.des Sourds Muets, 1851)

VERRILL (Addison E.), The Cephalopods of the Northeastern
Coast of North America (Part II, The Smaller
Cephalopods including the "squids" and the octopi,
with other allied forms) (*Trans. Connect. Academy of
Arts*, **5**, p. 258, 1879).

WILLIAMS (Woody), `Friend Octopus' (*Natural History*, New
York, **60**, (no 5), pp.210-215, May 1951).

9

PART 3
FAMILY ALBUM OF THE CEPHALOPODS

ALDROVANDI (Ulisse), *De reliquis animalibus exanguibus libri
quatuor, post mortem ejus editi, nemque de mollibus,
crustaceis, testaceis et zoophytis*. (Bononiae, apud
J.B. Bellagambam, 1606).

BARTSCH (Paul), `Pirates of the Deep and Stories of the
Squid and Octopus' (*Ann, Rep. Smithson.Institution*,
Wash.D.C., pp.347-375, 1917).

--- `The Octopuses, Squids and Their Kin' (*Smithson. Sci.
Ser.*, Wash.D.C., **10**, pp323-356, 1931).

BOULENGER (E.G.), *A Natural History of the Seas* (*op. cit.*).

BRELAND (Osmond P.), `Devils of the Deep' (*Science Digest*, Chicago, **32**, pp. 31-33, Oct. 1952).

DUNCAN (P.Martin), editor, *Cassell's Natural History* (*op. cit.*).

GOSLINE (John), and Edwin DEMONT, `La propulsion des calmars' (*Pour la Science*, Paris, pp. 31-37, mars 1985)

HEUVELMANS (Bernard), `Les Pieuvres' (*L'Aventure sous-marine*, Paris, 1er trimestre, 1955).

--- `Le monde inquiétant de ceux-qui-ont-les-pieds-à-la-tête (*Sciences et Avenir*, Paris, no 137, pp. 370-375, juillet 1958).

HEYERDAHL (Thor), *Kon-Tiki Ekspedisjonen* (Glydendal, Norsk Forlag, 1948).

--- *The Kon-Tiki Expedition* (London, George Allen & Unwin Ltd. 1950).

--- *L'Expédition du Kon-Tiki* (Paris, Albin Michel,1951)

LANE (Frank), *Kingdom of the Octopus: the Life History of the Cephalopoda* (London, Jarrolds, 1957).

10

LATIL (Pierre de), et Jean RIVOIRE, *A la recherche du monde marin* (Paris, Plon, 1954).

MANGOLD (Katherina), `Cephalopods' in P.P.Grassé, *Traité de Zoologie* (Paris, Masson, 1989).

MINER (Roy Waldo),`Marauders of the Sea' (*National Geographic Mag.*, Washington, D.C. **62**, (no 2), Aug 1935).
NIXON (Marion), and J.B. MESSENGER, editors, *The Biology of Cephalopods* (London, New York, San Francisco, Academic Press, 1977).

ORBIGNY (Alcide D'), *Voyage dans l'Amérique méridionale* (Paris, Strasbourg, T. V, part.3, Mollusques, pp. 1-6, 1835-1843).

BIBLIOGRAPHY

PARKER (G.H.), `The Power of Adhesion in the suckers of *Octopus bimaculatus*' (*J.Experiment. Zool.*,**33**, pp.391 -394, 1921).

PICKFORD (Grace Evelyn), `The Vampyromorpha, living fossil Cephalopoda' (*Trans.New York Acad. Sci.* **2,** (2), pp.169 -181, 1940).

RONDELET (Guillaume), (*op.cit.*).

SYKES (colonel W.H.),` On the power of leaping to a considerable height possessed by *Loligo (Ommastrephes) sagittata* Lam.'(*Proc.Zool.Soc.*London, **1,** p.90, 1833).

VARRO, *De Lingua latina*, (Paris, Collect. Nisard, 1845).

VERNE (Jules), *Vingt mille lieues sous les mers* (Paris, Hetzel, 1875)

--- *20,000 Leagues under the Sea* (Glasgow, William Collins & Sons, 1972).

WILLIAMS (Woody), `Friend Octopus' (*Natural History*, New York, **60**, (no 5), pp.210-215, May 1951).

On the systematics and biology of cephalopods, we advise the reader to consult, in addition to classical works, the more recent contributions of W.Adam, I.Akimushkin, F.A.Aldrich, A.Apellof, Z.M.Bacq, S.S,Berry, S.von Boletzky, G.Chun, M.R.Clark, R.Clark, W.G.Fields, G.Grimpe, F.G.Hochberg Jr., G.A.Jaeckel, L.Joubin, G.E. and N.McGinitie, A.Naef, G.Otterlind, G.Pfeffer, G.Pickford, A.Portmann, B.B.Rae, W.J.Rees, G.C.Robson, C.F.E.Roper, M.Sasaki, A.C.Stephen, J.Thiele, G.L.Voss, etc...

11

PART 4
GIANT POLYPS OF ANTIQUITY

AELIANUS (Claudius), (*op.cit.* bk.XIII, chap. VI).

ARISTOTLE (*op.cit.*, bk.IV, chap 1).

ATHENAEUS (*op.cit.*, bk.VII, t.III, p 199)

BROWNE (sir Thomas), *Pseudodoxia epidemica: Inquiry into*

vulgar error (London, E.Dod, 1646*)*.

COTTE (J.), *Poissons et animaux aquatiques du temps de Pline*
(Paris, P.Lechevalier, 1944).

FREDOL (A.) [pseudonym of Alfred MOQUIN-TANDON], *Le Monde de
la Mer* ((Paris, Hachette, 1865).

FULGOSIUS (Baptista), *De Dictis factisque memorabilibus
collectanea, Camillo Gilino latine facta* (bk. 1,
Mediolani, 1509).

GERVAIS (Paul), *Sur le Grand Calmar de la Méditerranée*
(Montpellier, s.d.) [1862].

--- *Des notions relatives aux Céphalopodes qui sont
consignées dans Aristote* (Mémoire lu le 8 avril 1863 à la Réunion
des Sociétés savantes tenue à Paris sous les
auspices de M. le ministre de l'Instruction publique).

HEPPEL (David), `Unveiled: Satan's murky origins'(*The
Independant*, p.13, Jul 1990).

HOMER, *L'Iliade* (XX, 147, trad. Leconte de Lisle, Paris.
Lemerre, 1802-1822).

--- *The Iliad*, (transl. R.Fagles, New York, Viking, 1990)

--- *L'Odyssée* (XII, pp. 73 ff).

--- *The Odyssey* (transl.W.Shewring, New York, Oxford Univ.
Press, 1980).

12

KOERNER (Otto), *Die homerische Tierwelt: ein Beitrag zur
Geschichte der Zoologie* (Berlin, Nicolaï, 1880).

KOLLMANN (J.),` Die Cephalopoden in der Zoologischen Station
des Dr. Dorn' (*Zeitsch. wiss. Zool.*, **26**, pp.1-23,1875).

LEY (Willy),`Scylla was a Squid' (*Natural History*, New York,
48, pp.11-13, Jun 1941).

MOULE (L.), `Etudes zoologiques et zootechniques dans la
littérature et dans l'art: la faune d'Homère' (*Mém. soc.*

zoolog. Paris, **22**, pp. 183 ff, 1909; **23**, pp.29-106, 1910.

PLINY(*op.cit*., bk. IX, chap. III and XXX; and bk XXXII, last chap.).

SALVERTE (Eusèbe Baconnière de), *Des Sciences occultes ou Essai sur la magie, les prodiges et les miracles* (p.32, Paris, Sédillot, 1829).

STRABO, *Geography* (bk.III) (London, G.Bell and Sons, Ltd., 1912-1916).

5

THE MEDIEVAL FABLE OF THE ISLAND-BEAST

AELIANUS (*op.cit.*).

ALBERTUS MAGNUS (*op.cit.*).

ALLATIUS (Leo), *S.Eustathii in hexahemeron commentarius* (Lyon, 1629).

AMBROISE, *Hexahemeron* (bk.V, chap. XI).

BASSETT (Fletcher S.), *Legends and Superstitions of the Sea and of Sailors in all Lands and all Times* (London, Sampson Low, 1885).

BELON (Pierre), *Histoire naturelle des estranges poissons marins* (Paris, 1551).

--- *De aquatilibus libri duo, cum eiconibus ad vivam ipsorum effigiem, quoad ejus fieri potuit, expressis* (Paris, Ch.Estienne, 1553)

13

--- *Nature et diversité des poissons.* (Paris, 1555).

BLOND (Georges), *La Grande Aventure des Baleines* (Paris, Amiot-Dumont, 1953).

--- *The Great Story of Whales* (transl. J.Cleugh, Garden City, N.Y., Hanover House, 1955).

BOISACQ (E.), *Dictionnaire étymologique de la langue grecque*
(2 ed. Paris, Klincksieck, 1923).

CARUS (Victor), *Histoire de la Zoologie, depuis l'Antiquité jusqu'au XIXe siècle* (Paris, J.B. Baillière et fils, 1880).

DENIS (Ferdinand), *le Monde enchanté: Cosmographie et Histoire naturelle fantastique du Moyen Age* (Paris, A.Fournier, 1843).

FRANKLIN (Alfred), *La Vie privée d'autrefois: les Animaux*
(t. 1er, Paris, Plon, 1897).

IZZI (Massimo), `L'Isola che vive'(*Abstracta*, Roma, **9,** (no 27), pp. 74-81, Guigno 1988).

MELVILLE (Herman), *Moby Dick of the White Whale* (London, Toronto, New York, Harper, 1851).

POUCHET (F.A.), *Histoires des sciences naturelles au Moyen Age, ou Albert le Grand et son époque considérés comme point de départ de l'école expérimentale* (Paris, J.B.Baillière, 1853).

PHYSIOLOGUS AND BESTIARIES

CAHIER (Charles) et Arthur MARTIN, *Mélanges d'Archéologie* (Paris, 1847-1856).

--- *Nouveaux Mélanges d'Archéologie* (Paris, 1874-1877).

GOLDSTAUB (M.), 'Der Physiologus und seine Weiterbildung' (*Philologus*, Supplementband VIII, p.339, 1901).

14

GUILLAUME LE CLERC, *Le Bestiaire divin avec une introduction sur les bestiaires, volucraires et lapidaires du Moyen Age, considérés dans leurs rapports avec la symbolique chrétienne, par C.Hippeau* (Caen, 1852).

HUGHES de SAINT-VICTOR, *De Bestiis et aliis rebus*, in *Opera* (t II, chap XXVI, 1526).

BIBLIOGRAPHY

LANGLOIS (Charles-Victor), *La Connaissance de la nature et du monde au Moyen Age d'après quelques écrits français à l'usage des laïcs* (Paris, Hachette, 1911).

LAUCHERT (Friedrich), *Geschichte der Physiologus,* (Strasbourg, 1889).

MALL (E.), *Li Cumpoz Philipe de Thaun, mit einer Einleitung* (Strasbourg, 1873).

MANN (M.Fr.), `Der Physiologus von Philipp von Thaun, und seine Quellen' (*Anglia*, Halle, **7**, p.420; **9**,p.391, 447).

THEOBALDUS (Episcopus), *Physiologus: a Metrical Bestiary of Twelve Chapters* (London, John & Edward Bumpus, 1928).

WHITE (T.M.), *The Book of Beasts, being a translation of a latin bestiary of the twelfth century* (London, Jonathan Cape, 1954).

SAINT BRENDAN AND SAINT MALO

****, *De la légende anglo-normande de Saint Brendan* (Halle, a.s. Eisleben, 1886).

GOEJE (J.de), *La Légende de Saint Brendan* (Leide, E.J.Brill, 1890).

JUBINAL (Achille), *La légende latine de S.Brandaines, avec une traduction inédite en prose et en poésie romanes* (Paris, 1836).

LARGILLIERE, *Mélange d'Hagiographie bretonne* (Brest, 1923).

LOT (Ferdinand), *Mélange d'histoire bretonne* (VIe-XIe siecles) (Paris, Champion, 1907).

MEDINA (Pedro de), *Libro de grandezas y cosas memorables de España, dirigida al serenisimo y muy esclarecido señior don Felipe principe de España nuestro senor* (Madrid, 1544).

15

SIGEBERTUS GEMBLACENCIS,`Vita S.Maclovii sive Machutii
　　Episcopi et confessoris'(*Acta Sanctorum*, apud Surium,
　　t.VI, chap VII, nov. die XV).

STOKES (Whitley), editor and translator, `Imram Maeldwin'
　　(*Revue Celtique*, Paris-London, **9**, pp. 447 ff and **10**,
　　pp. 50 ff, 1891)

TUFFRAU (Paul), *Le Merveilleux voyage de Saint Brandan à la
　　recherche du Paradis* (Paris, l'Artisan du Livre, 1925).

WAHLUND (Carl), *Die Altfranzösische Prosaübersetzung von
　　Brendans Meerfahrt* (Upsala, 1900).

WATERS (E.G.R.), *The Anglo-Norman voyage of St. Brendan,
　　Benedict, a Poem of the Early Twelfth Century* (Oxford,
　　1928).

ORIENT

ABBOTH (Nabia), in `Ninth Century "Thousand Nights"
　　Manuscript Found'(*Science Digest*, Chicago, April
　　1949).

BASSETT (Fletcher), *Legends and Superstitions of the Sea and
　　Sailors in All Lands and All Times* (London, Sampson
　　Low, Marston, Searle & Rivington, 1885).

CAMP (L.Sprague De) and Willy LEY, *Lands Beyond* (chap. IV,
　　New York, Toronto, Rinehart & Co, 1950).

GALLAND (Antoine), *Les Mille et une nuits* (Paris, Henri
　　Beziat, s.d.)

KASWINI, `Agiaib al-Makhloukat' *in* HERBELOT, *Bibliothèque
　　orientale* (p.64, Maestricht,1776).

LANDRIN (Armand), (*op.cit.*)

MAS'OUDI, *Les Prairies d'or* (texte et traduct. C.Barbier,
　　Paris, Imprimerie Imperiale, 1861-1877).

BIBLIOGRAPHY

NEARKOS, *in* GRONOVIUS (J.), *Les Sept livres de l'expédition d'Alexandre et l'Histoire de l'Inde par Arrien* (1704).

PLINY, (*op.cit.* bk.IX).

16

REMUSAT (Abel),`Sur l'état des Sciences naturelles chez les peuples de l'Asie orientale' (*Mém. de l'Acad. des Inscriptions*, Paris, **10**,pp.116-167,1833).

VAN DER LITH (P.A.) et L.M.DEVIC, *Livre des Merveilles de l'Inde* (Leiden, E.J.Brill, 1883-1886).

NORD

*** , *Kongs-skuggsjó (eller Speculum regale)* (udg. af Hafdan Einarsen, Sorö, 1768).

---- *The King's Mirror* (from the old Norwegian, transl. by Laurence Marcellus Larson, the American-Scandinavian Foundation, New York, 1917).

BARTHOLIN (Thomas), *Historiarum anatomicarum rariorum centuria IV* (hist.24, Hagae Comitum, ex typographia A.Vlacq, 1657).

BRINGSVAERD (Tor Åge), *Phantoms and Fairies from Norwegian Folklore* (Oslo, Johan Grunt Tanum, s.d.).

GESNER (Conrad), *Historiae animalium liber IV, qui est de piscium et aquatilium animantium natura* (ord. 12 de Cetis, Tiguru, 1560).

LEEMS (Knud), aka Canutus Leemius, *Beskrivelse over Finmarksenslappen* (KiØbenhavn, tryckt af G.C.Salikalk, 1767).

MAGNUS (Olaus), *Historia de gentibus septentrionalibus earum diversis statibus, conditionibus* (lib.XXI, chap.XXV and XXVI, Romae, 1555).

--- *Histoire des pays septentrionaux [...] en laquelle sont brièvement et clèrement déduites toutes les choses rares ou estranges qui se trouvent entre les Nations septentrionales* (Anvers, imp. de C.Plantin 1561).

NANSEN (Fridtjof), *In Northern Mists* (t.2, p. 234, London, William Heineman, 1911).

PAULLINUS (Christianus Franciscus), `De Singulari monstro marino' (*Miscellanea curiosa sive Ephemeridum medico -physicarum germanicarum Academiae naturae curiosorum*, I, anno 8, 1677, obs. LI, pp. 79-80, Norimbergae, Wolffg. Maur, Endteri. 1702).

17

--- *Observationes medico-physicae, rarae, selectae et curiosae, quatuor centuriis comprensae* (Cent.II, LXX, Lipsiae, apud viduam J.Heinrichii, 1706).

RINK (Heinrich Johannes), *Tales and traditions of the Eskimo, with a sketch of their habits, religion, language and other peculiarities* (Edinburgh, W.Blackwood & Sons, 1875).

WORM (Olaf), *Museum Wormianum, seu Historia rerum rariorum [...] adornata ab Olav Worm* (Chap. XIII, Lugduni Batavorum, apud J.Elsevirium, 1655).

AMBERGRIS

**** `On the Production of Ambergris: a Communication from the Committee of Council appointed for the Consideration of all Matters relating to Trade and Foreign Plantations, with a prefatory letter from William Fawkener, Esq. to Sir Joseph Banks, Bart.' (*Phil. Trans. Roy. Soc.*, London, *81*, 1791).

BEALE (Thomas), *The Natural History of the sperm whale, to which is added a sketch of a South sea whaling voyage* (2nd ed., London, J.Van Voort, 1839).

BOYLE (Robert), `A Letter of the Honorable Robert Boyle of Sept. 13, 1673, to the publisher, concerning Amber Greece, and its being a Vegetable Production.' (*Phil. Trans. Roy. Soc.*, London, **8**, p.6113, 1672).

BIBLIOGRAPHY

BOYLSTON (Dr.), `Ambergris found in Whales' (ibid. **33**, p.193, 1724).

CARDAN (Jerome), *De Subtilitate libri XXI, ab authore plusquam mille locis illustrati, nonnullis etiam cum additionibus* (Basileae, ex officine Petrina, 1560).

--- *Les Livres de Hiérome Cardanus [...] intitulés de la subtilitéet subtiles inventions, ensemble les causes occultes et raisons d'icelles traduit de latin en françois par Richard le Blanc* (Paris, G.le Noir, 1556).

CLARKE (Robert), `A great haul of ambergris' (*Nature*, London, 24 Jul 1954).

18

CLOQUET (Dr Hippolyte), *Faune des Médecins ou Histoire des Animaux et de leurs produits, etc...*(art. *Ambre*, t.1, p.329, Paris, Crochard,1822-1825).

CLUSIUS (Carolus), *Exoticorum libri X* (bk VII, 1605).

DIOSCORIDES, *Les Six livres de Pedacion Dioscoride d'Anazarbe, de la Matière médicinale, translatez du latin en françois [...] par Martin Mathée* (Lyon, T.Poyan, 1559).

--- *De medica materia, libri VI...* (Basle, 1532)

DUDLEY (Paul), `An Essay upon the Natural History of Whales, with a particular Account of the Ambergris found in the Sperma Ceti Whale' (*Phil.Trans.Roy.Soc.* London, **23,** p.256, 1725).

KAEMPFER (Dr Engelbert), *Histoire naturelle, civile et ecclésiastique de l'empire du Japon* (La Haye, P.Gosse & J.Neaulme, 1729)

KLOBIUS (Justus Fidus), *Ambrae historia* (Wittembergae, sumpt haered. D.T.Mevii et E.Schumacheri, 1666)

LAUFER (Berthold), `Arabic and Chinese trade in Walrus and Narwhal Ivory' (*T'oung Pao*, London, **14**, Oriental Printing Office, E.J.Brill, 1913).

MAGNUS (Olaus), (*op.cit.*).

MURPHY (Robert Cushmnan), `Floating Gold: the romance of Ambergris' (*Natural History*, New York, **33**, (2), pp.117 -130; (3), pp. 303-310, Mar-Apr,May 1933).

NIEREMBERG (J.E.), *Historia Naturae maxime peregrinae libris XVI distincta* (Antverpiae, 1635).

PIESSE (S.), *Des odeurs, des parfums et des cosmétiques.* (Paris, 1865).

POUCHET (F.A.), `Sur l'Ambre-gris' (*Centenaire de la fondation du Muséum d'Histoire naturelle: Volume commémoratif.* Paris, Impr. Nationale, 1843).

---- `Sur les calculs intestinaux du Cachalot (ambre-gris)' (*C.R. Acad. Sci.*, Paris, 20 juin 1892).

POUCHET (F.A.) et BEAUREGARD `Note sur l'ambre-gris' (*C.R.Acad.Sci.*, Paris, IX, **4** p.588, 24 juin 1892).

SANDERSON (Ivan T.), *Follow the whale* (*op.cit.*).

19

SCHOTT (Gaspard), *Physica curiosa, aucta et correcta, sive Mirabilia naturae et actis libris XII, comprehensa* (liv.X, chap X, Herbipoli, sumpt. J.A.Endteri et Wolfgangi,1667).

SCHWEDIAWER (Dr.Franz), `An account of Ambergris' (*Phil.Trans.Roy.Soc.*, London **73**, 1783).

---- `Recherches sur l'ambre-gris' (Observat. sur la physique, sur l'histoire naturelle et sur les arts, t. XXV, pp. 278-287, *Journ. de Physique*, Paris, juillet 1784).

THROUGHTON (Ellis D.), *Furred animals of Australia* (5 ad., pp.241-243, Sydney & London, Angus & Robertson, 1954).

WEGNER (A.M.R.) `What is amber, waarvoor dient het ?' (*Tropische Natuur*, Weltevreden, **33**, pp.78-81, 1953).

6

SCIENCE LOOKS AT THE KRAKEN

***, `Observationes Francisci Redi circa animalium viventia, quae reperiuntur in animalibus viventibus' (*Acta Eruditorum anno 1686 publicata*, Lipsiae, p.38, 1686).

AUBERT de La CHESNAYE DES BOIS (François Alex.), *Dictionnaire raisonné et universel des animaux* (art. *Microcosmus*, Paris, C.J.B.Bauche, 1759).

BARTHOLIN (Thomas), (*op.cit.*).

BERGEN (Carolus Augustus von), `De Microcosmo, bellua marina omnium vastissima' (*Nova Acta Physico-medica Academiae Caesarae Leopoldino-Carolinae Naturae Curiosorum exhibentia Ephemerides*, t.II, obc.38, p.143, Norimbergae, Imp. Wolfgangi Schwarzkopfii, 1761).

BLAINVILLE (Henri Marie DUCROTAY de), *Manuel de malacologie et de conchyliologie* (t.II, Strasbourg, 1825-1827).

DEBES (Lucas Jacobson), *Faeroe et Faeroea, deter: Faeröensis of Faoröeske Indyggers Beskrivelse, etc* (Kiöbenhaffn, tryckt aff M.Jorgensön, 1673).

20

--- *Faeroe et Faeroea reserata* (englished by John Sterpin, London, W.Iles, 1676).

HAPPEL (Eberhard Werner), *Mundus mirabilis tripartitus,oder wunderbare Welt in einer kurtzen Cosmographia furgestellt* (Tome I, book iv, chap.XX and XXI, Ulm , M.Wagner, 1687-1689).

LINNAEUS (Carl von), *Fauna suecica, sistens animalia Sueciae, quadrupedia, aves, amphibia, pisces, insecta, vermes distributa per classes et ordines, genera et species* (Lugduni Batavorum, apud C. et G.J. Wishoff, 1746).

--- *Systema naturae, sistens regna tria naturae in classes et ordines, genera et species reducta, secundum textum*

Stockholmiensem editionem (p.75, Lipsiae, imp. G.Kiesewetteri, 1748).

MAGNUS (Olaus), (*op.cit.*,bk.XXI, chap.V).

MOQUIN-TANDON (Alfred), (*op.cit.*, nom. FREDOL).

OKEN (Lorenz), *Lehrbuch der naturgschichte* (Dritten Theil (Zoologie), **I,** p.344, Jena, Aug. Schmid und Co, 1815).

PONTOPIDDAN (Erik Ludvigsen), *Det förste Forsög paa Norges Naturlige Historie* (part.II, chap. VIII, Kiöbenhavn, ved. L.H. (???) 1752-1753).

--- *Versuch einer naturlichen Historie von Norwegen, aus dem Danischen ubersetz von Joh.Ad.Scheiben* (Kopenhagen, F.Ch.Mumme, 1753-1754).

--- *The Natural History of Norway, translated from the Danish original* (London, A. Linde, 1755).

REDI (Francesco), *Osservazioni intorno agli animali viventi que se trovano negli animali viventi* (t.XXII, figs 1,4 and 5, Firenze, P. Batini, 1684).

SANDERSON (Ivan T.), *Follow the Whale* (*op.cit.*).

TORFAEUS (Thormod), *Historiae Rerum Norwegicarum, Pars Quarta* (Hafniae, ex typographes Joachimi Schmit, Jenii, 1711).

21

For a quick overview on the topic of the "deep scattering layer".

**** `Oceans' strange echo' (*Science News Letter*, Washington, D.C. 13 Jan 1951).

--- `Sea animals dive at dawn' (*ibid.*, 1 Sep 1951).

--- `Sea creatures seek level of deep shade' (*ibid.*, 7 Aug. 1954).

CARSON (Rachel), (*op.cit.*, pp. 55-59).

BIBLIOGRAPHY

MATTHEWS (Sam), `Mystery of the deep-sea echoes' (*Science Digest*, Chicago, Mar 1951).

7

THE FIRST PIECES OF EVIDENCE

***, *Abbildung eines erschroecklischen Meer-Wunders, so am Ende des 1661 Jahres in Holland zwischen Schevelingen und Cattwick auff der See gefanden worden* in JACOBAEUS seu LAUERENTZEN (*vide infra*).

ALDROVANDI (Ulisse) (*op.cit.*, liv.V, chap. XVII).

BJÖRN of SKARSDA, *Annalar Björns a Skarsda* (II, p.238, 1639).

BOULENGER (E.G.), *A Natural History of the Seas* (*op.cit.*).

COOK (James), *Account of a Voyage round the World in 1769-71* (vol.II, London, 1773).

FERUSSAC (André Etiénne d'Audebard, baron de), et Alcide d'ORBIGNY, *Histoire naturelle générale et particulière des Céphalopodes acétabuliferes* (pp.49 and 52, Paris, 1835-48).

GESNER (Conrad), (*op.cit.*,liv.IV, p.1543).

GRAY (John Edward), *Spicilegia zoologica* (p.3, London, Treuttel, Wurtz & Co., 1830).

22

JACOBAEUS (Oliger), *Museum regium, seu Catalogus rerum tam naturalium quam artificialium quae in basilica bibliotheca [...] Friderici IV. Havniae asservantur.* (S.17, tab. VI, fig.1, Hafniae, literis J.Schmetgen, 1696).

LAUERENTZEN (Johannes), *Museum regium [etc.], ab Oligero Jacoboeo. Med.et.Phil.Prof.Regio, quandam descriptus, [...], accurante J.Lauerentzen, assessore Consistorii Havniensis Regio* (Part.I, sec.III, tab.III, no 40, Havniae, ex Reg. Majest.et Universit. typographeo, 1710).

LÖNNBERG (Einar), `Ofversigt öfver Sveriges Cephalopoder'
(*Bihang till K.Svenska Vet. Akad. Handlingar*,
Stockholm, **17**, Afd. IV, no 6, p.36, 1891).

MOLINA (Juan-Ignacio), *Saggio sulla Historia naturale des
Chile* (Bologna, 1782).

MORE (A.G.), `Notice of a gigantic Cephalopod (*Dinoteuthis
proboscideus*) which was stranded at Dingle in Kerry,
two hundred years ago' (*Zoologist*, London, 2 ser. no
118, p 4536, Jul 1875).

OLAFSSON (Eggert), *Vice-lovmand Eggert Olafsens og
Langphysici Biarne Povelsens Reisen igiennen Island,
foranstalter of Videnskabernes Saelskab i Kiöbenhavn og
beskreven og forbemeldte Eggert Olafsen* (II, p.716,
Sorö. J.Lindgrens Enke, 1772).

--- *Voyage in Islande fait par ordre de Sa Majeste danoise,
traduit du danois par Gauthier de LaPeyronie [et
Biornered]*, (vol, IV, p.186, Paris, Levrault, 1802).

*OLEARIUS (Adam), Die Göttorfische Kunst-Kammer, worinnen
allerhand ungemeine Sachen So theils die Natur theils
kunstliche Hande hervor gebracht* (S.45, tab.XXVI, fig 2
and 3, Schlesswig, gedruckt durch J.Holwein, 1666)

PENNANT (Thomas), *British Zoology: Crustacea, Mollusca,
Testacea* (London, print. for Benjamin White, 1777).

PERNETTY (Dom Antoine-Joseph), *Histoire d'un voyage aux
isles Malouines, fait en 1763 et 1764 avec des
observations sur le détroit de Magellan et sur les
Patagons* (t.II, p.76, Paris, Saillant et Nyon, 1770).

SCHWEDIAWER (Dr Franz), (*op.cit.*)

23

SMITH (colonel A.), *in* Sir Richard OWEN, (*Brit. Assoc. for Advanc.
Science*, Plymouth, 2nd session, 1841).

BIBLIOGRAPHY

SPERM WHALES VERSUS CEPHALOPODS

*** , `Deep-swimming whales tangle with cables' (*Science News Letter*, Washington D.C., 19 Apr 1958).

BOURLIERE (Dr.F.), *Vie et moeurs des mammifères* (pp. 20-23, Paris, Payot, 1951).

BULLEN (Frank), (*op. cit.*, ch. XII, "About the Kraken").

GROUT (Jack), *Dans le sillage de la flibuste* (Paris, Arthaud, pp. 133-134, 1970).

IRVING (L.), `Respiration in diving animals' (*Physiol. Reviews*, **19**, pp. 112-134, 1939).

--- `The action of heart and circulation during diving' (*Trans. New York Acad. Sci.*, **7**, pp. 11-16, 1942).

IRVING (L.), SCHOLANDER (P.J.) and GRINNEL (S.W.), ` The respiration of the Porpoise *Tursiops truncatus*' (*Jour. Cell. Comparat. Physiol.*, **27**, pp.145-163, 1941).

KÜKENTHAL (Willy), `Untersuchungen an Walen' (Zweiter Teil, *Jenaische Zeitschr. f. Naturwiss.*, vol. LIV, 1914).

PORTIER (Paul), *Physiologie des animaux marins* (Paris, Flammarion, Bibl. de Phil. scientifique, 1938).

ROBIN (capitaine Max P.), `Combat de monstres' (*Le Chasseur français*, Saint-Etienne, no 663, p.707, novembre 1952).

SHALER (N.S.), `Notes on the Right and Sperm whale' (*Amer. naturalist*, Salem Mass., **7**, pp.1-4, 1873).

24

8

PIERRE DENYS DE MONTFORT, OUTLAWED MALACOLOGIST

BOSC (Louis Augustin), *Histoire naturelle des vers, contenant leur description et leurs moeurs*, in *Histoire naturelle de Buffon et suites* (t. LXIII, p.36, Paris,

Déterville, an X [1801]).

--- *Nouveau dictionnaire d'histoire naturelle appliquée aux arts, etc.* (t.XXX, p.462, Paris, Deterville, 1816 -1819).

BUCKLAND (Frank), *Log-book of a fisherman and zoologist* (London, Chapman & Hall, 1891).

CARRINGTON (Richard), *Mermaids and Mastodons* (pp.51-53, London, Chatto & Windus, 1957).

CUVIER (Georges), *Le Règne animal* (2 ed, t.III, p. 392, Paris, 1830).

DAWSON (G.M.), in TRYON (George Washington, *op.cit.*).

DENYS DE MONTFORT (Pierre), *Aux citoyens français professeurs et administrateurs du Muséum national d'histoire naturelle* (13 nivose an VIII [janvier 1800], Paris).

--- *La Vie et les aventures politiques de Nadir-Mirza Schah prince de Perse* (Paris, H.J.Jansen, an IX [1800-1801]).

--- *Histoire naturelle des mollusques*, in BUFFON, *Histoire naturelle générale et particulière* (nouvelle édition par C.S.Sonnini, vol. LIII-LV, Paris, Dufart, an X, [1801-1802]).

--- *Conchyliologie systématique* (Paris, Schoell, 1808-1810).

--- *Ruche à trois récoltes annuelles, fortifiée, économique, et son gouvernement, ou moyen de mettre les abeilles à couvert contre les attaques de leurs ennemis* (Paris, Audibert, 1813).

--- *Petit vocabulaire à l'usage des Français et des Alliés [...] renfermant les noms d'une partie des choses les plus essentielles à la vie en plusieurs langues, etc.* (Paris, Plancher, 1815).

25

--- [Divers articles,] in *Bibliothèque physico-économique, instructive et amusante* (1e année à 14e année, Paris, an XI, 1816)

OKEN (Lorenz), *Lehrbuch der Naturgeschichte* (Dritter Theil,
Zoologie) (vol 1, p.343, Jena, Aug. Schmid u. Co, 1815)

QUERARD (Joseph Marie), *La France litéraire, ou Dictionnaire
bibliographique des savants, etc.* (art. Denys de
Montfort, t. II, p.80, Paris, Firmin Didot, 1828).

LANDRIN (Armand), (*op.cit.*)

REES (W.J.), `On the habits of the octopus'(*The Times*,
London, August 12, 1950).

VAILLANT (Léon), `Quelques mots sur Denys de Montfort à
propos d'une brochure parue en 1815' (*Bulletin du Mus. Nat. d'Hist.
nat. de Paris*, **17**, pp. 83-86, 1911).

VALMONT de BOMARE (Jacques Christophe), *Dictionnaire
raisonné universel d'histoire naturelle* (Lyon, Bruynet,
an VIII [1800]).

9

THE MONSTER IS IDENTIFIED

CHENU (Dr. Jean-Charles), *Encyclopédie d'histoire naturelle:
Crustacés, Mollusques et Zoophytes* (en collaboration
avec M.Eugène Desmarest) (Paris, Marescq, & Co,
1858).

COOK (James), (*op.cit.*).

FERUSSAC and d'ORBIGNY (*op.cit.*).

GERVAIS (Paul), `Remarques au sujet des grands céphalopodes
décabrachides constituant le genre *Architeuthis*'
(*Journal de Zoologie*, Paris, **4**, no 2, pp. 88-90, 1875).

HARTING (Pieter), `Description de quelques fragments de deux
Céphalopodes gigantesques' (*Verhandel. Koningl. Akad.
Wetensch.*, **9**, Amsterdam, C.G. Van der Post, 1860).

JOHNSTON (George), *Einleitung in die Konchyliologie* [trad.
of *Introduction to Conchyology*] (Stuttgart, J.B.Muller,
1853).

26

KENT (William Saville), `Note on a Gigantic Cephalopod from
        Conception Bay, Newfoundland' (*Proc. Zool. Soc. London*,
        32, p. 178, 1874).

--- `A further communication upon certain Gigantic
        Cephalopods recently encountered off the Coast of
        Newfoundland' (*Ibid.* **32**, pp. 489-494, 1874).

LEY (Willy), *The Dodo, the Lungfish and the Unicorn*
        (chap.VI, *the Story of the Kraken*, New York, Viking
        Press, 1948).

OLAFSSON (Eggert), (*op.cit.*).

OLEARIUS (Adam), (*op.cit.*).

OWEN (sir Richard), `Description of some new and rare
        Cephalopoda' (*Transact, Zool. Soc.London,* **11**, pp. 13
        -170, 1881).

PACKARD Jr (A.S.), `Colossal cuttle fishes' (*American
        Naturalist*, Salem, Mass. **7**, pp.87-93, 1873).

PAULSSON (Svein), *Dagböger* (for February 1792, pp. 76-77) in
        STEENSTRUP (1847).

PERON (François), *Voyage de découvertes aux terres australes
        par M.F.Péron et M.C.A.Lesueur* (t. 1, p.216, Paris,
        Imprimerie Imperiale,1807).

POSSELT (Henri I.), `Blacksprutterne' (*Studentersamfundets
        Museumskrifter*, Kjobenhavn, p.37, 1895).

QUOY (Jean Rene Constant) et Joseph Paul GAIMARD in
        FREYCINET (Louis Claude Desaulces de), *Voyage autour du
        monde exécuté sur l'Uranie et la Physicienne: Zoologie*
        (t.1, 2e partie, p. 411, Paris, 1824-1844).

RANG (Paul Karel Sander), *Manuel de l'histoire naturelle des
        mollusques et de leur coquilles* (p.86, Paris, 1829).

REVOIL (Benedict Henry), (*op.cit.*)

RONDELET (Guillaume), *Libri de piscibus marinis* (*op.cit.*, p. 392).

--- *Histoire entière des poissons* (*op.cit.*, p.361).

SMITH (colonel A.), `Sur les Sépiaires gigantesques' (*L'Institut*, Paris, **10**, (no 428), pp. 85-86, 1842).

27

STEENSTRUP (Johan Japetus), `Meddelelse om tvende Kjaempestore Blaecksprutter, opdrevne 1639 og 1790 ved Islands Kyst , og om nogle andre nordiske Dyr' (*Förhandlinger ved de Skandinaviske Naturfoskeres*, Kjöbenhavn, **5**, pp. 950-957, 1847) (publ. in 1849)

--- `Om den i Kong Christian III^S Tid i Öresundet fangne Havmand (Sömunken kaldet)' (*Dansk Maanedskrift*, pp. 63 -96, 1855).

--- `Ueber den Zeit Christian's III in Oeresund gefangenen Meerman' (*Die Natuur*, pp. 337, 355, 382, 398, Halle, 1858).

--- `Kjäber af en kolossal Blaecksprotte' (*Overs. K.D. Vid. Selsk. Förhandl.*, pp. 190-200, 1855).

--- `Oplysninger om Altanterhavets kolossale Blaecksprutter' (*Förhandl. Skand. Naturforsk.*, Christiania, pp. 182 -185, 1856) (publ. in 1857).

--- `De Ommatostrephagtige Blaecksprutters indbyrdes Forhold: en Orientering' (*Overs. K.D. Vid. Selsk. Förhandl.*, 1880).

VERANY (Jean-BAptiste), *Mollusques méditerranéens* (Gênes, impr. des Sourds-Muets, 1851).

10

THE INADMISSIBLE SQUID OF THE ALECTON

BERTHELOT (Sabin) et le capitaine Fréderic BOUYER, `Poulpe géant observé entre Madère et Téneriffe' (*C.R.Acad. Sc.*, Paris, **53**, (no 27), pp. 1263 ff. 1862).

BOUYER (F.) `Poulpe géant dans les eaux de Téneriffe' (*L'Illustration*, Paris, **39**, p.140, 2 mars 1862).

CROSSE (Henri) et Paul FISCHER, `Nouveaux documents sur les Céphalopodes gigantesques' (*Journal de Conchyliologie*, Paris, 3e serie, **10**, pp. 124-140, 1862).

FIGUIER (Louis), *Zoophytes et Mollusques* (p.467, Paris, Hachette, 1866).

GIBSON (John), *Monsters of the sea* (London, Thomas Nelson, 1887).

28

HARTING (Pieter), (*op.cit.*).

JEFFREYS (John Gwyn), *British Conchology or an Account of the Mollusca which now inhabit the British Isles and the surrounding seas* (art. Cephalopoda, vol. V, p.124, London, J.van Voort, 1862-1869).

LANDRIN (Armand), (*op.cit.*).

MANGIN (Arthur). *les Mystères de l'Océan* (Tours, A.Mame et fils, 1864).

MOQUIN-TANDON (Alfred), (*op.cit.*).

PACKARD Jr.(A.S.), (*op.cit.*).

REVOIL (Benedict-Henry), (*op.cit.*).

SCOTT (Walter), *The Pirate* (vol. 1, chap.II, Paris, Galignani, 1826).

11

ARCHITEUTHIS GALORE

***. `Giant Cuttlefishes' (*The Scotsman*, p.3, May 22, 1882).

*** [Giant squid of the Bancals Rocks, off St. Honorat, Cannes, 1889] (*Standard*, London, Aug 15, 1889).

BIBLIOGRAPHY

BLANCHERE (de La), `Les Calmars géants de Conception Bay (Terre Neuve)' (*Nature*, 1er sem., p.196, 1874).

CARSON (Rachel), (*op.cit.*).

DALL (W.H.), `Aleutian Cephalopods' (*Amer.Naturalist.*, Salem, **7,** pp.484-485, 1873).

GERVAIS (Paul), `Remarques au sujet des grands céphalopodes décabrachides, etc.' (*op.cit.*).

29

HARVEY (Rev.M), `Gigantic Cuttlefishes in Newfoundland' (*Annals & MAg. of Natural History*,**13**,p.74, 1874). See also: *American Sportsman* (Dec. 6, 1873), *The Field* (Dec. 13, 1873) and *Amer. Jour. Sci.* (**7,** p.460, vol. VIII, p.158, 1874).

--- [*Letter*], (*Boston Traveller*, 30 Jan, 1879).

--- `How I discovered the great devil-fish' (*Wide World Mag.*, London, **2,** pp.732, 740, 1899).

HILGENDORF (F.), *Mittheilungen der deutschen Gesellschaft für Natur und Volkerkunde Ostasiens* (Herausgegeben von dem Vorstande) (1ste Heft, p.21, Yokohama, Japan, 1873)

--- `Ueber einen riesigen Tintenfish aus Japan, *Megateuthis martensii* n.g.n.sp.'(*Sitz. naturf. Fr.*, Berlin, pp. 65 -67, 1880).

KENT (William Saville), (*op.cit.*).

KIRK (T.W.), `On the Occurence of Giant Cuttlefish on the New Zealand Coast' (*Trans.& Proc. of the N.Z.Institute*, **12,** pp.310-313, 1879).

--- `Description of New Cephalopoda' (*ibid.*, **14,** p.283, pl. XXXVI, 1881).

--- `Brief description of a new species of large Decapod (*Architeuthis longimanus*)'(*ibid.*, **20,** p.34, pl.VII-IX, 1887).

MASSY (Anne L.), `The Cephalopoda Dibranchiata of the coasts of Ireland' (*Fisheries Ireland Sci. Investigations 1907*, **1**, pp. 1-39, 3 pl. 1909).

MITSUKURI & IKEDA, `Notes on a Gigantic Cephalopod' (*Zool. Mag.*, Tokyo, **7**, no 77, pp.39-50, tab.10, 1895).

MORE (A.G.), `Some account of the gigantic squid (*Architeuthis dux*) lately captured off Boffin Island (Connemara), (*Zoologist*, 2nd ser., **10**, pp. 4569-71, 1875).

--- `Gigantic squid on the west coast of Ireland' (*Ann. & Mag. of Nat. Hist.*, 4th ser, **16**, pp.123-124, Aug. 1875)

MURRAY (Alexander), `Capture of a Gigantic Squid at Newfoundland' (*Amer. Naturalist*, Salem, **7**, no 2, p.120, Feb 1874).

30

NORMAN (Rev. Cannon A.M.), `Revision of British Mollusca' (*Ann. & Mag. of Nat. Hist.*, (6), **5**, p.479, 1890).

O'CONNOR (Thomas), `Capture of an enormous cuttle-fish off Boffin Island, on the coast of Connemara' (*Zoologist*, 2nd ser, **10**, pp4502-4503, 1875.

ORBIGNY (Alcide Dessalines d'), *Mollusques vivants et fossiles, contenant: 1° une étude générale des mollusques; 2° la monographie complète des céphalopodes acétabulifères* (Paris, A.Delahays, 1855).

PACKARD Jr (A.S.) (*op.cit..*)

PFEFFER (George), *Ergebnisse der in dem Atlantischen Ozean [...] Plankton Expedition der Humboldt Stiftung: Die Cephalopoden der Plankton Expedition, zugleich eine monographische Uebersicht der Oegopsiden Cephalopoden* (Kiel u. Leipzig, Verlag von Lipsius u. Tischer, 1912).

ROBSON (C.W.),`On a new species of giant Cuttlefish, stranded at Cape Campbell, June 30th, 1886' (*Trans.New Zealand Inst.*, **9**, pp.155-157, 1886).

BIBLIOGRAPHY

ROBSON (G.C.), `On *Architeuthis clarkei*, a new species of Giant Squid, with observations on the genus' (*Proc. Zool. Soc.*, **3**, pp.681-697, Sep 1933).

SARS (G.O.), *Bidrag til Kundskaben om Norges arktishe Fauna: I. Mollusca regionis articae Norvegiae* (p.337, Christiania, 1878).

STEENSTRUP (Johan Japetus), *Spolia atlantica: Kolossale Blaeksprutter fra det nordlige Atlanterhav* (Kjöbenhavn, F.Dreyer, 1898).

STEWART (Alexander), *Twixt Ben Nevis and Glen Coe* (Edinburgh, William Paterson, 1885).

TAYLOR (Rev. Richard), *Te Ika a Maui, or New Zealand and its inhabitants etc*,(p.118, London, 1855).

TRYON (George Washington), (*op.cit.*,**1**, p. 89 and 184).

VELAIN (Charles), `Observations générales sur la faune des deux iles Saint Paul et Amsterdam' (*Arch.Zool. Exp.*, **6**, p.83, 1877). See also (*C.R.Acad. Sci.* Paris, **90**, p.1002, 19 avril 1875).

31

VERRILL (Addison E.), `Occurence of gigantic Cuttlefishes on the coasts of Newfoundland' (*Amer.Journ.Sci.*, **7**, (3), pp.158-167, 1874) and (*Ann.Mag.Nat.Hist.*, **13**, (4), pp.255-258, 1874).

--- `The Giant Cuttle-Fishes of Newfoundland and the common squids of the New England coast' (*Amer. Naturalist*, Salem, **8**, pp. 167-174, 1874).

--- `The Colossal Cephalopods of the Western Atlantic' (*ibid.*, **9**, pp. 21-36 and 78-86, 1875).

--- `The Gigantic Cephalopods of the North Atlantic' (*Amer. Journ. Sci.*, **9**, (3), pp. 123-130, 177-185, 1875).

---`Notice on the occurence of another gigantic Cephalopod (*Architeuthis*) on the coast of Newfoundland in December 1874, (*ibid.*, **10**, (3), pp. 213 ff, 1875).

--- `Notice on Gigantic Cephalopods: a Correction' (*ibid.*,
    12, (3), p. 236, 1876).

--- `Occurence of another Gigantic Cephalopod on the coast of
    Newfoundland' (*ibid.*,**14**,(3), pp.425-426, 1877).

--- `The Cephalopods of the North-Eastern Coast of America'
    (*Transact. Connect. Academy of Arts, 5,* pp. 177-257,
    1879).

--- `Report on the Cephalopods of the North-Eastern Coast of
    America' (*Report U.S.Comm.Fish.for 1879*, pp. 211-260,
    1882.)

--- `Second Catalogue of Mollusca recently added to the Fauna
    of the New England Coast (Cephalopoda: pp. 140-146, pp.
    243-249) (*Transact. Connect. Academy of Arts.*, **6**,
    1884).

VERRILL (A.Hyatt), *Moeurs étranges des Mollusques* [*Strange
    sea shells*] (Paris, Payot, 1952).

PART 5
THE SHADOW SIDE

12

FINAL UNCERTAINTIES

*** `The most macabre episode of the war at sea: marine
    monsters devour men adrift on a raft' (*Illustrated
    London News*, **199**, pp. 564-565, Nov 1, 1941).

32

*** `*Architeuthis clarkei*, sp. nov.' (*Naturalist*, **926**, pp.57
    -59, 1934).

*** `A Successor to the Sea-Serpent' (*Times*, London, Jul 4,
    1874).

BIBLIOGRAPHY

*** `The Case for the Sea-Serpent' [Stranding at Baven-on-Sea, 1924] (*Wide World Magazine*, London, p.451, March 1925).

ALBERT I, Prince of Monaco, `La Carrière d'un navigateur: la mort d'un cachalot' (*La Nouvelle Revue*, Paris, avril 1896).

--- *La Carrière d'un Navigateur* (Paris, Plon et Nourrit, 1902).

ALDRICH (Frederick A.), `Newfoundland's Giant Squids' (*Animals*, London, **10**, (no 1), pp. 20-21, May 1967).

--- `The Giant Squid in Newfoundland' (*Newfoundland Quarterly*, **65**, no 3, pp. 4-8, 1967).

--- `The Distribution of Giant Squids (Cephalopoda, Architeuthidae) in the North Atlantic and particularly about the shores of Newfoundland' (*Sarsia*, **34**, pp. 393 -398, 1968).

ALLAN (Joyce), `A rare giant squid' (*Austr. Mus.Mag.*, Sydney, **9**, no 9, pp.306-308, 1946).

--- `The Kraken, legendary of the seas' (*ibid.*, **11**, (no 9), pp. 275-278, 1955).

ASHLEY (Clifford W.), *The Yankee Whaler* (Boston, Houghton, Mifflin Co, 1926).

BENEDEN (P.J.van), `Histoire naturelle des Delphinidés des mers d'Europe' (*Mém. cour. etc., Ac.Roy.de Belgique*, **43**, 1889).

BRELAND (Osmond), `Devils of the Deep' (*Science Digest*, Chicago, Oct. 1952).

BRINKMANN (August), `Kjaempeblaekspruten (*Architeuthis dux* Stp) i Bergens Museum' (*Naturen*, 1916).

BROCH (Hj.), `Blekksprut' (*Fauna*, no4, 1954).

33

BULLEN (Frank) (*op.cit.*)

--- *Creatures of the Sea* (London, 1904).

BURTON (Maurice), `Mass death at sea' (*Illustrated London News*, Jul 10, 1954).

CADENAT (J.), `Note sur la première capture du Céphalopode géant, *Architeuthis nawaji*' (*C.R.Assoc.Franc. Av. Sci.*, **59**, p.513, 1935).

--- `Note sur un Céphalopode géant (*Architeuthis harveyi* Verrill) capturé dans le golfe de Gascogne' (*Bull. Mus. Hist. nat.*, Paris (2), **8**, (no 3), pp. 277-285, 1936).

CARRINGTON (Richard), *Mermaids and Mastodons* (London, Chatto and Windus; New York, Rinehart, 1957).

CHICHESTER (Francis), *Along the Clipper Way* (London, Hodder & Stroughton, 1966).

CLARKE (Malcolm R.), `A Review of the Systematics and Ecology of Ocean Squids' (*Advances in Marine Biology*, **4**, pp. 91-300, 1966).

CLARKE (Robert), `A Giant Squid Swallowed by a Sperm Whale; En kjempeblekksprut slukt av en spermhval' (*Norsk Hvalfanst-Tidende*, *40*, (no 10), pp. 589-593, Oct 1955).

---- `Sperm Whales of the Azores' (*Discovery Reports*, **28**, pp. 237-298, Dec 1956).

CLARKE (W.J.) and G.C.ROBSON, `Notes on the stranding of giant squids on the N.E. Coast of England' (*Proc. Malacol. Soc.*, London, **18**, pp.154-158, 1929).

CLARKE (W.J.), `Giant squid (new to science) at Scarborough' (*Naturalist*, London, **918**, pp. 157-158, 1933)

--- `Giant squids near Scarborough (Yorkshire)' (*Journ. Conch.*, London, **21**,(6), pp.163-164, 1939).

--- `A giant squid swallowed by a sperm-whale' (*Norsk

BIBLIOGRAPHY

Hvalfangst Tidende, **44**, no 10, pp.589-593, 1955).

COLEMAN (Loren), `The Plum Island Squid' (*Boston*, **73**, pp.63
  65,Apr 1981).

--- `A Birthday Gift for Jules Verne' (*Fortean Times*,
  London.no 34, pp 37-38, Winter 1981).

34

COWEN (Robert C.), *Frontiers of the Sea: the Story of*
  Oceanographic Exploration (New York, Doubleday and Co.
  1960).

DANIEL (R.), *Animal life in the sea* (London, Hodder &
  Stoughton, 1925).

DELL (R.K.) `A Specimen of the Giant Squid *Architeuthis* from
  New Zealand ' (*Records of the Dominion Museum*,
  Wellington, N.Z. **7**, (no 4) p 25-36, Sep 15, 1970).

DUNCAN (David D.) `Fighting Giants of the Humboldt' (*Nat.*
  Geogr.Mag., Washington, D.C., **79**, pp. 373-400, 1942).

DYSON (John), `Mysterious Monster from the Deep' (*Readers'*
  Digest, Pleasantville, N.Y. [Canadian edition], pp.
  107-114, Nov 1982).

FICHTER (G.S.), `Tentacles of Terror' (*Internat. Wildlife*,
  Washington, D.C. **10**, pp.12-16, Jan-Feb 1930).

FOURNIER (R.P.Georges), *Hydrographie, contenant la théorie*
  et la practique de toutes les parties de la navigation

  (Paris, M.Soly, 1643; 2e ed. augm. Paris, J.Dupuis,
  1667).

FRIDRICKSSON (Arni), `Remarks on the age and the growth of
  the squid' (*Greinar*, **2**, p.2, 1943).

FROST (Nancy), `Notes on a giant squid (*Architeuthis* sp.)
  captured at Dildo, Newfoundland in December 1933'
  (*Rep. from the Newfoundland Research Commission*, St.
  John's, **2**, (no 2), p. 100-113, 1934).

---- (*ibid*, **2**, (no 5), pp.89-95, 1936).

GIRARD (Albert A.) *Les Céphalopodes des iles Açores et de l'ile de Madère* (Lisbonne, 1890) (Abstracted by CROSSE and FISCHER in: *Journ. de Conchyl.*, Paris, **40**,p.365, 1892).

---- *Révision des Céphalopodes du Muséum de Lisbonne* (t.II, pp.210-220, Lisbon 1892).

--- `Céphalopodes des iles Açores' (*Jorn.Sc.math.phys.e natur*, Lisboa, **2**,1892).

GOSS (Michael), `Giant Squids on the Attack' (*Fate*, Highland Park, Ill. **38**, (7), pp. 34-41; **38**, (8) pp.78-82, Jul Aug 1985).
35

GRIMPE (G.), `Die Cephalopoden des arktischen Gebietes' (in ROMER u. SCHAUDINN, *Fauna Arctica*, Jena, **6**, pp.489-514, 1933).

GRONNINGSAETER (Arne), `Sjöormen-blekkspruten'(*Naturen*, nr 12, 1946).

HAMILTON (J.E.), `Belmullet Whaling Station: Report to the Committee' (*Rep. Brit. Ass.*, 125-161, 1954).

HARDY (Alister C.), *The Open Sea: the World of Plankton* (chap. XIV: *Squids, Cuttlefish and Kin*, London, Collins, 1956).

HEYERDAHL (Thor), (*op.cit.*).

JOUBIN (Louis), `Céphalopodes recueillis dans l'estomac d'un Cachalot recueilli aux iles Açores' (*C.R.Acad. Sci.*, Paris, **121**, p1172, 1895).

--- `Contribution à l'étude des Céphalopodes de l'Atlantique nord' (in *Résultats des campagnes scientifiques accomplies sur son yacht par Albert Ier, prince de Monaco* (Fasc. **9**, Monaco, 1895; fasc. **67**, Monaco, 1924).

--- `Céphalopodes provenant des campagnes de la "Princesse Alice' (1891-1894) (2e sér.) et 1898-1910) (3e sér.)' (*ibid*, fasc. **17**,Monaco, 1900; fasc.**54**, Monaco, 1920).

BIBLIOGRAPHY

KEIL (Albert), `Riesentintenfish aus dem Pottwal-Magen'
(*Natur und Museum*, Frankfurt a/Mein, **93**, (8), pp 319
323, 1 Aug 1963).

KELLY (William J.), `The Kraken Mystery' (*Explorer*,
Cleveland, Ohio, **27**, (no 2), pp. 27-31, Summer 1985).

KJENNERUD (Johannes), `Kjempeblekksprut' (*Universitetet i
Bergen smaa godbiter fra samlingene*, Ser. 2, nr 2,
1951).

LANE (Frank.W.), (*op.cit.*)

LEBLOND (Paul H.) and John SIBERT, *Observations of Large
Unidentified Marine Animals in British Columbia and
Adjacent Waters* (Vancouver, Institute of Oceanography,
University of British Columbia, Manuscript Report No
28, June 1973).

36

LEY (Willy), *The Dodo, the Lungfish and the Unicorn*
(*op.cit.*).

---- `Scylla was a Squid' (*Natural History*, New York, **48**,
(no 1), pp. 11-13, 1941).

MACGINITIE(G.E.) and Nettie MACGINITIE, *Natural History of
Marine Animals* (New York, McGraw Hill, 1949).

--- (*op.cit.*, 2nd ed. 1968).

MACKAL (Roy P.) *Searching for Hidden Animals* (Garden City,
N.Y., Doubleday, 1980).

MASSY (A.L.), `The Cephalopods of the Irish Coast'
(*Proc.R.Irish. Academy*, **38**, pp.25-37, 1928).

MATTHEWS (L.Harrison), `The Sperm Whale, *Physeter catodon*'
(*Discovery Reports*, Cambridge, **17**,pp 96-168, 1938).

MURRAY (Sir John) and Johan HJORT, *The Depths of the Ocean:
a general account of the modern scicence of
Oceanography based largely on the scientific researches
of the Norwegian steamer "Michael Sars"* (London,
Macmillan, 1912).

MYKLEBUST (Björn), `Et nytt funn av kjempeblekksprut i Romsdal (*Naturen*,nr 12, 1946).

NORDGAARD (O.), `The Cephalopoda dibranchiata observed outside and in the Trondhjem Fjord' (*Det. Kgl. Norske Vid. Selsk.Skr.*, no 5, 1922).

--- `Faunistic Notes on Marine Invertebrates III (*Det. Kgl. Norske Vid. Forhandling*, Bd.I, nr 26, pp 70-72, 1928) (publ. 1929).

OTTERLING (Gunnar), `Bläkfisk och fiske i Skandinavien' (*Faunistik Revy*, nr 3, 1954).

PFEFFER (Georg), (*op.cit.*).

QUOY (J.R.G.) and F.P.GAIMARD (*op.cit.*).

RAE (Bennet B.), `Description of a giant Squid, stranded near Aberdeen' (*Proc. Malac. Soc.*, London, **28**, pp. 103 -167, pls 20-21, 1950).

RANCUREL (Paul), `Les Calmars géants existent-ils en Nouvelle-Calédonie ?' (*Bull. du Pacifique Sud*, **24**, (no 3) pp. 5-7, 1974).

37

REES (W.J.), `On a giant squid *Ommastrephes caroli* Furtado, stranded at Looe, Cornwall' (*Bull.of the Brit. Museum (Natural History) (Zoology)*, London, vol I, no 2, 1950).

--- `Giant squid: the quest for the kraken' (*Illustr. London News*, **215**, p.826, 1949).

REES (W.J.) and MAUL (G.E.), `The Cephalopods of Madeira' (*Bull. Brit. Mus. (Nat. Hist.) (Zool.)*, London, **3**, pp. 259-281, 1956).

RICHARD (J.) and H.NEUVILLE, `Sur quelques cétaces observés pendant les campagnes du yacht "Princesse Alice".' (*Mém. Soc. Zool.*, Paris, **18**, pp.100-109, 1897).

RITCHIE (James), `Occurence of a giant squid (*Architeuthis*) on the Scottish coast' (*Scott. Nat.*, Edinburgh, pp.

BIBLIOGRAPHY

133-139, 1918).

--- Giant squid cast ashore North Uist, Outer Hebrides'
(*ibid.*, p. 57, 1920).

--- `Giant squids on the Scottish coast' (*Rep.Brit.Ass.*,
p.423, 1921).

ROBSON (G.C.), *A Monograph of the Recent Cephalopoda, Based
on the Collection of the British Museum* (London,
British Museum, (Nat. Hist.) 1929-1932).

--- `A Giant Squid from the North Sea' (*Natur. Hist. Mag.*,
2, p.6, 1929).

--- `On *Architeuthis clarkei*, a new species of Giant Squid'
(*op. cit.*).

ROE (H.S.J.), `The food and the feeding habits of the Sperm
Whale (*Physeter catodon* L.) taken off the west coast
of Iceland' (*J.Cons. int. Explor.mer*, Copenhagen,
33,(no 1), pp. 93-102, Nov. 1969).

RONDELET (Guillaume) (*op.cit.*).

RUSSELL (Eric Frank), *Great World Mysteries* (pp 143-144,
London, Dennis Dobson, 1957).

SANDERSON (Ivan T.), *Follow the Whale*, (*op.cit.*).

SASAKI (Madoka), `A Monograph of the dibranchiate
cephalopods of the Japanese and adjacent waters'
(*J.Fac.Sci.Univ.Tokyo, Zool.*, **20**, pp. 1-357, pl. I-XXX,
Suppl. 1929).

38

SHALER (N.S.), (*op.cit.*)

SIVERTSEN (Erling), `Blekksprut' (*Aarbok 1954 for Det Kgl.
Norske Vidensk. Slesk Museet*)

STEPHEN (A.C.), `Notes on Scottish Cephalopods' (*Scot. Nat.*,
Edinburgh, **227**, pp. 131-132, 1939).

--- `Giant Squid *Architeuthis* in Shetland (*ibid*, **62**, pp. 52

-53, 1950).

--- `Rare invertebrates recently found in the Scottish
Area' (*ibid.* **65**, pp. 120-121, 1953).

--- `The Cephalopods of Scottish and adjacent waters'
(*Trans. Roy. Soc.*, Edinburgh, **61**, pp. 247-270,
1944).

STORM (V.), `Om 2 udenfor Trondhjhemsforden fundne kjempe-
blekkspruter' (*Naturen*, p.97, 1897).

SUTER (Henry), *Manual of the New Zealand Mollusca*
(Wellington, J. Mackay, 1913).

TAGO (Katsuya), `Cetacea found in Japanese waters' (*Int.
Congr. Zool.*, **21**, pp. 2192-2228, 1937).

TAMBS-LYCHE (Hans), `Et nytt funn av kjempeblekksprut ved
Norskekysten' (*Naturen*, nr 9, 1946).

TAYLOR (M.R.), `Great squids' (*Trans. Suffolk Nat. Soc.*,
Norwich, **2**, (1), pp. 1-3, 1932).

THOMPSON (d'Arcy Wentworth), `Note on a Dolphin showing
traces of an encounter with a cuttlefish' (*Ann. Mag. Nat.* (7), **7**, pp.
503-505, 1901).

THURSTON (Harry), `Quest for the Kraken' (*Equinox*, **46**, pp.
50-55, Jul-Aug 1939).

VERE (Maximilian Schele DE), *Wonders of the Deep* (London,
Guilford, 1871; New York, 1869).

VOSS (Gilbert L.), `Hunting Sea Monsters' (*Sea Frontiers*,
Miami, FL, **5,** (3), pp. 134-146, Aug 1959).

WALKER (N.W.Gregory), `Sperm Whale and Squid' (*Discovery*,
London, pp. 308-310, Oct 1937).

39

WRIGHT (Bruce S.), `How to Catch a Giant Squid - Maybe!'
(*Atlantic Advocate*, Fredericton, **51**,pp. 17-23, Aug
1961).

--- *Wildlife Sketches* (Fredericton, Brunswick Press, 1962).

YONGE (C.M.) and T.E.THOMPSON, *Living Marine Molluscs* (London, Collins, 1976).

13

THE INCOMPLETE STORY OF THE COLOSSAL OCTOPUS.

***, `Una grossa piovra, con tentacoli che superano i dieci metros de lunghezza (*Corriere della Sera*, Milano, p.4, 6. Marzo 1986).

*** `The Sea Monster that came ashore on the Florida coast' (*The Grit*, Williamsport, Pa. **15**, (no 1), p.1, Dec 13, 1896).

*** `Big Octopus on the beach' (*Florida Times-Union,* Jacksonville, Fla. p.3, Dec 1, 1896).

***`An Octopus in the hand' (*INFO Journal*, College Park,Md., **2**, (no 4), pp 1-3, 1972).

***`San Diego Membership Meeting' (*I.S.C.Newsletter*, **4**, (no 2), p.6, Summer 1985).

*** `Giant octopus blamed for deep-sea fishing disruption' (*ibid.*, **4**, (no 3), pp. 1-6, Autumn 1985).

*** `Blaming the Giant Octopus for the Cyclops mystery' (*Literary Digest*, New York, **60**, (no 10), p.92, Mar 8, 1919).

*** `Scraps from Serials' (*Natural Science*, London, **10**, p.130, Feb 1897).

***`News of the Universities' (*ibid.*, **10**, p. 285, Apr 1887).

*** `Some new books' (*ibid.*, **10**, p.207, Mar 1897).

*** `Views and News' (*ibid.*, **10**, p.356, May 1897).

40

***`Un octopode gigantesque dans les mers du Japon' (*La Nature*, Paris, **2**, (no 47), p.332, 25 avril 1874).

*** `Was not an Octopus' (*New Haven Evening Register*, New Haven, Conn., p. 1, Feb 24, 1897).

*** `Scientists say Sea Creature of 1896 was Monster Octopus' (*New Haven Journal and Courier*, New Haven, Conn., p. 23, Apr 10, 1971).

***`Last of this Sea Serpent' (*New York Herald*, New York, p.6, Dec 2, 1896).

*** `Gigantic Octopus: the largest known member of the family washed ashore in Florida' (*ibid.*, 2 nd ed., p.18, Jan 3, 1897).

***`Octopuses kill fishermen' (*New Zealand Herald*, Auckland, sect 1, p. 20, Sep 10, 1984).

***`Combat d'un scaphandrier contre une pieuvre' (*Petit Journal Illustré,* Paris, 24 janvier 1909).

*** `Un scaphandrier attaqué par une pieuvre' (*ibid.*, 16 juin 1912).

***`Aux prises avec une pieuvre géante' (*Petit Var*, Toulon, 1 fevrier 1912).

*** `Divers debunk the Sea Monster' (*Science Digest*, Chicago, Nov 1953).

*** `Killed by giant octopus' (*Star*, Kuala Lumpur, Malaysia, Jul 18, 1977).

*** `Octopus kills tot' (*Sun*, London, p.5, Dec 27, 1989).

*** `Kraken zogen zwei Fischer in die Tiefe' (*Die Welt*, 11 Sep 1984).

AKIMUSHKIN (Igor), *Cephalopods of the Sea of the USSR* (Jerusalem, Israel Program for Scientific Translations, 1965.)

BIBLIOGRAPHY

BARRATT (P.H.J.), *Grand Bahama* (p.25, London, David and Charles, 1972).

BARTSCH (Paul), (*op.cit.*)

BAVENDAM (Fred), `Eye to eye with the Giant Octopus' (*National Geographic*, Washington, D.C., **179**, (no 3), pp.86-87, Mar 1991).
41

BENJAMIN (George J.), `Diving into the Blue Holes of the Bahamas' (*ibid.*,**138**, pp. 347-352, Sep 1970).

BERGE (Victor), *Danger is my life* (pp. 31-34, *op.cit.*)

BERRILL (N.J.), (*op.cit.*)

BOLETZKY (Sigurd von), `Petits et grands "poulpes à oreilles" ' (*L'Univers du Vivant*, Paris, no 3, septembre 1985).

BOULENGER (E.G.), *A Natural History of the Seas* (*op. cit.*).

BRELAND (Osmond P.), `Harmless or deadly?' (*op.cit.*).

BRIGHT (Michael), `The Big Blob, or what grabbed the crab traps?' (*B.B.C. Wildlife*, **3**, (no 9), pp. 431-433, Sep 1985).

CAHILL (Tim) & Bob WALLACE, `The Devil and the Deep Blue Sea' (*Rolling Stone*, New York, pp. 34-39, Mar 8, 1979).

CAZARD (Paul), *Aux quatre coins des océans, souvenirs d'un consul* (Aurillac, Cantal, Imprimerie Moderne, 1938).

CHENU (Jean-Charles), (*op.cit.*).

CHICHESTER (sir Francis), *Along the Clipper Way* (London, Hodder & Stoughton, 1966).

COUSTEAU (Jacques-Yves) and Philippe DIOLÉ, *Pieuvres, la fin d'un malentendu* (pp.216-217, Paris, Flammarion, 1973).

DALL (W.H.),`Note on *Octopus punctatus*, Gabb (*Proc. California Acad. Nat. Sciences*, **3**, pp. 243-244, 1866).

--- `Aleutian Cephalopods' (*Amer. Nat.*, Salem, Mass., **7**,
    pp . 484-485, 1873).

---`The arms of the octopus, or devil fish' (*Science,*
    Lancaster, Pa.,**6**, (no 145), p.432, 1885).

DENYS DE MONTFORT (Pierre), *Histoire naturelle des
    mollusques (op.cit.)*

DRAGESCO (Jean), *(op.cit.)*

FANNING (James), `A Touch of Humour' (*Nat. History*, New
    York, **80**,(no 5), p.92, May 1971).

FOURNIER (R.P.Georges), *(op.cit.)*.

42

FRANCE (Raoul Heinrich), *Lebenswunder der Tierwelt* (p.51,
    Berlin, Deutscher Verlag, 1940).

GABB (M.),`Description of two species of Cephalopods' (*Proc.
    Calif. Acad. Nat. Sciences*, Apr 7, 1862).

GENNARO (Joseph F.), `The Creature Revealed' (*Natural
    History*, New York, p.24, p. 84, Mar 1971).

--- `*Octopus giganteus*, largest creature in the world'
    (*Argosy*, New York, **376**, (no 3), pp. 30-32, Mar 1973).

GEORGE (J.S.) *in* Prof. B.J.WILDER, `A Colossal Octopus'
    (*Amer. Natur.*, Salem, Mass., **6**,p. 772, 1872).

GILL (Rev. William Wyatt), *Myths and Songs of the South
    Pacific* (p.146, London, H.S.King,1876).

GNAEGY (Charles), `Deadly Mystery of the Bahamas Blue Holes'
    (*Saga*, New York, pp.10-13, 72-76, Jul 1973).

HIGH (William L.), `The Giant Pacific Octopus' (*Marine
    Fisheries Review*, Seattle, **38**, (no 9), pp 27-22, Sep
    1976).

HOCHBERG (F.G.) and Gordon FIELDS, `Cephalopods: the Squids
    and Octopuses' in MORRIS, ABBOTT & HADERLIE *Intertidal
    Invertebrates of California* (Stanford, Stanford

BIBLIOGRAPHY

University Press, 1989).

HOYLE (William E.), `Report on the Cephalopods collected by *HMS Challenger* during the years 1873-76' in *Report on the Voyage of the Challenger* (Zoology, **16**, (44), pp.56 60, 66, 100) (London, Longmans and Co, 1886).

KIRCHER (R.P.Athanasius), *Mundus subterraneus in XII libros digestus* (t.I, chap. 25, p.97, Amsterdam, 1678).

KOTZEBUE (Otto von), *Entdeckungs-Reise in die Sud-See und nach der Berings-Strasse zur Erforschung einer nordostlicher Durchfahrt, unternommen in den Jahren 1815, 1816, 1817 und 1819* (t.II, p.108, Weimar, Gebruder Hoffman, 1821).

KUSCHE (Lawrence David), *La Solution du mystère* (Montréal, l'Etincelle, 1976).

43

LUCAS (Frederick Augustus), `The Florida Monster' (*Science*, Lancaster, Pa., N.S. **5**, (116), p.467, Mar 19, 1897).

--- `Some mistakes of scientists' (*Natural History*, New York, *28*, (no 2), pp. 169-174, Mar-Apr 1928).

MACGINITIE (G.E. & Nettie), (*op.cit.*)

MACKAL (Roy P.), `Biochemical analyses of preserved *Octopus giganteus* tissue' (*Cryptozoology*, **5**, pp.55-62, 1986).

MANGIACOPRA (Gary S.), `*Octopus giganteus* Verrill: a new species of Cephalopod' (*Of Sea and Shore*, Port Gamble, Washington, D.C. **6**, (no 1), pp 3-10, Spring 1975).

--- `*Octopus giganteus* Verrill' (*Pursuit*, Columbia, R.D.1, N.J., **8**,pp. 95-101, Oct 1975).

--- `*Octopus giganteus* Verrill: Giant Octopus or Whale?' (*Fate*, Highland Park, Ill., *29*, (no 7), pp. 78-85, Jul 1976).

--- `The Great Ones' (*Of Sea and Shore*, Port Gamble, Washington, D.C., **7**, (no 2), pp 51-52, Summer 1976).

--- `Monster on the Florida Beach' (*INFO Journal*, no 17, pp. 2-6, May; no 18, pp 2-6, Jul 1976).

--- `Fisherman Fables' (*Of Sea and Shore*, Port Gamble, Washington, D.C. **9**,(no 4), pp. 207-208, Winter 1978 -79).

MOFFITT (Donald), `A 200-foot Octopus washed up in Florida, two scientists claim' (*Wall Street Journal*, New York, **177**, (no 68), pp. 1, 33, Apr 8 1973).

PACKARD Jr (A.S.), (*op.cit.*)

PALMER (Robert), `*The Blue Holes of the Bahamas*' (pp. 72 -73,   London, Jonathan Cape, 1985).

---- `Ecology beneath the Bahama Banks' (*New Scientist*, London, **110**, (no 1507), pp. 45-48, May 8, 1986).

--- `In the Lair of the Lusca' (*Natural History*, New York, **96**, (no 1), pp. 42-47, Jan 1987).

PARKER (T.Jeffery), `On the size and the external sexual characters of the New Zealand Octopus (*O. maorum*, Hulton)'(*Nature*, London, p.586, Oct 15, 1885).

44

PICKFORD (Grace), `*Octopus dofleini* (Wulker), the giant octopus of the North Pacific' (*Bull. of the Bingham Oceanographic Collection*, Cambridge Mass., **19**, pp.5-70, 1964).

PRACHAN (Jean), *Le Triangle des Bermudes* (p.180, Paris, Belford, 1978).

RAYNAL (Michel), `L'incroyable dossier des poulpes géants' (*Amazone*, St Julien, no 1, pp.42-46, octobre 1982)

---- `Poulpes géants, d'autres témoignages' (*ibid.*, no 2, pp 28-31, janvier 1983).

--- `Le Poulpe colossal des Caraïbes' (*Soc. Ed. Sc. nat. Beziers*, Beziers, publ. spec. pp 1-18, 1985).

RAYNAL (Michel) et Michel DETHIER, `Le "Monstre de Floride"

de 1896: Cétacé ou Poulpe Colossal?' (*Bull. Soc. neuchâteloise des Sc.nat.,* Neuchatel, **114**, pp 105-115, 1991).

RÉVOIL (Benedict-Henry), (pp. 237-247, Paris, L.Hachette, 1863).

TASSI (Franco), `La Prova dell'Octopus' (*Il Messagero*, Roma, p.16, 11 Jul 1986).

TRÉGOUBOFF (G.) [Communication personnelle, lettre du 8 juillet 1956].

VERANY (J.B.), (*op.cit.*).

VERRILL (Addison E.), `The Cephalopods of the North-Eastern Coast of America' (*op.cit.*).

--- `A Gigantic Cephalopod on the Florida Coast' (*Amer. Journ. Sci.*, New Haven, Conn. 4th ser., **3**, p.79, Jan 1897).

---`Additional information concerning the Giant Cephalopod of Florida' (*ibid.*, 4th ser., **3**, pp.162-163, Feb 1897).

---`The Florida Monster' (*Science*, Lancaster, Pa., N.S., **5**, (no 114), p.392, Mar 5, 1897).

`The Supposed Great Octopus of Forida: Certainly Not a Cephalopod' (*Amer. Journ.Sci.*, New Haven, Conn., 4th ser., **3**, pp. 355-356, Apr 1897).

--- `The Florida Sea-Monster' (*Science*, Lancaster, Pa., N.S., **5**, (no 116), p.476, 1897).

45

--- `What is this Creature?' (*New York Herald*, New York, 2nd ed., p.13, Mar 7, 1897).

---`The Florida Sea-Monster' (*Amer. Naturalist*, Salem, Mass., **31**, pp. 304-307, Apr 1897).

VERRILL (Alpheus Hyatt), `An Octopus for Yale' (*Hartford Daily Courant*, Hartford, Conn., p.11, Feb 18, 1897)

--- `Octopus for New Haven' (*New Haven Evening Register*, New Haven, Conn., Feb 14, 1897).

--- *The Ocean and its Mysteries* (pp. 118-119, New York, Duffield and Co., 1916).

--- *Moeurs étranges des Mollusques* (pp. 128-133, Paris, Payot, 1952).

WEBB (DeWitt), `A Large Decapod' (*The Nautilus*, Philadelphia, Pa., **10**, (no 9), p.108, Jan 1897).

WILES (Doris C.) `DeWitt Webb' (*El Escribano*, St Augustine, Fla. **3**, (no 4), p. 14, Oct 1966).

WILLOCK (Colin), `In Search of the Hairy-handed Monster' (*T.V.Times*, London, Sep 9-15, 1972).

WOOD (Forrest G.), `Stupefying Colossus of the Deep' (*Natural History*,New York, **80**, (3), pp.15-24, Mar 1971)

--- `In Which Bahamian Fishermen Recount their Adventures with the Beast' (*ibid.*, pp.84-87, Mar 1972).

WOOD (Gerald L.), *Guinness Book of Animal Facts and Feats* (Enfield, Middlesex, Guinness Superlatives Ltd., 2nd ed. pp 208-211, 1976; 3d ed., pp 191, 194-197, 1982).

WRIGHT (Bruce S.), `The lusca of Andros' (*Atlantic Advocate*, Fredericton, **57**,pp. 32-39, Jun 1967).

WÜLKER (Gerhard), `Über Japanische Cephalopoden. Beitrage zur Kenntnis der Systematik und Anatomie der Dibranchiaten' *in* DOFLEIN (F.), `Beitrage zur Naturgeschichte Ostasiens' (*Abh. d. II Kl. d.K.Acad.d.Wiss.*, München, III, Suppl. Bd., 72 pp., 1910).